Zhong Li

Fuzzy Chaotic Systems

Studies in Fuzziness and Soft Computing, Volume 199

Editor-in-chief
Prof. Janusz Kacprzyk
Systems Research Institute
Polish Academy of Sciences
ul. Newelska 6
01-447 Warsaw
Poland
E-mail: kacprzyk@ibspan.waw.pl

Further volumes of this series can be found on our homepage: springer.com

Vol. 183. Larry Bull, Tim Kovacs (Eds.)
Foundations of Learning Classifier Systems, 2005
ISBN 3-540-25073-5

Vol. 184. Barry G. Silverman, Ashlesha Jain, Ajita Ichalkaranje, Lakhmi C. Jain (Eds.)
Intelligent Paradigms for Healthcare Enterprises, 2005
ISBN 3-540-22903-5

Vol. 185. Spiros Sirmakessis (Ed.)
Knowledge Mining, 2005
ISBN 3-540-25070-0

Vol. 186. Radim Bělohlávek, Vilém Vychodil
Fuzzy Equational Logic, 2005
ISBN 3-540-26254-7

Vol. 187. Zhong Li, Wolfgang A. Halang, Guanrong Chen (Eds.)
Integration of Fuzzy Logic and Chaos Theory, 2006
ISBN 3-540-26899-5

Vol. 188. James J. Buckley, Leonard J. Jowers
Simulating Continuous Fuzzy Systems, 2006
ISBN 3-540-28455-9

Vol. 189. Hans Bandemer
Mathematics of Uncertainty, 2006
ISBN 3-540-28457-5

Vol. 190. Ying-ping Chen
Extending the Scalability of Linkage Learning Genetic Algorithms, 2006
ISBN 3-540-28459-1

Vol. 191. Martin V. Butz
Rule-Based Evolutionary Online Learning Systems, 2006
ISBN 3-540-25379-3

Vol. 192. Jose A. Lozano, Pedro Larrañaga, Iñaki Inza, Endika Bengoetxea (Eds.)
Towards a New Evolutionary Computation, 2006
ISBN 3-540-29006-0

Vol. 193. Ingo Glöckner
Fuzzy Quantifiers: A Computational Theory, 2006
ISBN 3-540-29634-4

Vol. 194. Dawn E. Holmes, Lakhmi C. Jain (Eds.)
Innovations in Machine Learning, 2006
ISBN 3-540-30609-9

Vol. 195. Zongmin Ma
Fuzzy Database Modeling of Imprecise and Uncertain Engineering Information, 2006
ISBN 3-540-30675-7

Vol. 196. James J. Buckley
Fuzzy Probability and Statistics, 2006
ISBN 3-540-30841-5

Vol. 197. Enrique Herrera-Viedma, Gabriella Pasi, Fabio Crestani (Eds.)
Soft Computing in Web Information Retrieval, 2006
ISBN 3-540-31588-8

Vol. 198. Hung T. Nguyen, Berlin Wu
Fundamentals of Statistics with Fuzzy Data, 2006
ISBN 3-540-31695-7

Vol. 199. Zhong Li
Fuzzy Chaotic Systems, 2006
ISBN 3-540-33220-0

Zhong Li

Fuzzy Chaotic Systems

Modeling, Control, and Applications

Springer

Dr. Zhong Li
Faculty of Electrical
and Computer Engineering
FernUniversität in Hagen
58084 Hagen
Germany
E-mail: zhong.li@fernuni-hagen.de

Library of Congress Control Number: 2006923229

ISSN print edition: 1434-9922
ISSN electronic edition: 1860-0808
ISBN-10 3-540-33220-0 Springer Berlin Heidelberg New York
ISBN-13 978-3-540-33220-6 Springer Berlin Heidelberg New York

Springer is a part of Springer Science+Business Media
springer.com

Printed in The Netherlands

Typesetting: by the author and techbooks using a Springer LaTeX macro package

Printed on acid-free paper SPIN: 11353348 89/techbooks 5 4 3 2 1 0

To Juan and Yifan

Preface

Bringing together the two seemingly unrelated concepts, *fuzzy logic* and *chaos theory*, is primarily motivated by the concept of *soft computing* (SC), initiated by Lotfi A. Zadeh, the founder of fuzzy set theory. The principal constituents of SC are fuzzy logic (FL), neural network theory (NN) and probabilistic reasoning (PR), with the latter subsuming parts of belief networks, genetic algorithms, chaos theory and learning theory. What is important to note is that SC is not a melange of FL, NN and PR. Rather, it is an integration in which each of the partners contributes a distinct methodology for addressing problems in their common domain. In this perspective, the principal contributions of FL, NN and PR are complementary rather than competitive.

SC differs from conventional (hard) computing in that it is tolerant of imprecision, uncertainty and partial truth. In effect, the role model for soft computing is the human mind. From the general SC concept, we extract FL and chaos theory as the object of this book to study their relationships or interactions.

Over the past few decades, fuzzy systems technology and chaos theory have received ever increasing research interests from, respectively, systems and control engineers, theoretical and experimental physicists, applied mathematicians, physiologists, and other communities of researchers. Especially, as one of the emerging information processing technologies, fuzzy systems technology has achieved widespread applications around the globe in many industries and technical fields, ranging from control, automation, and artificial intelligence (AI) to image/signal processing and pattern recognition. On the other hand, in engineering systems chaos theory has evolved from being simply a curious phenomenon to one with real, practical significance and utilization. We are now standing at the threshold of major advances in the control and synchronization of chaos with new applications across a broad range of engineering disciplines, where chaos control promises to have a major impact on novel time- and energy-critical engineering applications.

Notably, studies on fuzzy systems and chaos theory have been carrying out separately. Although there have been some efforts on exploring the interactions between fuzzy logic and chaos theory, it is still far away from fully understanding their mutual relationships. Intuitively, fuzzy logic resembles human reasoning in its use of approximate information and uncertainty to generate decisions, and chaotic dynamics could be a fundamental reason for human brain to produce and process massive information instantly. It is believed that they have a close relationship in human reasoning and information processing, and it has a great potential to combine fuzzy systems with chaos theory for future scientific research and engineering applications.

This book does not intend to – in fact, is not able to – give an in-depth explanation of the interactions or intrinsic relationships between fuzzy logic and chaos theory, but tries to provide some heuristic research achievements and insightful ideas to attract more attention on the topic. Although this book may raise more questions than it can provide answers, We hope that it nevertheless contains seeds for future brooming research.

More precisely, this book is written in the following way: it starts with the fundamental concepts of fuzzy logic and fuzzy control, chaos theory and chaos control, as well as the definition of chaos on the metric space of fuzzy sets, followed by fuzzy modeling and (adaptive) fuzzy control of chaotic systems, all based on both Mamdani fuzzy models and Takagi-Sugeno (TS) fuzzy models; then, it discusses some other topics, such as synchronization, anti-control of chaos, intelligent digital redesign, all for TS fuzzy systems, and spatiotemporal chaos and synchronization in complex fuzzy systems; finally, it ends with a practical application example of fuzzy-chaos-based cryptography.

I am very grateful to Prof. Wolfgang A. Halang and Prof. Guanrong Chen for their long-term support and friendship in various aspects of my work and life, without which this book would not have been completed. Special thanks are given to Dr. Hojae Lee for providing me with the materials presented in Chapters 6, 9, and 12. Thanks also go to the following individuals who provided some original figures or helpful comments: Oscar Calvo, Federico Cuesta, Patrick Grim, Z.H. Guan, K.-Y. Lian, Domenico M. Porto, M. La Rosa, Hua O. Wang, and H.B. Zhang. I am very appreciative of the discussions and collaborations with Peter Kloeden, Zongyuan Mao, Ping Ren, Bo Zhang, Yaobin Mao, Wenbo Liu, Shujun Li, Martin Skambraks, Jutta Düring, Ping Li, and Hong Li. I owe my deepest thanks to my parents for their fostering and education. Finally, I wish to express my appreciation to Prof. Janusz Kacprzyk for recommending this book to the Springer series, Studies in Fuzziness and Soft Computing, and the editorial and production staff of Springer-Verlag in Heidelberg for their fine work in producing this new monograph.

Hagen, Germany *Zhong Li*
February 2006

Contents

1

Introduction

At first glance, fuzzy logic and chaos theory may seem two totally different areas with merely marginal connections to each other. In this introduction, after reviewing the evolution of fuzzy set theory and chaos theory, respectively, we explain briefly the ideas why we bring them together, and we shall show that the understanding of the interactions between fuzzy systems and chaos theory lays a solid foundation for better applications of the two promising new technologies, and their integration offers a great number of interesting possibilities in their interplay and future developments.

1.1 Fuzzy Logic and Fuzzy Control Systems

1.1.1 Fuzzy Logic

The precision of mathematics owes, to a high extent, its success to the efforts of Aristotle and the philosophers who preceded him. In their efforts to devise a concise theory of logic, and later mathematics, the so-called "Laws of Thought" were posited [1]. One of those, the "Law of the Excluded Middle," states that every proposition must either be True or False. Even when Parminedes proposed the first version of this law (around 400 B.C.) there were strong and immediate objections: for example, Heraclitus proposed that things could be simultaneously True and not True.

It was Plato who laid the foundation for what would become fuzzy logic, by indicating that there was a third region (between True and False) where these opposites appeared. Other more modern philosophers echoed his sentiments, notably Hegel, Marx, and Engels. But it was Lukasiewicz who first proposed a systematic alternative to the bi-valued logic of Aristotle [2].

In the early 1900's, Lukasiewicz described a three-valued logic, along with the mathematics to accompany it. The third value he proposed can best be translated as the term "possible", and he assigned it a numeric value between

Zhong Li: *Fuzzy Chaotic Systems*, StudFuzz **199**, 1–11 (2006)
www.springerlink.com

True and False. Eventually, he proposed an entire notion and axiomatic system from which he hoped to derive modern mathematics.

Later, he explored four-valued logic, five-valued logic, and then declared that, in principle, there was no difficulty in deriving an infinite-valued logic. Lukasiewicz felt that three- and infinite-valued logics were the most intriguing, but he ultimately settled on a four-valued logic, because it seemed to be the most easily adaptable to the Aristotelian logic.

In about the same time, Knuth proposed a three-valued logic similar to that of Lukasiewicz, from which he speculated that mathematics would become even more elegant than that in traditional bi-valued logic. His insight, apparently missed by Lukasiewicz, was to use the integral range $[-1, 0, +1]$ rather than $[0, 1, 2]$. Nonetheless, this alternative failed to gain acceptance, and has fallen into relative obscurity.

It was not until relatively recently that the notion of an infinite-valued logic was brought forward. In 1965, Lotfi A. Zadeh published his seminal work "Fuzzy sets" [3, 4], which described the mathematics of what is called fuzzy set theory today. This theory proposed a membership function (or the values False and True) to operate over the range of real numbers $[0.0, 1.0]$. New operations for the calculus of logic were formulated, and showed to be, in principle, a generalization of classic logic.

Since then, fuzzy set theory has received tremendous interest in research and applications. This is mainly due to the immense success of Japanese consumer products that made use of fuzzy technology in the 1980s, and some theoretical breakthrough on the stability analysis of fuzzy systems in the 1990s.

Fuzzy systems, including fuzzy logic and fuzzy set theory, are an alternative to traditional notions of set theory and logic that have their origins in ancient Greek philosophy, and applications at the leading edge of Artificial Intelligence. The main idea of fuzzy systems is to extend the classical two-valued modeling of concepts and attributes like *tall*, *fast* or *old* in a sense of partial truth. This means that a person is not just viewed as *tall* or *not tall*, but as *tall* to a certain degree between 0 and 1.

Classical models usually try to avoid *vague*, *imprecise* or *uncertain* information, because it is considered to have a negative influence in an inference process. Fuzzy systems, on the other hand, deliberately make use of this kind of information. This usually leads to simpler, more suitable models, which are easier to handle and are more familiar to human beings.

Zadeh's proposal of modeling the mechanism of human thinking, with linguistic fuzzy values rather than numbers [5], led to the introduction of fuzziness into systems theory and to the development of a new class of mathematical systems called *fuzzy systems*. In general, we shall refer to fuzzy systems as those *resulted from fuzzification of a conventional system*. A central characteristic of fuzzy systems is that they are based on the concept of fuzzy coding (partitioning) of information. Fuzzy systems operate with fuzzy sets instead of numbers. Each fuzzy set has more expressive power than a single number. The use of fuzzy sets permits a generalization of information. This generalization is associated with the introduction of imprecision. In many real problems the imprecision is admissible, even useful, because the categories of human thinking are vague ideas which are very hard to quantify. In essence, the representation of the information in fuzzy systems imitates the mechanism of approximate reasoning performed in the human mind. The precision of conventional systems theory is obtained as a limiting case in the continuity of varying levels of abstraction.

Fuzzy system models basically fall into two categories, which differ fundamentally in their abilities to represent different types of information. The first category includes linguistic models, which have been referred to so far as Mamdani fuzzy models. They are based on collections of IF-THEN rules with vague predicates and use fuzzy reasoning [6, 7]. In these models, fuzzy quantities are associated with linguistic labels, and a fuzzy model is essentially a qualitative expression of the underlying system. Models of this type form a basis for qualitative modeling that describes the system behavior by using natural language [8]. A corresponding fuzzy logic controller is a prototypical example of such a linguistic model, in which its rules give a linguistic expression of the control strategy in a common sense.

The second category of fuzzy models is based on the Takagi-Sugeno (TS) method of reasoning [9, 10, 11]. These models are formed by logical rules that have a fuzzy antecedent part and a functional consequent. They are combinations of fuzzy and nonfuzzy models. Fuzzy models based on the TS method of reasoning integrate the ability of linguistic models for qualitative knowledge representation with great potential for expressing quantitative information. In addition, this type of fuzzy models permits a relatively easy application of various powerful learning techniques for system identification from data and controller design.

Because of the linear dependence of each rule on the input variables of the underlying system, the TS method is capable of acting as an interpolating supervisor of multiple linear controllers that are to be applied, respectively, under different operating conditions of a dynamic nonlinear system. For example, the performance of an aircraft may change dramatically with altitude and Mach number. Linear controllers, though easy to compute and well-suited to any given flight condition, must be updated regularly and smoothly to keep up with the changing state of the airplane. A TS fuzzy inference model is extremely well suited to the task of smoothly interpolating the linear control gains that would be applied across the input space; it is a natural and efficient gain scheduler. Similarly, a TS fuzzy system is suited to model nonlinear systems by interpolating between multiple linear models.

Since a TS fuzzy system is a more compact and computationally efficient representation than a Mamdani fuzzy system, TS fuzzy system typically uses adaptive techniques for model construction. These adaptive techniques can be used to customize the membership functions so that the fuzzy system can best fit the data available.

Here are some general comparisons of the two different fuzzy models:

Advantages of the Mamdani fuzzy model:

i) Intuitive;
ii) Widespread acceptance;
iii) Well-suited to human linguistic input.

Advantages of the TS fuzzy model

i) Computationally efficient;
ii) Works well with linear techniques (e.g., PID control);
iii) Works well with optimization and adaptive techniques;
iv) Guaranteed continuity of the output surface;
v) Well-suited to mathematical analysis.

1.1.2 Fuzzy Control Systems

Fuzzy control is the most successful and active branch of fuzzy system technology, in terms of both theoretical analysis and practical applications. The primary thrust of this novel control paradigm, created in the early 1970's, was to utilize the knowledge and experience extracted from a human control operator to intuitively construct controllers so that the resulting controllers were able to emulate human control behavior to a certain extent. Compared to the traditional control paradigm, the advantages of the fuzzy control paradigm are twofold. First, a precise mathematical model of the system to be controlled is not required; second, a satisfactory nonlinear controller can often be developed empirically without using complicated mathematics [12].

Industrial automation and commercial production have successfully been developed worldwide using fuzzy control. In this regard, Japan has led the way. Its success includes Hitachi automated train operation of the Sendai subway system that has been in daily operation since 1987. The trains, controlled by fuzzy predictive controllers, consume less electric energy, and ride more comfortably than the ones controlled by conventional controllers. Another Hitachi product is the group fuzzy control operation for elevators. The waiting time and idle time of the elevators are both reduced during the rush hours; and riding and stopping are smoother.

In the late 1980s, a real-time fuzzy control drug delivery system was successfully developed and clinically implemented to regulate blood pressure in postsurgical open-heart patients at cardiac surgical intensive care units [13]. This is the world's first real-time fuzzy control application in medicine. Fuzzy systems have also been applied to the control of muscle immobility and hypertension during general anesthesia, assessment of cardio-vascular dynamics during ventricular assistance, diagnosis of artery lesions and coronary stenosis, support for seriodiagnosis, intelligent medical alarms, and multineuron studies. Other successful medical applications are the detection of coronary artery disease, classification of tissues and structures in electrocardiograms, and classification of normal and cancerous tissues in brain magnetic resonance images.

The theory of fuzzy systems has advanced significantly along with the rapid increase of their practical applications. In early days, most fuzzy controllers were used as black-box controllers in that their internal mathematical structures were unknown. Since the late 1980s, significant progress has been made to mathematically explore the analytic structures of various types of fuzzy controllers [14]. Fuzzy control has been related to PID control, sliding mode control, adaptive control, relay control, etc. in conventional control systems, resulting in insightful understanding of fuzzy control within the context of classical control. The advances have also been used to analyze some important aspects of fuzzy control systems, including system stability and control performance, and to better design fuzzy controllers.

System modeling and system control are two closely related problems. Fuzzy modeling is a new modeling paradigm, and fuzzy models are nonlinear dynamic models. Comparing with the conventional black-box modeling techniques that can only utilize numerical data, the uniqueness of the fuzzy modeling approach lies in its ability to utilize both qualitative and quantitative information. This advantage is practically important and even crucial in many circumstances. Fuzzy models are often intuitive, as fuzzy sets, fuzzy logic and fuzzy rules are intuitive and linguistic. However, fuzzy models are not as simple as those models that can be expressed in mathematical formulae. In general, fuzzy models are black-box models. Nevertheless, under certain conditions, analytical structures of some fuzzy models can be derived and, hence, can be used conveniently.

A system capable of uniformly approximating any continuous function on a compact set is called a *functional approximator* or a *universal approximator*. The issue of universal approximation is crucial to fuzzy systems. In the context of control, the question is whether a fuzzy controller can always be constructed to uniformly approximate a desired continuous nonlinear control solution with sufficient accuracy. For modeling, the question is whether always a fuzzy model can be established, which is capable of uniformly approximating any continuous nonlinear physical system. Recent theoretical work has led to affirmative answers to these qualitative questions for both Mamdani fuzzy models and TS fuzzy models.

1.2 Chaos and Chaos Control

1.2.1 Chaos

Chaos, defined in the *Encyclopaedia Britannica*, originates from the Greek $\chi\alpha o\zeta$, which means "the primeval emptiness of the universe before things came into being of the abyss of Tartarus, the under world.... In the later cosmology, chaos generally designated the original state of things, however conceived. The modern meaning of the word is derived form Ovid, who saw chaos as the original disordered and formless mass, from which the maker of the Cosmos produced the ordered universe" [15]. Yet, there is another interpretation of chaos in the ancient Chinese vocabulary, where chaos generally refers to an unusual phenomenon that is disorderly and irregular. A modern dictionary definition of chaos provides two meanings: (i) utter disorder and confusion, and (ii) the unformed original state of the universe. In the modern scientific terminology, however, chaos has a rather precise but fairly complicated definition by means of the dynamics of a generally nonlinear system.

Historically, at the turning of last century (around 1900), French mathematician Henry Poincaré had already observed that fully deterministic dynamics do not necessarily imply explicit predictions on the evolution of a dynamical system. This can be considered as a milestone in the approach to the study of dynamical chaos. Although in the 1930s and 1940s strange behavior was observed in many physical systems, the notion that this phenomenon was inherent to deterministic systems was never suggested. Even with the powerful results of Steve Smale in the 1960s, complicated behavior of deterministic systems remained no more than a mathematical curiosity. Scientifically, chaos implies the existence of undesirable randomness, but the self-organization concept at the edge of chaos denotes the order out of chaos. The American essayist and historian Henry Adams (1858-1918) expressed the scientific meaning of "chaos" succinctly: "Chaos often breeds life, when order breeds habit" [16].

Not until the late 1970s, with the advent of fast and powerful computers, was it recognized that chaotic behavior was prevalent in almost all domains of science and technology. In the development of chaos theory, the first evidence of physical chaos seems to be the discovery of Edward Lorenz, the Lorenz attractor or "butterfly effect", in 1963. The first underlying mechanism within chaos was observed by American physicist Mitchell Feigenbaum, who in 1976 discovered that "when an ordered system begins to break down into chaos, a consistent pattern of rate doubling occurs". Later on, the works of Michel Hénon and Carl Heiles [17] and Boris Chirikov [18] provided new insights into the origin of chaotic behaviors in dissipative as well as in conservative systems. To that end, the term *chaos* was first formally introduced into mathematics by Li and Yorke in their paper "period-3 implies chaos" [19]. Ever since then, there have been several alternative but closely related definitions of chaos, among which that of Devaney is perhaps the most popular one [20]. Nevertheless, a unified, universally accepted, and rigorous definition of chaos is not yet available in the current scientific literature [21].

Devaney states that a map $f : S \to S$, where S is a set, is *chaotic* if

(i) f is *transitive* on S: for any pair of nonempty open sets U and V in S, there is an integer $k > 0$ such that $f^k(U) \cap V$ is nonempty;

(ii) f has *sensitive dependence on initial conditions*: there is a real number $\delta > 0$, depending only on f and S, such that in any nonempty open subset of S there is a pair of points, whose eventual iterates under f are separated by a distance of at least δ;

(iii) the periodic points of f are *dense* in S.

Another definition requires the set S to be compact, but drops condition (iii) from the above [22]. It is even believed that only the transitive property is essential in this definition. Although a precise and rigorous mathematical definition of chaos does not seem to be available any time soon, some fundamental features of chaos are well accepted, which can be used to signify or identify chaos in most cases. A fundamental property of chaos is its extreme sensitivity to initial conditions. Other features of chaos include the embedding of a dense set of unstable periodic orbits in its strange attractor, positive leading Lyapunov exponent, finite Kolmogorov-Sinai entropy or positive topological entropy, continuous power spectrum, positive algorithmic complexity, ergodicity, Arnold's cat map, Smale horseshoe map, a statistically oriented definition of Šhil'nikov, and some other unusual limiting properties [21, 25]. In other words, chaotic motion is an unstable bounded stationary motion (i.e. locally unstable but globally bounded). This definition unfolds the two aspects of chaotic motion: instability and boundedness. Or, in other words, chaotic motion is a bounded stationary motion without equilibrium, periodicity and almost-periodicity [27].

In fact, our real world is essentially nonlinear, thus chaos is ubiquitous. Besides the above mathematical descriptions of chaos, some scientists' views may help understand the meaning of chaos from various perspectives. For instance, Physicist Paul C. W. Davies said that reality without chaos is very rare, yet scientists have developed their understanding of the world by studying only a small piece of orderly and predictable reality, but ignoring chaos or dismissing it as "noise" in an otherwise well-defined system, even though chaos is the rule much more than the exception. Biochemist Linda Jean Shepherd believes that "chaos science is shifting how we see the world", because it is changing the central metaphor by which science understands things. Instead of the metaphor of the Newtonian clock with its cogwheels and levers, all being rational and predictable, chaos theory suggests metaphors that are much more indeterminate, unpredictable and random, e.g., turbulent rivers, weather, or smoke, and closer to how the world really works. Shepherd believes that chaos theory describes a real universe and demonstrates that precise prediction is impossible in complex systems [23]. Physicist Stephen H. Kellert agrees and believes that chaos theory offers a way to understand this unpredictability and randomness, allowing us to see how randomness arises, what it looks like, and how, at large scale and long term, randomness shows a complex non-repetitive pattern. Ultimately, physicist M. Mitchell Waldrop sees chaos theory as widening our view of order in the world. No longer will scientists have to focus only on the small portion of reality that is predictable; now they can also study the great wildness beyond simplistic traditional order. So to speak, chaos theory reverts the primary face of the real world.

Mathematician James Gleick describes three essential features of chaos. First, chaos is characterized by sensitive dependence on initial conditions, or what has become known as the "butterfly effect" – the flap of a butterfly's wings in a Brazilian rainforest may produce a tornado in Texas next week. In chaotic systems, like weather, turbulence, or smoke, small disturbances can have dramatic effects due to the underlying webs of interconnectedness that can amplify small changes. Second, chaotic systems are aperiodic or never undergo a regular repetition of values. They never settle into a precise pattern, because nothing repeats in any systematic way. They are "no-repeat" systems. And third, chaotic systems are focused on "strange attractors" – attractors that not only attract but also repel. An attractor is the central organizer of a system, but the attractor in a chaotic system is affected by the butterfly effect and the no-repeat characteristic; so, the attraction is paradoxically a strange attraction toward butterfly unpredictability and no-repeat randomness that create a complex, non-repetitive pattern which can only be described in long term and in large scale.

Chaos is considered together with relativity and quantum mechanics as one of the three monumental discoveries of the twentieth century. For physicists and philosophers, relativity and quantum mechanics may rank above chaos for their impacts on the way we view the world. As for science in general, it is not clear that these theories have had any distinct effect on medicine, biology, and geology. Yet, chaotic dynamics are having an important impact on all of the fields, particularly, chemistry, physics, economics and sociology [26].

1.2.2 Chaos Control

Understanding chaos has long been the main focus of research in the field of nonlinear dynamics. Although chaos is a very attractive subject for study, due to its intrinsic topological complexity it was once believed to be neither predictable nor controllable. However, recent research efforts have shown that not only (short term) prediction but also control of chaos are possible. It is now well known that most conventional control methods and many special techniques can be used for chaos control [25, 28].

For many years, the feature of the extreme sensitivity to initial conditions made chaos undesirable, and most experimentalists considered such characteristic as something to be definitely avoided. Besides this feature, chaotic systems have two other important properties. Firstly, there is an infinite number of unstable periodic orbits embedded in the underlying chaotic attractor. In other words, the skeleton of a chaotic attractor is a collection of an infinite number of unstable periodic orbits. Secondly, the dynamics in the chaotic attractor are ergodic, which implies that during its temporal evolution the system ergodically visits a small neighborhood of every point in each one of the unstable periodic orbits embedded within the chaotic attractor [29].

In their attempt of controlling chaos, physicists and mathematicians brought about some fresh ideas and novel techniques that utilize the very nature of chaos for controls. For instance, the so-called OGY control method [30] employs the classical feedback control idea, which might be understood as a kind of pole-placement method and is technically quite simple, but virtually it takes advantage of chaos itself in the sense of using its structural stability and its basic property of having a dense set of periodic orbits near a saddle. Such special properties are not available for non-chaotic systems; therefore, such a control methodology was not first recommended by control theorists and engineers who usually – if not always – try to use "brute force" to regulate and stabilize unstable dynamics. When "brute force" type of controls is not allowed, e.g., in fragile and microscopic biological control systems such as human brain and heart regulations, new control methods utilizing the extreme sensitivity of chaos to tiny variations are very desirable. This usually leads to some non-conventional approaches. Today, there are some non-traditional control ideas and methods developed in the field of chaos control, including system parameter tuning, bifurcation monitoring, entropy reduction, state pinning, phase delay, weak oscillation input, disorder input, and some special-purpose feedback and adaptive controls, to list just a few [31].

In contrast, recent research has shown that chaos can actually be useful under certain circumstances, and there is growing interest in utilizing the very nature of chaos. Today, the traditional trend of understanding and analyzing chaos has evolved to ordering and utilizing chaos. A new research direction in the field of applied chaos technology not only includes controlling chaos, which means to weaken or completely suppress chaos when it is harmful, but also includes chaotification, called anti-control of chaos, which refers to enhancing existing chaos or purposely creating chaos when it is useful and beneficial.

For example, fluid mixing is not only useful but actually important in the situation that two fluids are to be thoroughly mixed by consuming minimal energy. The fluid mixing problem is a curiosity in your coffee cup, a minor problem when baking in the kitchen, but a $20-billion problem for the U.S. chemical industry: Why does it take so long for stirred liquids to mix? Part of the answer is that regular stirring causes some streams to return to their starting points. The path that the stream travels may be highly complex, but eventually it meets up with itself, forming "regular islands" – essentially unmixed pockets of liquid woven intricately through the mixture. "Mixing is one of the nicest applications of exploiting chaos," Julio M. Ottino therefore said, "in many cases, you would like to get rid of chaos, but in the case of mixing, we would often like to enhance it, and we would certainly like to understand it" [24]

Scientists have always been trying to unravel the mechanism of how our brains endow us with inference, thoughts, perception, reasoning and, most fascinating of all, emotion such as happiness and sadness. The fundamental reason for the ability of human brains to process massive information instantly may lie in the controlled chaos of the brains. Other potential applications of chaos control in biological systems have reached out from the brain to elsewhere, such as the human heart. In contrast to the common belief that healthy heartbeats are completely regular, a normal heart rate may fluctuates in an erratic fashion, even at rest, and may actually be somewhat chaotic [47].

Chaos control and anti-control methodologies promise to have a major impact on many novel, time- and energy-critical applications, such as high-performance circuits and devices (e.g., delta-sigma modulators and power converters), liquid mixing, chemical reactions, biological systems (e.g., in the human brain, heart, and perceptual processes), crisis management (e.g., in jet-engines and power networks), secure information processing (e.g., chaos-based encryption), and decision-making in critical events. This new and challenging research area has become a scientific inter-discipline, involving engineers in controls, systems, electronics, and mechanics, as well as applied mathematicians, theoretical and experimental physicists, physiologists, and above all, nonlinear dynamics specialists [25].

1.3 Interactions between Fuzzy Logic and Chaos Theory

Although the relationship between fuzzy logic and chaos theory is not yet completely understood at the moment, the study on their interactions has been carried out for more than two decades, at least with respect to the following aspects: fuzzy control of chaos [32, 33], adaptive fuzzy systems from chaotic time series [34], theoretical relations between fuzzy logic and chaos theory [35, 36], fuzzy modeling of chaotic systems with assigned properties [37, 38], chaotifying Takagi-Sugeno (TS) fuzzy models [39, 40, 41, 42], and fuzzy-chaos-based cryptography [43].

At about the same time, fuzzy logic and chaos theory entered into the ken of science. Fuzzy logic was originally introduced by Lotfi Zadeh in 1965 in his seminal paper "fuzzy sets" [3], while the first evidence of physical chaos was Edward Lorenz's discovery in 1963 [44], although the study of chaos can be traced back to hundreds of years ago to some philosophical pondering [15] and to the work of the French mathematician Jules Henri Poincaré at the beginning of the last century [45]. Is it only a coincidence?

Roughly speaking, fuzzy set theory resembles human reasoning using approximate information and inaccurate data to generate decisions under uncertain environments. It is designed to mathematically represent uncertainty and vagueness, and to provide formalized tools for dealing with imprecision in real-world problems. On the other hand, chaos theory is a qualitative study of unstable aperiodic behavior in deterministic nonlinear dynamical systems. Research reveals that it is due to the drastically evolving and changing chaotic dynamics that the human brain can process massive information instantly. "The controlled chaos of the brain is more than an accidental by-product of the brain complexity, including its myriad connections, but rather, it may be the chief property that makes the brain different from an artificial-intelligence machine" [46]. Therefore, it is believed that both fuzzy logic and chaos theory are related to human reasoning and information processing. It is also believed that to understand the complex information processing within the human brain, fuzzy data and fuzzy logical inference are essential, since precise mathematical descriptions of such models and processes are clearly out of question with today's scientific knowledge.

Based on the above-mentioned observations, the study on the interactions between fuzzy logic and chaos theory allows a better understanding of their relation, and may provide a new and promising, although challenging, approach for theoretical research and simulational investigation of human intelligence.

1.4 About This Book

This book is organized as follows: Chapters 2– 4 introduce the fundamental concepts of fuzzy logic and fuzzy control, chaos theory and chaos control, as well as the definition of chaos on the metric space of fuzzy sets, respectively. Then, fuzzy modeling and (adaptive) fuzzy control of chaotic systems, all based on both Mamdani fuzzy models and Takagi-Sugeno (TS) fuzzy models, will be given in Chapters 5– 9. A very important topic, synchronization of fuzzy systems, will be discussed in Chapter 10. In Chapter 11, systematic anti-control approaches will be studied. Chapter 12 discusses intelligent digital redesign for fuzzy systems. More complicated spatiotemporally chaotic phenomena and synchronization in complex fuzzy systems will be investigated in Chapter 13. Finally, an application of fuzzy-chaos-based control method in cryptography will be illustrated in Chapter 14.

With this book, we try to bridge the apparent gap between fuzzy logic and chaos theory. It is expected that the studies within the book can be useful for a wide range of challenging applications and that the ideas derived from the studies would lay a foundation on which further research in this promising area can be based.

This book can serve as a reference book for researchers working in the interdisciplinary areas related to, among others, both fuzzy logic and chaos theory.

2

Fuzzy Logic and Fuzzy Control

Over the past few decades, there has developed a tremendous amount of literature on the theory of fuzzy set and fuzzy control. This chapter attempts to sketch the contours of fuzzy logic and fuzzy control for the readers, who may have no knowledge in this field, with easy-to-understand words, avoiding abstruse and tedious mathematical formulae.

2.1 Introduction

Fuzzy logic is in nature an extension of conventional (Boolean) logic to handle the concept of partial truth – truth values between "completely true" and "completely false". It was introduced by Lotfi Zadeh in 1965 as a means to model the uncertainty of natural language.

Fuzzy logic in the broad sense, which has been better known and extensively applied, serves mainly as a means for fuzzy control, analysis of vagueness in natural language and several other application domains. It is one of the techniques of soft-computing, i.e. computational methods tolerant to suboptimality and impreciseness (vagueness) and giving quick, simple and sufficiently good solutions [48, 49, 50, 51].

Fuzzy logic in the narrow sense is a symbolic logic with a comparative notion of truth developed fully in the spirit of classical logic (syntax, semantics, axiomatization, truth-preserving deduction, completeness, and propositional and predicate logic). It is a branch of many-valued logic based on the paradigm of inference under vagueness. This fuzzy logic is a relatively young discipline, not only serving as a foundation for the fuzzy logic in a broad sense but also of independent logical interest, since it turns out that strictly logical investigation of this kind of logical calculi can go rather far [52, 53, 54, 55].

Fuzzy logic has several unique features that make it a particularly good choice for many control problems:

i) It is inherently robust, since it does not require precise, noise-free inputs and can, thus, be fault-tolerant if a feedback sensor quits or is destroyed.

Zhong Li: *Fuzzy Chaotic Systems*, StudFuzz **199**, 13–29 (2006)
www.springerlink.com

The output control is a smooth control function despite a wide range of input variations.

ii) Since a fuzzy logic controller processes user-defined rules governing the target control system, it can be modified and tweaked easily to improve or drastically alter system performance. New sensors can easily be incorporated into the system simply by generating appropriate governing rules.

iii) Fuzzy logic is not limited to a few feedback inputs and one or two control outputs, nor is it necessary to measure or compute rate-of-change of parameters. It is sufficient with any sensor data to provide some indication of a system's actions and reactions. This allows the sensors to be inexpensive and imprecise thus keeping overall system cost and complexity low.

iv) Because of the rule-based operation, any reasonable number of inputs can be processed (1–8 or more) and numerous outputs (1–4 or more) generated, although defining a rule-base quickly becomes complex if too many inputs and outputs are chosen for a single implementation, since rules defining their interrelations must be defined, too. Then, it would be better to break the control system into smaller chunks and use several smaller fuzzy logic controllers distributed on the system, each one with more limited responsibilities.

v) Fuzzy logic can control nonlinear systems that would be difficult or impossible to model mathematically. This opens doors for control systems that would normally be deemed unfeasible for automation.

In summery, fuzzy logic was conceived as a better method for sorting and handling data, and has proven to be an excellent choice for many control system applications, since it mimics human control logic. It can be built into anything from small, hand-held products to large computerized process control systems. It uses an imprecise but very descriptive language to deal with input data more like a human operator. It is very robust and forgiving of operator and data input, and often works when first implemented with little or no tuning.

2.2 Fuzzy Set Theory

2.2.1 Crisp Sets and Fuzzy Sets

Normal sets in classic set theory are called *crisp sets* as compared with the fuzzy sets in fuzzy set theory. Let C be a crisp set defined on the universe of discourse U, then for any element x of U, either $x \in C$ or $x \notin C$. For a crisp set C its *characteristic function* $\mu_C : U \to \{0, 1\}$ is defined as [56, 57]:

Definition 2.1. $\mu_C : U \to \{0, 1\}$ *is a characteristic function of the set* C *iff for all* x *(iff stands for "if and only if"):*

$$\mu_C(x) = \begin{cases} 1, \ when \ x \in C, \\ 0, \ when \ x \notin C. \end{cases} \tag{2.1}$$

In fuzzy set theory this property is extended. Thus, in a *fuzzy set* F, it is not necessary that either $x \subset F$ or $x \not\subset F$. The characteristic function is extended to a so-called *membership function*, which assigns to each $x \in U$ a value from the unit interval $[0, 1]$ instead of from the two-element set $\{0, 1\}$ in classic set theory. The set defined with such an extended membership function is called a *fuzzy set.*

Definition 2.2. *The membership function* μ_F *of a fuzzy set* F *is a function*

$$\mu_F : U \rightarrow [0, 1]. \tag{2.2}$$

Therefore, each element x in U has a *membership degree* $\mu_F(x) \in [0, 1]$. F is completely determined by the set of tuples

$$F = \{(x, \mu_F(x)) | x \in U\}. \tag{2.3}$$

It should be noticed that a fuzzy set is actually a generalized subset of a classical set, and a universe of discourse is never fuzzy [58].

Example 2.3. Membership functions of three fuzzy sets, namely, "slow", "medium" and "fast", for the speed of a car are shown in Fig. 2.1. Here, the universe of discourse is all speeds of the car, i.e., $U = [0, V_{max}]$, where V_{max} is the maximum possible speed of the car. At the speed of 45 km/h, for instance, the fuzzy set "slow" has membership degree 0.5, namely, $\mu_{slow}(45) = 0.5$, the fuzzy set "medium" has membership degree 0.5, namely, $\mu_{medium}(45) = 0.5$, and the fuzzy set "fast" has membership degree 0, namely, $\mu_{fast}(45) = 0$ [59].

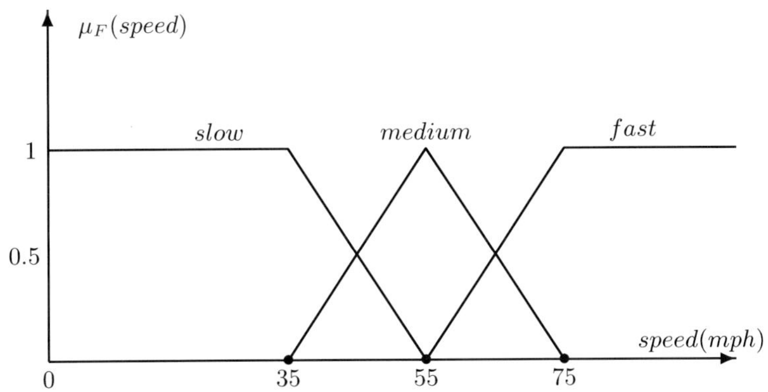

Fig. 2.1. Membership functions of three fuzzy sets, namely, "slow", "medium" and "fast", for the speed of a car

A fuzzy set can be expressed as:

1. Discrete case (when the universe of discourse is finite): Let the universe of discourse U be $U = \{u_1, u_2, \ldots, u_n\}$. Then, a fuzzy set F on U can be represented as follows:

$$F = \mu_F(u_1)/u_1 + \mu_F(u_2)/u_2 + \cdots + \mu_F(u_n)/u_n = \sum_{i=1}^{n} \mu_F(u_i)/u_i. \quad (2.4)$$

2. Continuous case (when the universe of discourse is infinite): When the universe of discourse U is an infinite set, a fuzzy set F on U can be represented as follows:

$$F = \int_u \mu_F(u_i)/u_i. \quad (2.5)$$

Membership functions of fuzzy sets may have various forms. The most important one is the triangular shape, like the fuzzy set "medium" in Fig. 2.1. Other common forms are the trapezoidal as shown in Fig. 2.2, the Gaussian membership functions in Fig. 2.3, as well as the irregularly shaped and arbitrary membership functions.

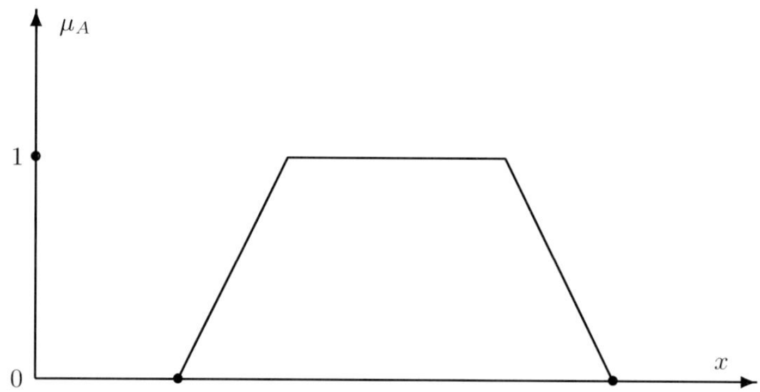

Fig. 2.2. Trapezoidal type membership function

Each kind of membership function has its own merits and disadvantages. For instance, the triangular membership function is extensively used in practice for their simplicity, because it is defined with a minimal amount of information (data). It can easily be modified with their parameters, and its linear segments ease the derivation of a model's input/output map. On the other hand, it lacks continuous differentiability. The Gaussian function is expressed as

$$\mu_F(x) = exp\left[-\left(\frac{x-b}{a}\right)^2\right]. \quad (2.6)$$

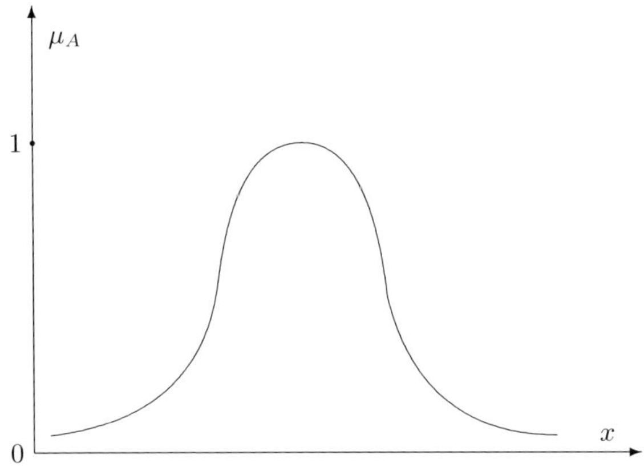

Fig. 2.3. Gaussian membership function

It is determined by the two parameters, a and b, where b is the modal or central value, and a determines its width. The Gaussian function is of any grade of smooth and continuous differentiability, which eases the theoretical analysis of fuzzy systems.

The actual shape of a fuzzy set depends completely on the semantics of the concept intended to be represented. In other words, there are no universal or pre-defined fuzzy sets [57]. A fuzzy set makes no sense without the context of a system or model, which means that certain shapes are representative of particular classes of knowledge.

The shape of a fuzzy set is almost always a continuous line from left to the right edge of the set. The contours of a fuzzy set represent the semantic properties of the underlying concept, so the closer the set surface maps to the behavior of a physical or conceptual phenomenon, the better our model will reflect the real world. When we turn to building fuzzy models, fuzzy systems are tolerant of approximations not only in their problem spaces but also in the representation of fuzzy sets. This means that they perform well even when a fuzzy set does not map exactly with its model concept.

2.2.2 Fundamental Operations of Fuzzy Sets

Union of fuzzy sets A and B, $A \cup B$, is a fuzzy set defined by the membership function:

$$\mu_{A\cup B}(x) = \mu_A(x) \vee \mu_B(x), \tag{2.7}$$

where

$$\begin{aligned}\mu_A(x) \vee \mu_B(x) &= \begin{cases} \mu_A(x), & \mu_A(x) \geq \mu_B(x) \\ \mu_B(x), & \mu_A(x) < \mu_B(x) \end{cases} \\ &= \max\{\mu_A(x), \mu_B(x)\}.\end{aligned}$$

Intersection of fuzzy sets A and B, $A \cap B$, is a fuzzy set defined by the membership function:

$$\mu_{A \cap B}(x) = \mu_A(x) \wedge \mu_B(x), \tag{2.8}$$

where

$$\begin{aligned}\mu_A(x) \wedge \mu_B(x) &= \begin{cases} \mu_A(x), & \mu_A(x) \leq \mu_B(x) \\ \mu_B(x), & \mu_A(x) > \mu_B(x) \end{cases} \\ &= \min\{\mu_A(x), \mu_B(x)\}.\end{aligned}$$

Complement of fuzzy set A, $\bar{A}$, is a fuzzy set defined by the membership function:

$$\mu_{\bar{A}}(x) = 1 - \mu_A(x). \tag{2.9}$$

It should be noted that union, intersection, and complement of crisp sets are special cases of union, intersection, and complement of fuzzy sets, respectively.

The above definition of the classical operators was introduced by Lotfi Zadeh, which is only one possible choice of operations for union, intersection, and complement, as it is by no means unique. More general terms, *t-norm* and *t-conorm*, are defined as follows:

A *t-norm* is a function $t : [0,1] \times [0,1] \longrightarrow [0,1]$ that satisfies the following four conditions:

i) Boundary condition: $t(0,0) = 0, t(x,1) = t(1,x) = x$;
ii) Commutativity: $t(x,y) = t(y,x)$;
iii) Monotonicity: $t(x,y) \leq t(\xi,\upsilon)$, if $x \leq \xi$ and $y \leq \upsilon$;
iv) Associativity: $t(t(x,y),z) = t(x,t(y,z))$.

Besides the classical min operator proposed by Lotfi Zadeh, other *t-norms* tend to compensate for the strict upper bound of the minimum, such as those:

i) Algebraic product: $t_a : (x,y) \rightarrow xy$;
ii) Bounded product: $t_b : (x,y) \rightarrow \max(0, x+y-1)$;
iii) Drastic product: $t_d : (x,y) \rightarrow \begin{cases} x, \text{ if } y = 1 \\ y, \text{ if } x = 1 \\ 0, \text{ otherwise} \end{cases}$.

A *t-conorm* (or *s-norm*) is a function $s : [0,1] \times [0,1] \longrightarrow [0,1]$ that satisfies the following four conditions:

i) Boundary condition: $s(1,1) = 1, s(x,0) = s(0,x) = x$;
ii) Commutativity: $s(x,y) = s(y,x)$;
iii) Monotonicity: $s(x,y) \leq s(\xi,\upsilon)$, if $x \leq \xi$ and $y \leq \upsilon$;
iv) Associativity: $s(s(x,y),z) = s(x,s(y,z))$.

Besides the above given classical max operator, other *t-conorms* include:

i) Algebraic sum: $s_a : (x,y) \rightarrow x + y - xy$;

ii) Bounded sum: $s_b : (x, y) \to \min(1, x + y)$;

iii) Drastic sum: $s_d : (x, y) \to \begin{cases} x, \text{ if } y = 1 \\ y, \text{ if } x = 1 \\ 0, \text{ otherwise} \end{cases}$.

2.2.3 Properties of Fuzzy Sets

Let A, B, and C be fuzzy sets on the universe of discourse U [60].

i) Properties valid for both fuzzy and crisp sets.
Idempotency law:

$$A \cup A = A, A \cap A = A; \tag{2.10}$$

Commutativity law:

$$A \cup B = B \cup A, A \cap B = B \cap A; \tag{2.11}$$

Associativity law:

$$A \cup (B \cup C) = (A \cup B) \cup C, A \cap (B \cap C) = (A \cap B) \cap C;$$

Distributivity law:

$$A \cup (B \cap C) = (A \cup B) \cap (A \cup C), A \cap (B \cup C) = (A \cap B) \cup (A \cap C);$$

The law of double negation:

$$A = \bar{\bar{A}}; \tag{2.12}$$

De Morgan's law:

$$\overline{A \cup B} = \bar{A} \cap \bar{B}, \overline{A \cap B} = \bar{A} \cup \bar{B}.$$

ii) Properties valid for crisp sets, but in general not for fuzzy sets.
The law of excluded middle:

$$A \cup \bar{A} = U; \tag{2.13}$$

The law of contradiction:

$$A \cap \bar{A} = \emptyset, \tag{2.14}$$

where $\emptyset$ means an empty set.

2.2.4 Some Other Fundamental Concepts of Fuzzy Sets

Here, we introduce some concepts of fuzzy sets, which may not be applicable in classical set theory [61].

i) A fuzzy set F of U is called *normal* if there exists at least one element $x \in U$ such that $\mu_F(x) = 1$. A fuzzy set that is not normal is thus called *subnormal.*
Note that all crisp sets except for the null set are normal.
ii) The *height* of a fuzzy set F is the largest membership degree of any element in F. It is denoted by $height(F)$, hence, $height(F) = \max_x\{\mu_F(x)\}$.
iii) The *support* of a fuzzy set F, denoted by $Supp(F)$, is the crisp set of all points $x \in U$ such that $\mu_F(x) > 0$.
iv) The *center* of a fuzzy set F is the point(s) $x \in U$ at which $\mu_F(x)$ achieves its maximum value.
v) If the support of a fuzzy set F is a single point in U at which $\mu_F = 1$, the F is called a *fuzzy singleton.*
vi) Assume A and B are two fuzzy sets of U. A is said to be a *subset* of (or contained in) B, denoted by $A \subset B$, if $\mu_B(x) \geq \mu_A(x)$ for each $x \in U$.
vii) Assume A and B are two fuzzy sets of U. A and B are said to be *equal,* denoted by $A = B$, if $A \subset B$ and $B \subset A$, or $\mu_A(x) = \mu_B(x)$ for each $x \in U$.
viii) *α-cuts* of the fuzzy set F on the universe of discourse U is defined as:

$$\text{strong } \alpha\text{-cut: } F_\alpha = \{x|\mu_F(x) > \alpha\}, \alpha \in [0,1),$$
$$\text{weak } \alpha\text{-cut: } F_\alpha = \{x|\mu_F(x) \geq \alpha\}, \alpha \in (0,1].$$

Weak α-cuts are also called α-level sets.

2.2.5 Extension Principle

Under many circumstances we can only characterize and deal with numeric information imprecisely. For instance, we use such terms as, *about 3*, *near zero*, and *more or less than 15*. These are examples of the so-called *fuzzy numbers.* In fuzzy set theory, we represent fuzzy numbers as fuzzy sets of real numbers. Fuzzy numbers are used to perform arithmetic operation in intelligent systems, where the extension principle plays a fundamental role in extending any point operations to operations involving fuzzy sets.

Definition 2.4 (Extension Principle). *Let U and V be two universes of discourse, and f be a map from U to V. For a fuzzy set A on U, the extension principle defines a fuzzy set B on V by*

$$\mu_B(v) = \begin{cases} \sup_{v=f(u)} \mu_A(u), & \text{if } f^{-1}(v) \neq \emptyset, \\ 0 & \text{if } f^{-1}(v) = \emptyset. \end{cases} \tag{2.15}$$

When f is a one-to-one map, it is recast as:

$$\mu_B(v) = \mu_A(u). \tag{2.16}$$

2.3 Fuzzy Relations and Their Compositions

Fuzzy relations can be explained as extensions of relations in classical set theory, and the compositions of fuzzy relations allow us to perform fuzzy reasoning, which plays a key role in clustering, pattern recognition, inference, and control. They are also applied in the so-called "soft sciences", such as psychology, medicine, economics and sociology.

2.3.1 Fuzzy Relations

Ambiguous relationships in daily conversations like "x and y are almost equal" and "x is much more beautiful than y" are difficult to be expressed in terms of ordinary relations. Fuzzy relations are what makes it possible to express those frequently used ambiguous relationships.

Definition 2.5 (Fuzzy relation). *Let U and V be two universes of discourse. A fuzzy relation R is a fuzzy set in the product space $U \times V$, which is characterized by a membership function μ_R:*

$$\mu_R : U \times V \to [0, 1]. \tag{2.17}$$

Especially when $U = V$, R is known as a fuzzy relation on U.

As a generalization of fuzzy relations, the *n-ary fuzzy relation* R in $U_1 \times U_2 \times \ldots \times U_n$ is defined by

$$R = \int_{U_1 \times U_2 \times \ldots \times U_n} \mu_R(x_1, x_2, \ldots, x_n)/(x_1, x_2, \ldots, x_n), x_i \in U_i, \tag{2.18}$$

where $\mu_R : U_1 \times U_2 \times \ldots \times U_n \to [0, 1]$.

Note that when $n = 1$, R is an unary fuzzy relation, and a fuzzy set in U_1; when $n = 2$, it is the fuzzy relation defined in (2.17).

The *converse fuzzy relation* of fuzzy relation R, R^{-1}, is defined as

$$\mu_{R^{-1}}(y, x) = \mu_R(x, y). \tag{2.19}$$

The following are the most basic fuzzy relations: For any $x, y \in U$

Identity relation

$$I \Leftrightarrow \mu_I(x, y) = \begin{cases} 1; \, x = y \\ 0; \, x \neq y \end{cases}. \tag{2.20}$$

Zero relation

$$O \Leftrightarrow \mu_O(x, y) = 0. \tag{2.21}$$

Universe relation

$$E \Leftrightarrow \mu_E(x, y) = 1. \tag{2.22}$$

Example 2.6. The following are examples of these three relations.

$$I = \begin{bmatrix} 1\,0\,0 \\ 0\,1\,0 \\ 0\,0\,1 \end{bmatrix}, O = \begin{bmatrix} 0\,0\,0 \\ 0\,0\,0 \\ 0\,0\,0 \end{bmatrix}, E = \begin{bmatrix} 1\,1\,1 \\ 1\,1\,1 \\ 1\,1\,1 \end{bmatrix}.$$

2.3.2 Operations of Fuzzy Relations

Since fuzzy relations are expressed by fuzzy sets on a Cartesian product space, we can apply the same operations on fuzzy relations as on fuzzy sets.

Let R and S be fuzzy relations on $U \times V$. The operations are defined as follows:

Union of fuzzy relations: $R \cup S$

$$\mu_{R \cup S}(x, y) = \mu_R(x, y) \vee \mu_S(x, y). \tag{2.23}$$

Intersection of fuzzy relations: $R \cap S$

$$\mu_{R \cap S}(x, y) = \mu_R(x, y) \wedge \mu_S(x, y). \tag{2.24}$$

Complement of fuzzy relations:

$$\mu_{\bar{R}} = 1 - \mu_R(x, y). \tag{2.25}$$

Inclusion of fuzzy relations: $R \subseteq S$

$$R \subseteq S \Leftrightarrow \mu_R(x, y) \leq \mu_S(x, y), \forall x \in U, \forall y \in V. \tag{2.26}$$

2.3.3 Composition of Fuzzy Relations

Let R and S be fuzzy relations on $U \times V$ and $V \times W$, respectively. The so-called *Sup-Star composition* of R and S, $R \circ S$, is a fuzzy relation on $U \times W$ defined as:

$$R \circ S \Leftrightarrow \mu_{R \circ S}(x, z) = \sup_{y \in V}\{\mu_R(x, y) \star \mu_S(y, z)\}, \tag{2.27}$$

where $x \in U$ and $z \in W$, and $\star$ can be any t-norm operator. It is possible that S is just a fuzzy set in V; in this case, the $\mu_S(y, z)$ in (2.27) becomes $\mu_S(y)$, then $\mu_{R \circ S}(y, z)$ becomes $\mu_{R \circ S}(y)$, and the others remain the same.

2.4 Fuzzy Reasoning

Fuzzy reasoning is sometimes called fuzzy inference or approximate reasoning. It is used in a fuzzy rule to determine the rule outcome from the given rule input information. Fuzzy rules represent control strategy or modeling knowledge/experience. When specific information is assigned to input variables in the rule antecedent, fuzzy inference is needed to calculate the outcome for output variables in the rule consequence.

2.4.1 Generalized Modus Ponens and Modus Tollens

In classical logic, reasoning is based on "modus ponens" and "modus tollens", which are complementary. In modus ponens, when the statement "If A, then B" is true, we infer "If A is true, then B is true". This reasoning can be written as

$$\begin{array}{ll} \text{premise 1:} & A \longrightarrow B \\ \text{premise 2:} & A \\ \hline \text{consequence:} & B \end{array}$$

where A and B are crisp sets, and "$\longrightarrow$" means implication.

It can also be recast in an IF-THEN form as

$$\begin{array}{ll} \text{premise 1:} & \text{If } x \text{ is } A \text{ Then } y \text{ is } B \\ \text{premise 2:} & x \text{ is } A \\ \hline \text{consequence:} & y \text{ is } B. \end{array}$$

On the other hand, in modus tollens, when the statement "If A, then B" is true, we infer "If B is not true, then A is not true". The reasoning process is described as

$$\begin{array}{ll} \text{premise 1:} & A \longrightarrow B \\ \text{premise 2:} & \text{not } B \\ \hline \text{consequence:} & \text{not } A. \end{array}$$

Fuzzy reasoning is, however, based on the *generalized modus ponens* (GMP), which can be written in IF-THEN form as

$$\begin{array}{ll} \text{premise 1:} & \text{If } x \text{ is } A \text{ Then } y \text{ is } B \\ \text{premise 2:} & x \text{ is } A' \\ \hline \text{consequence:} & y \text{ is } B' \end{array}$$

where A, A', B, and B' are fuzzy sets.

It is noted that other than modus ponens based on crisp sets, in the GMP the fuzzy sets A in premise 1 and A' in premise 2 can be different. On the other hand, even though in the GMP the fuzzy sets A in premise 1 and A' in premise 2 do not have to be the precisely the same, we can still infer the conclusion "y is B'" from the premise "y is B" by their similarity. Thus, fuzzy reason is sometime called "approximately reasoning".

Accordingly, the *generalized modus tollens* is described as the following inference procedure:

$$\begin{array}{ll} \text{premise 1:} & \text{If } x \text{ is } A \text{ Then } y \text{ is } B \\ \text{premise 2:} & y \text{ is } B' \\ \hline \text{consequence:} & x \text{ is } A' \end{array}$$

which, when $B' = \text{not } B$ and $A' = \text{not } A$, reduces to modus tollens.

2.4.2 Fuzzy Implications

Definition 2.7 (Fuzzy implications). *Let A and B be fuzzy sets on U and V, respectively. A fuzzy implication, denoted by $A \to B$, is a special kind of fuzzy relation on $U \times V$ satisfying the same conditions known as t-norm, such as:*

Mamdani's method

$$\mu_{A\to B}(x,y) = \mu_A(x) \wedge \mu_B(y)$$

Algebraic product

$$\mu_{A\to B}(x,y) = \mu_A(x) \cdot \mu_B(y)$$

Bounded product

$$\mu_{A\to B}(x,y) = 0 \vee (\mu_A(x) + \mu_B(y) - 1)$$

Drastic product

$$\mu_{A\to B}(x,y) = \begin{cases} \mu_A(x), \ \textit{if } \mu_B(y) = 1 \\ \mu_B(y), \ \textit{if } \mu_A(x) = 1 \\ 0, \qquad \textit{otherwise} \end{cases}$$

Zadeh's method (Lukasiewicz's implication)

$$\mu_{A\to B}(x,y) = 1 \wedge (1 - \mu_A(x) + \mu_B(y))$$

Boolean logic implication

$$\mu_{A\to B}(x,y) = (1 - \mu_A(x)) \vee \mu_B(y)$$

Gödel logic implication

$$\mu_{A\to B}(x,y) = \begin{cases} 1, & \mu_A(x) \le \mu_B(y) \\ \mu_B(y), & \mu_A(x) > \mu_B(y) \end{cases}$$

Goguen's implication

$$\mu_{A\to B}(x,y) = \begin{cases} 1, & \mu_A(x) \neq \mu_B(y) \\ \mu_B(y)/\mu_A(x), & \mu_A(x) > \mu_B(y) \end{cases}$$

2.4.3 Fuzzy Rule Base

Fuzzy reasoning is performed with inference rules, which are expressed in IF-THEN format, called fuzzy IF-THEN rules. A *fuzzy rule base* is composed of a collection of fuzzy IF-THEN rules. The most popular fuzzy rules are Mamdani fuzzy rules and Takagi-Sugeno (TS) fuzzy rules, which are our concern throughout this book.

Mamdani Fuzzy Rules

A general Mamdani fuzzy rule, for either fuzzy control or fuzzy modeling, can be expressed as,

$$R_M^l : \text{IF } x_1 \text{ is } F_1^l \text{ and } \cdots \text{ and } x_n \text{ is } F_n^l, \text{ THEN } y \text{ is } G^l, \tag{2.28}$$

where F_i^l and G^l are fuzzy sets, $x = (x_1, \ldots, x_n)^T \in U$ and $y \in V$ are input and output linguistic variables, respectively, and $l = 1, 2, \ldots, q$. This kind of fuzzy IF-THEN rules provides a convenient framework to incorporate human experts' knowledge.

Example 2.8. A Mamdani fuzzy rule used to describe the air conditioner is:

IF room temperature is "a little high" AND humidity is "quite high"
THEN air conditioner setting is "high",

where "room temperature" and "humidity" are input variables and "air conditioner setting" is an output variable; "a little high", "quite high", and "high" are fuzzy sets.

Takagi-Sugeno (TS) Fuzzy Rules

Instead of using fuzzy sets in the consequence part of (2.28), TS fuzzy rules adopt linear functions, which represent input-output relations.

A TS fuzzy rule is described as,

$$\begin{aligned} R_T^l : &\text{IF } x_1 \text{ is } F_1^l \text{ and } \cdots \text{ and } x_n \text{ is } F_n^l, \\ &\text{THEN } y^l = c_0^l + c_1^l x_1 + \cdots + c_n^l x_n, \end{aligned} \tag{2.29}$$

where F_i^l are fuzzy sets, c_i are real-valued parameters, y^l is the system output, and $l = 1, 2, \ldots, q$. That is, in the TS fuzzy rules, the IF part (premise) is fuzzy but the THEN part (consequence) is crisp – the output is a linear combination of input variables.

Example 2.9. The preceding air conditioner rule can be written as:

IF room temperature x is "about 20°C" AND humidity y is "about 80%"
THEN air conditioner setting $z = 0.2x + 0.05y$.

The equation means that the room temperature has four times weight as that of humidity. In general, it is difficult to determine the linear equations for the consequence part empirically. Therefore, it is assumed that the rules are obtained with modeling techniques by using input-output data.

2.4.4 Fuzzy Inference Engine

In a fuzzy inference engine, the fuzzy IF-THEN rules in the fuzzy rule base are converted to a map from a fuzzy set in $U = U_1 \times \cdots \times U_n$ to fuzzy sets in V. Mamdani fuzzy rules and TS fuzzy rules use different fuzzy inference approaches, which will be introduced in the sequel, respectively.

Mamdani Fuzzy Inference Method

The reasoning method uses fuzzy relations and their composition. A fuzzy IF-THEN rule (2.28) is interpreted as a fuzzy implication $F_1^l \times \cdots \times F_n^l \to G^l$ on $U \times V$. Let a fuzzy set A' on U be the input to a fuzzy inference engine; then each fuzzy IF-THEN rule (2.28) determines a fuzzy set B' on V using the sup-star composition (2.27), which writes

$$\mu_{B'}(y) = \sup_{x \in U}\{\mu_{F_1^l \times \cdots \times F_n^l \to G^l}(x, y) \star \mu_{A'}(x)\}. \tag{2.30}$$

Denote $F_1^l \times \cdots \times F_n^l = A$ and $G^l = B$, the rule (2.28) is therefore denoted by $A \to B$. Thus, the fuzzy implications given in Definition 2.7 can be applied.

Takagi-Sugeno (TS) Inference Method

The Mamdani fuzzy model is a general framework in which linguistic information from human experts is quantified and dealt with. Its main disadvantage is that its inputs and outputs are fuzzy sets, whereas in many engineering systems the inputs and outputs of a system are real-valued variables. To overcome this disadvantage, the TS fuzzy model proposes a solution with real-valued inputs and outputs.

For TS fuzzy model (2.29), fuzzy reasoning is carried out by the weighted mean,

$$y(x) = \frac{\sum_{l=1}^{M} \omega^l y^l}{\sum_{l=1}^{M} \omega^l}, \tag{2.31}$$

where the weight ω^l is the overall truth value of the premise of rule R_T^l for the input, and is given by

$$\omega^l = \prod_{i=1}^{n} \mu_{F_i^l}(x_i), \tag{2.32}$$

where $\mu_{F_i^l}(x_i)$ is the membership degree of the fuzzy set F_i^l.

The advantage of the TS inference method lies in its compact system equation (2.32). Therefore, parameters can be estimated and order can be determined with systematic approaches. On the other hand, its disadvantage is also the no-fuzzy THEN part, which does not provide a natural framework to incorporate fuzzy rules from human experts.

In order to deal with real-valued inputs and outputs for fuzzy systems, it is straightforward to add a fuzzifier to the input and a defuzzifier to the output. The fuzzifier maps crisp points in U to fuzzy sets in U, and the defuzzifier maps fuzzy sets in V to crisp points in V.

2.4.5 Fuzzifier

A fuzzifier performs a map from a crisp point $x = (x_1, \ldots, x_n)^T$ into a fuzzy set A in U [59]. There are (at least) two possible choices of this map:

Singleton fuzzifier: A is a fuzzy singleton with support x, i.e., $\mu_A(x') = 1$ for $x' = x$ and $\mu_A(x') = 0$ for all other $x' \in U$ with $x' \neq x$.

Nonsingleton fuzzifier: $\mu_A(x) = 1$ and $\mu_A(x')$ decreases from 1 as x' moves away from x, for example, $\mu_A(x') = \exp\left[-\frac{(x'-x)^T(x'-x)}{\sigma^2}\right]$, where σ^2 is a parameter characterizing the shape of $\mu_A(x')$.

It seems that only the singleton fuzzifier has been applied in practice. However, the nonsingleton fuzzifier can be very useful in cases where the input data are corrupted by noise.

2.4.6 Defuzzifier

A defuzzifier performs a map from fuzzy sets in V to a crisp point in V. Here, we adopt the most popular centroid defuzzifier (also called center-average defuzzifier), which maps the fuzzy set $A \circ R$ in V to a crisp point,

$$y = \frac{\sum_{l=1}^{M} y_l \mu_{A \circ R}(y_l)}{\sum_{l=1}^{M} \mu_{A \circ R}(y_l)}, \tag{2.33}$$

where y_l is a point in V at which $\mu_{G^l}(y)$ achieves its maximum value, and μ_{B^l} is given by (2.30).

When $A = A_x$ is a fuzzy singleton, i.e., $\mu_{A_x}(x) = 1$ and $\mu_{A_x}(x') = 0$ for $x' \neq x$, (2.33) can be expressed as,

$$y = y(x) = \frac{\sum_{l=1}^{M} y_l \mu_{A_l}(x)}{\sum_{l=1}^{M} \mu_{A_l}(x)}. \tag{2.34}$$

2.5 Fuzzy Control

Based on the descriptions above, a fuzzy control system can thus be constructed as shown in Fig. 2.4. It is composed of the following four elements:

A fuzzy rule base: It contains a fuzzy logic quantification of the expert's linguistic description of how to achieve good control.
A fuzzy inference engine: It emulates the expert's decision-making in interpreting and applying knowledge about how to control the plant.
A fuzzifier: It converts controller inputs into information that the inference mechanism can easily use to activate and apply rules.
A defuzzifier: It converts the conclusions of the inference mechanism into actual outputs for the process.

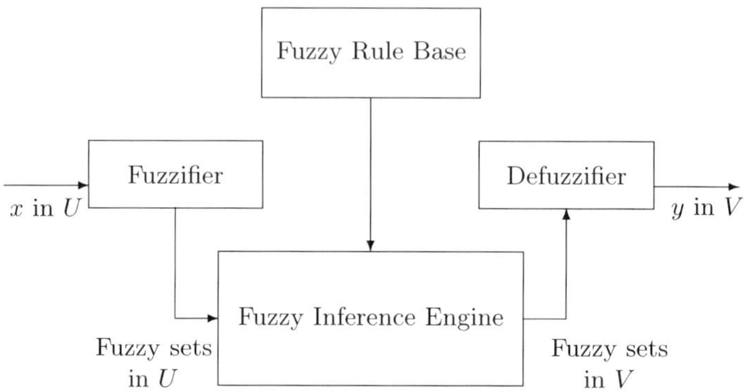

Fig. 2.4. Block diagram of fuzzy logic control systems

In summary, a fuzzy controller incorporates knowledge of human experts in a form of logical inference rules, enabling it to act in a human-like fashion.

2.6 Fuzzy Systems as Universal Approximators

Fuzzy logic systems have very strong functional capabilities, which means that if properly constructed, they can perform very complex operations (e.g., much more complex than those performed by a linear map) [59, 62]. Actually, it is known that fuzzy logic systems possess the universal approximation property.

Suppose that a fuzzy logic system $f(x)$ adopts center-average defuzzification, product for the premise and implication, and Gaussian membership functions. Then, the universal approximation theorem as given below holds.

Theorem 2.10 (The universal approximation theorem [59]). *For any given real continuous function $\psi(x)$ defined on a compact set $U \subset \Re^n$ and an arbitrary $\epsilon > 0$, there exists a fuzzy system $f(x)$ in the form of (2.33) such that*

$$\sup_{x \in U} |f(x) - \psi(x)| < \epsilon. \tag{2.35}$$

It is remarked that this theorem just guarantees that there exists a way to define a fuzzy logic system, which can uniformly approximate any given function to arbitrary accuracy. It does not, however, say how to find such a fuzzy logic system, which is indeed very difficult. Moreover, the more accurate it is, the larger is the number of rules the fuzzy logic system needs. It is also remarked that the theorem is only a justification for using the fuzzy logic systems. But what we need and are concerned about is the capability of fuzzy logic systems to incorporate linguistic information in a natural and systematic way, constituting their unique advantage.

3

Chaos and Chaos Control

For the abstruse and vast nonlinear dynamical and control systems it is difficult, if not impossible, to cover all the concepts within one chapter. In this chapter, through exploring the simplest logistic map, we sketch some basic but important concepts and some related essential ones in the theory of nonlinear dynamical and control systems, as well as review some now popular methodologies of chaos control.

3.1 Logistic Map

We begin with the logistic map to introduce some basic concepts as it possesses bifurcations, stable and unstable periodic orbits, periodic windows, ergodic and mixing behaviors, homoclinic connections, chaotic orbits and some kinds of universality.

The logistic map is denoted as a difference equation in the form of

$$x_{n+1} = f(\mu, x_n) = \mu x_n(1 - x_n), \tag{3.1}$$

which was discovered by Robert May and Mitchell Feigenbaum in 1976 as a population model. It represents the population as a fraction, x_n, of the maximum population that is supported by a habitat. μ represents the population growth ratio of the species. Thus, $x_n \in [0,1]$ and $0 < \mu \leq 4$. It is easy to see that, if $\mu > 4$, the map (3.1) does not map the interval $[0,1]$ into itself [26, 66, 67].

Solving the fixed-point equation

$$x = f(\mu, x), \tag{3.2}$$

we can derive the map's fixed point(s). Now, let us investigate the stability of the fixed points. When $0 < \mu < 1$, $x_1 = 0$ is the unique fixed point ($x_2 = (\mu - 1)/\mu$ is out of $[0,1]$). Because of $|f'(\mu, x_1)| = |f'(\mu, 0)| = \mu < 1$, the fixed point $x_1 = 0$ is stable and is an attracting point, thus, all points

Zhong Li: *Fuzzy Chaotic Systems*, StudFuzz **199**, 31–52 (2006)
www.springerlink.com

map into 0 under the iterations of f no matter where the point starts, see Fig. 3.1 (a). Further, when $\mu \geq 1$, there are two fixed points $x_1 = 0$ and $x_2 = \frac{(\mu - 1)}{\mu}$. Due to $|f'(\mu, 0)| = \mu \geq 1$, the fixed point x_1 becomes unstable just as the second fixed point x_2 is born. The second fixed point x_2 is stable for $1 < \mu < 3$, because $|f'(\mu, x_2)| = |f'(\mu, (\mu - 1)/\mu)| = |1 - \mu| < 1$, then all points converge to x_2, see Fig. 3.1 (b). Thereafter, what happens when both fixed points become unstable for $\mu > 3$?

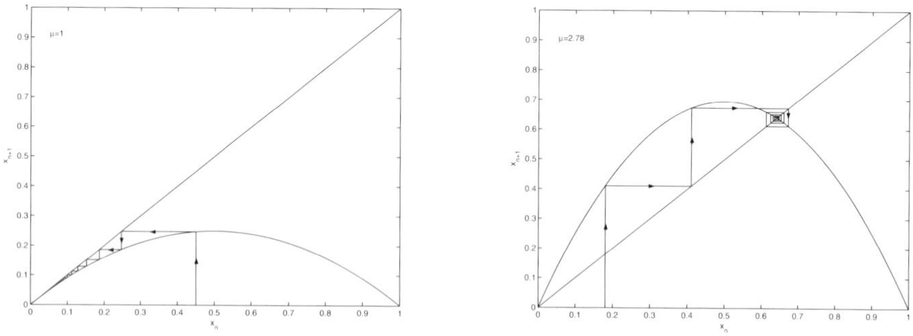

(a) (b)

Fig. 3.1. The logistic map with various parameter values of μ

It is clear that the stability of the fixed points changes as the parameter μ varies. To see this, we plot the attracting set of f as a function of μ with a value $\mu = 2.8$ where the fixed point x_2 is still an attracting (or stable) fixed point. This numerical simulation is shown in Fig. 3.2, which can display some characteristic properties of the asymptotic solution of the logistic map, allowing one to see at a glance, where qualitative changes in the asymptotic solution occur. Such changes are termed as *bifurcation.*

We can see that initially the attracting set consists of a single point that bifurcates into two at $\mu = 3.0$. Subsequently, these points bifurcate again into four points, which bifurcate into eight, and so on. The interval in μ between bifurcations decreases until eventually what looks like a chaotic set appears. The chaotic region appears interspersed with bands, which consist of only a small number of points termed *periodic windows.*

Let us now look closely at the mechanism behind these phenomena. As noted above, the first bifurcation occurs at $\mu = 1$, with which the fixed point $x_1 = 0$ becomes unstable and the fixed point $x_2 = (\mu - 1)/\mu$ is born and becomes stable. The second bifurcation takes place at $\mu = 3$, where the fixed point x_2 becomes unstable and an attracting 2-cycle is born, as shown in Fig. 3.3. Similar to deriving the fixed points of the logistic map, the 2-cycle points x_1^* and x_2^* can be obtained by solving the equations $f(\mu, f(\mu, x)) = x$. This is a fourth order equation:

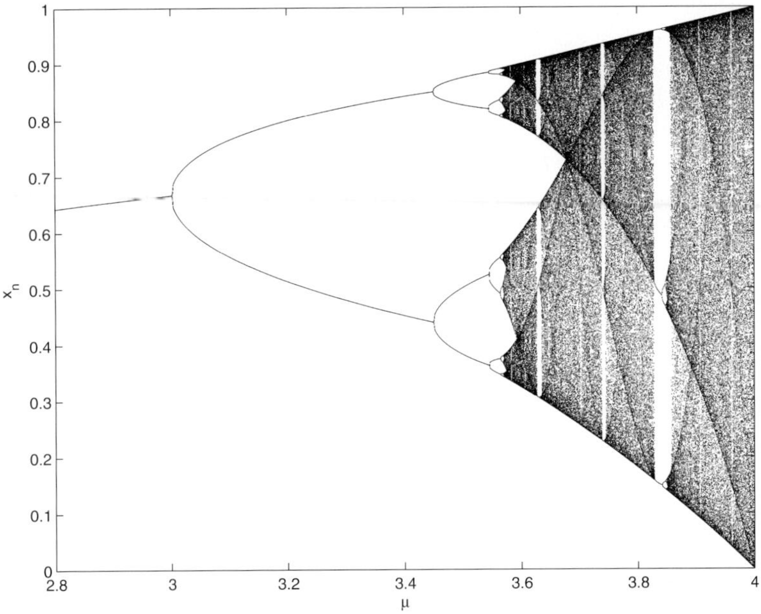

Fig. 3.2. Bifurcation diagram of the logistic map for $2.8 < \mu < 4$

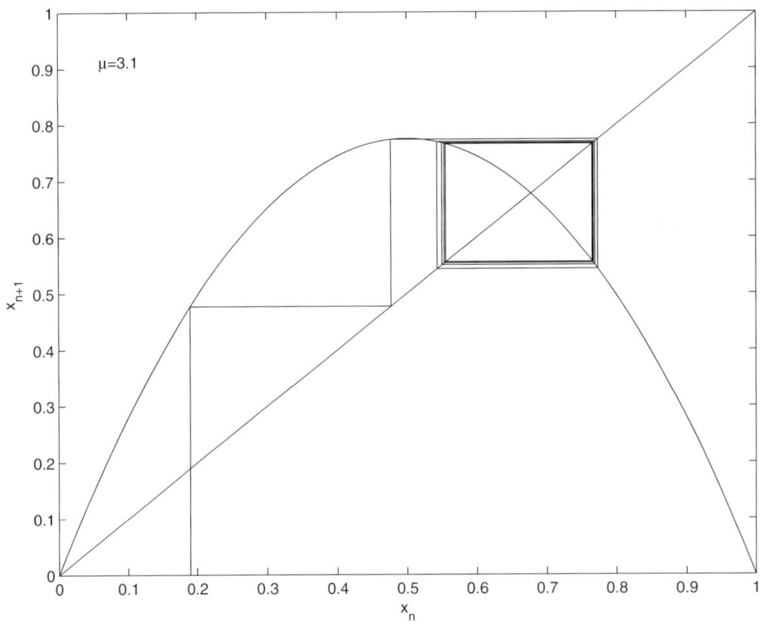

Fig. 3.3. Periodic-2 orbits of the logistic map at $\mu = 3.1$

$$-\mu^3x^4 + 2\mu^3x^3 - (\mu^2+\mu^3)x^2 + (\mu^2-1)x = 0. \tag{3.3}$$

Already knowing the two fixed points of the iteration, (3.3) can be first reduced to a third-order equation with $x_1 = 0$,

$$-\mu^3x^3 + 2\mu^3x^2 - (\mu^2+\mu^3)x + (\mu^2-1) = 0, \tag{3.4}$$

and further dividing (3.4) by $(x-(\mu-1)/\mu)$ yields:

$$-\mu^3x^2 + (\mu^2+\mu^3)x - (\mu^2+\mu) = 0. \tag{3.5}$$

The solutions of (3.5) are

$$x^*_{1,2} = \frac{1+\mu \pm \sqrt{\mu^2-2\mu-3}}{2\mu}, \tag{3.6}$$

which form a 2-cycle for the logistic map f with each a fixed point of $f^2 = f \circ f$.

Let the points $x_1,\ x_2,\ \ldots,\ x_p$ denote the points of a p-cycle of the logistic map, where $f(x_i) = x_{i+1}$ and $f(x_p) = x_1$. Each of the points x_i is a fixed point of f^p. By the chain rule of differentiation, one has

$$\left.\frac{df^p}{dx}\right|_{x_i} = f'(x_1)f'(x_2)\cdots f'(x_p) = \prod_{j=1}^{p} f'(x_j), \text{ for all } i = 1,\ldots,p, \tag{3.7}$$

and thus

$$\left|\left.\frac{df^p}{dx}\right|_{x_i}\right| = \prod_{j=1}^{p} |f'(x_j)|\,, \text{ for all } i = 1,\ldots,p. \tag{3.8}$$

Applying the stability criterion for the case $f^2(\mu, x)$ gives

$$-1 < \mu^2(1-2x_1)(1-2x_2) < 1. \tag{3.9}$$

Substituting the two points $x^*_{1,2}$ given in (3.6) for x_1 and x_2 into (3.9) yields $3 < \mu < 1+\sqrt{6} \approx 3.44949$. At $\mu = 1+\sqrt{6}$, the 2-cycle points lose their stability and bifurcate to a 4-cycle, and so on. The sequence of bifurcations is infinite. It is rather tedious to compute them analytically.

Let μ_n be the parameter value at which a period 2^n-cycle appears, and denote the converged value by μ_∞. Assuming geometric convergence, the difference between the values μ_∞ and μ_n is denoted by

$$\mu_\infty - \mu_n = \frac{c}{\delta^n}, \tag{3.10}$$

where c and $\delta > 1$ are constants. Little algebraic transformation yields

$$\delta = \frac{\mu_n - \mu_{n-1}}{\mu_{n+1} - \mu_n}. \tag{3.11}$$

We can calculate the onsets of bifurcations occurring at $\mu_1 = 3.0$, $\mu_2 = 3.449490\ldots$, $\mu_3 = 3.544090\ldots$, $\mu_4 = 3.564407\ldots$, $\mu_5 = 3.568759\ldots$, $\mu_6 = 3.569692\ldots$, $\mu_7 = 3.569891\ldots$, and $\mu_8 = 3.569934\ldots$, which can roughly be estimated to converge to $\mu_\infty = 3.5699456\ldots$ by (3.10). Using these values of μ, we obtain an approximate value for $\delta = 4.669201609102990\ldots$, which was named as Feigenbaum constant after its discoverer. From (3.10) we can also find that $c = 2.637\ldots$. Note that the Feigenbaum constant is universal, i.e., it appears in any dynamical system that approaches chaotic behavior via *period-doubling bifurcation*, and has a single quadratic maximum.

It is clearly seen that the interval between μ_∞ and 4 in Fig. 3.2 contains a large number of small windows, where the attracting set is a stable periodic cycle. The biggest window corresponds to a stable period 3-cycle. It appears at about $\mu = 3.828427\ldots$ and is stable for a relative large interval of μ. Outside these windows the map looks chaotic [26].

3.2 Bifurcations

As discussed above, a bifurcation, the Latin word for "forking into two", is a qualitative change in the dynamics of a system that occurs when a control parameter is varied. Typically, the bifurcation occurs when an attractor becomes unstable. Bifurcations are classified according to how stability is lost. In this section, we discuss the types of bifurcations such as *saddle node*, *pitchfork* and *transcritical* [63].

We take maps in the form $f : \Re \times \Re \to \Re$, where $(\mu, x) \mapsto f(\mu, x) \in \Re$. The stability of the fixed points is determined by the value of $\partial f/\partial x$. The value of μ for which the fixed point becomes unstable leading to a bifurcation is called a bifurcation value of μ.

Here, we give a few examples to distinguish some typical bifurcations.

Example 3.1 (Transcritical bifurcation). The one-dimensional system

$$\dot{x} = f(\mu, x) = \mu x - x^2 \tag{3.12}$$

has two fixed points $x_1 = 0$ and $x_2 = \mu$. If μ is varied, then there exist two fixed-point curves as shown in Fig. 3.4, where the solid curves refer to stable equilibria and the dashed curves refer to the unstable ones. Since the Jacobian, J, of the one-dimensional system is simply $J = \partial f/\partial x|_{x_1} = \mu$, we see that for $\mu < \mu_0 = 0$, the fixed point $x_1 = 0$ is stable, but for $\mu > \mu_0 = 0$ it changes to be unstable. Thus, we call $(x_1, \mu_0) = (0, 0)$ a bifurcation point. Similarly, (x_2, μ_0) is also a bifurcation point. Owing to the nature of the bifurcation point $(0, 0)$ in the $x - \mu$ plane, it is called a *transcritical bifurcation.*

Example 3.2 (Saddle-node (fold, tangent) bifurcation). The one-dimensional system

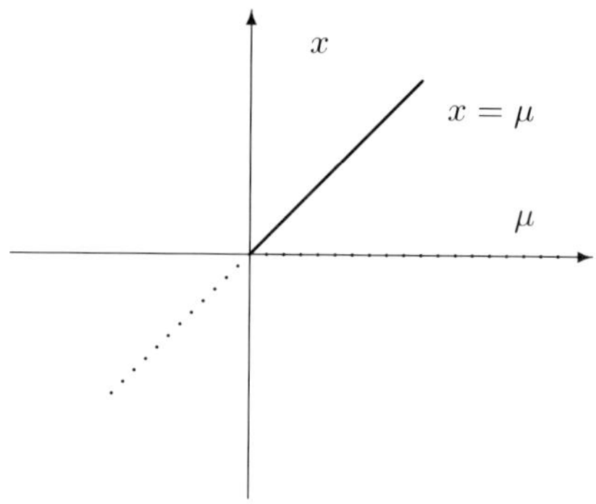

Fig. 3.4. A transcritical bifurcation

$$\dot{x} = f(\mu, x) = \mu - x^2 \tag{3.13}$$

has a fixed point $x_1 = 0$ at $\mu_0 = 0$, and fixed points $x_2 = \pm\sqrt{\mu}$ at $\mu > 0$, where $x_2 = \sqrt{\mu}$ is stable and $x_2^* = -\sqrt{\mu}$ is unstable for $\mu > 0$. Hence, by the nature of the bifurcation point $(x, \mu) = (0, 0)$ in the $x - \mu$ plane, as shown in Fig. 3.5, it is called a *saddle-node bifurcation.*

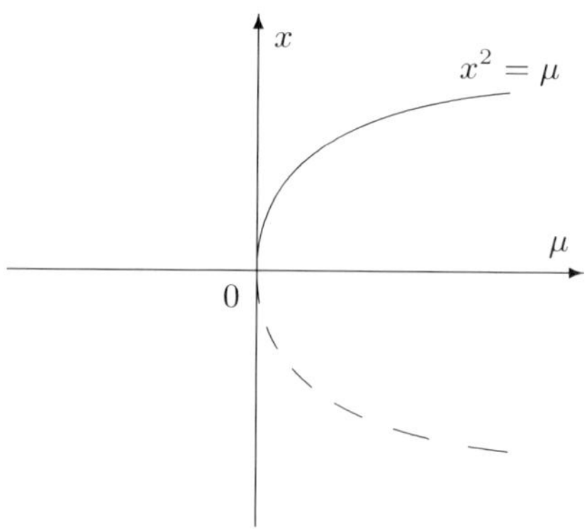

Fig. 3.5. A saddle-node bifurcation

Example 3.3 (Pitchfork bifurcation). The one-dimensional system

$$\dot{x} = f(\mu, x) = \mu x - x^3 \tag{3.14}$$

has a fixed point $x_1 = 0$ at $\mu_0 = 0$, and fixed point $x_2 = \pm\sqrt{\mu}$ at $\mu > 0$. Since $x_0 = 0$ is unstable for $\mu > \mu_0 = 0$ and stable for $\mu < \mu_0 = 0$, and $x_2 = \pm\sqrt{\mu}$ is stable for $\mu > 0$ due to the fact that the Jacobian $J = -2\mu$, this case is called a *pitchfork bifurcation*, as shown in Fig. 3.6.

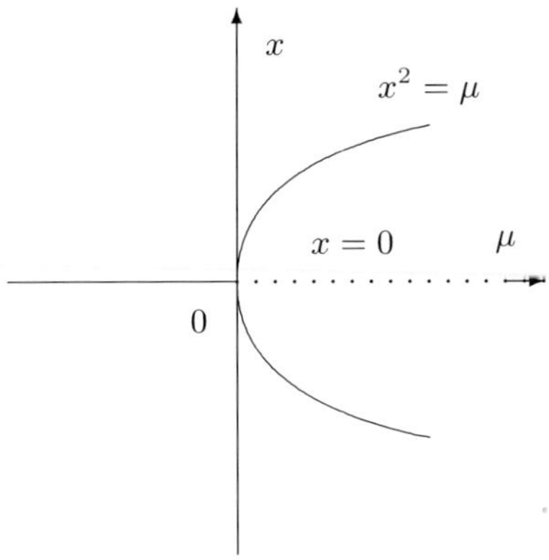

Fig. 3.6. A pitchfork bifurcation

3.3 Hopf Bifurcation of Higher-dimensional Systems

In higher-dimensional systems or maps, the situation is much more complicated than that for one-dimensional parametrized nonlinear maps. Here, an additional bifurcation phenomenon occurs – the *Hopf bifurcation* [63].

A Hopf bifurcation occurs as the parameter μ is varied to pass a critical value μ_0, and the system Jacobian has one pair of complex conjugate eigenvalues moving from the left-half plane to the right, crossing the imaginary axis, while all other eigenvalues remain stable. At the moment of crossing, the real parts of the two eigenvalues become zero, and the stability of the existing equilibrium changes from stable to unstable.

Theorem 3.4 (Poincaré-Andronov-Hopf [64]).
Suppose that the two-dimensional system,

$$\begin{cases} \dot{x} = f(x, y; \mu) \\ \dot{y} = g(x, y; \mu), \end{cases} \tag{3.15}$$

has a zero equilibrium, $(x^*, y^*) = (0, 0)$, and that its associate Jacobian has a pair of purely imaginary eigenvalues, $\lambda(\mu)$ and $\bar{\lambda}(\mu)$. If

$$\left. \frac{dR\{\lambda(\mu)\}}{d\mu} \right|_{\mu=\mu_0} > 0,$$

for some μ_0, *then*

i) $\mu = \mu_0$ *is a bifurcation point of the system;*
ii) for close enough values $\mu < \mu_0$, *the zero equilibrium is asymptotically stable;*
iii) for close enough values $\mu > \mu_0$, *the zero equilibrium is unstable;*
iv) for close enough values $\mu \neq \mu_0$, *the zero equilibrium is surrounded by a limit cycle of magnitude* $O(\sqrt{|\mu - \mu_0|})$.

For the discrete-time setting, consider a two-diementional parametrized system:

$$\begin{cases} x_{k+1} = f(x_k, y_k; \mu) \\ y_{k+1} = f(x_k, y_k; \mu), \end{cases} \tag{3.16}$$

with a real variable parameter $\mu \in \Re$ and an equilibrium point (x^*, y^*), satisfying $x^* = f(x^*, y^*; \mu)$ and $y^* = g(x^*, y^*; \mu)$ simultaneously for all μ in a neighborhood of $\mu^* \in \Re$. Let $J(\mu)$ be its Jacobian at this equilibrium, and $\lambda_{1,2}(\mu)$ be its eigenvalues, with $\lambda_2(\mu) = \bar{\lambda}_1(\mu)$. If

$$|\lambda_1(\mu^*)| = 1, \qquad \left.\frac{\partial|\lambda_1(\mu)|}{\partial\mu}\right|_{\mu=\mu^*} > 0, \tag{3.17}$$

the system undergoes a *Hopf bifurcation* at (x^*, y^*, μ^*), in a way analogous to the continuous-time setting [65, 163].

3.4 Lyapunov Exponents

A qualitative discussion has been given above. It is known that a hallmark of chaos is the extreme sensitivity of the system dynamics to initial conditions (or parameters), in the sense that two trajectories starting close to one another in phase space will move exponentially away from each other. Other hallmarks of chaos include the existence of a dense set of unstable periodic orbits in its regime [66], positive Lyapunov exponents or finite Kolmogorov-Sinai entropy [68], continuous power spectrum [68, 69], ergodicity [68, 70], mixing [70], as well as some other limiting properties.

Among all the features of chaos, positive Lyapunov exponent is the easiest one to verify, which is used to quantitatively measure chaotic behavior. The quantitative test for chaotic behavior can sometimes distinguish it from noisy behavior due to random, external influences. With the quantitative measure of the degree of chaoticity, we can see how chaos changes as the parameters are varied.

Starting from two close initial values x_0 and y_0, we have

$$x_k = f(\mu, x_{k-1}) = \cdots = f^k(\mu, x_0) \text{ and } y_k = f^k(\mu, y_0).$$

If x_k and y_k are separated exponentially fast in the iterations, then

$$|y_k - x_k| = |y_0 - x_0|e^{k\lambda} \quad (\lambda > 0)$$

and

$$\frac{1}{k}\ln|y_k - x_k| \to \lambda \text{ as } k \to \infty.$$

In the case that the motion is within a bounded region, this exponential separation phenomenon can not occur unless the two initial values are close enough to each other. Thus, we let $|y_0 - x_0| \to 0$ before we take the limit $k \to \infty$, so as to define the constant

$$\begin{aligned}
\lambda &= \lim_{k\to\infty}\frac{1}{k}\lim_{|y_0-x_0|\to 0}\ln\left|\frac{y_k - x_k}{y_0 - x_0}\right| \\
&= \lim_{k\to\infty}\frac{1}{k}\lim_{|y_0-x_0|\to 0}\ln\left|\frac{f^k(\mu, y_0) - f^k(\mu, x_0)}{y_0 - x_0}\right| \\
&= \lim_{k\to\infty}\frac{1}{k}\ln\left|\frac{df^k(\mu, x_0)}{dx_0}\right| \\
&= \lim_{k\to\infty}\frac{1}{k}\sum_{i=0}^{k-1}\ln\left|\frac{df(\mu, x_i)}{dx_i}\right|, \qquad (3.18)
\end{aligned}$$

which is called the Lyapunov exponent of the trajectory $x_k = f^k(x_0)$, $k = 0, 1, \ldots$.

In the case of the logistic map, we have

$$\begin{aligned}
\lambda(x_0) &= \lim_{k\to\infty}\frac{1}{k}\sum_{i=0}^{k-1}\ln|\mu - 2\mu x_i| \\
&= \ln\mu + \lim_{k\to\infty}\frac{1}{k}\sum_{i=0}^{k-1}\ln|1 - 2x_i|. \qquad (3.19)
\end{aligned}$$

Especially, note that since $x = 0$ is a fixed point, we have $\lambda(0) = \ln\mu$ (and also $\lambda(1) = \ln\mu$).

Figure 3.7 shows the Lyapunov exponent for the logistic map as a function of μ. The Lyapunov exponent is independent on the initial value x_0. We can see that the values of μ for which λ becomes negative in Fig. 3.7 correspond to the regions of periodic behavior evident in Fig. 3.2.

3.5 Routes to Chaos

The manner of bifurcation introduced in Section 3.1 is called *period-doubling bifurcation*, which states that as μ increases, a sequence of period-doubling bifurcations occurs. The period-doubling solution appears in the bifurcation

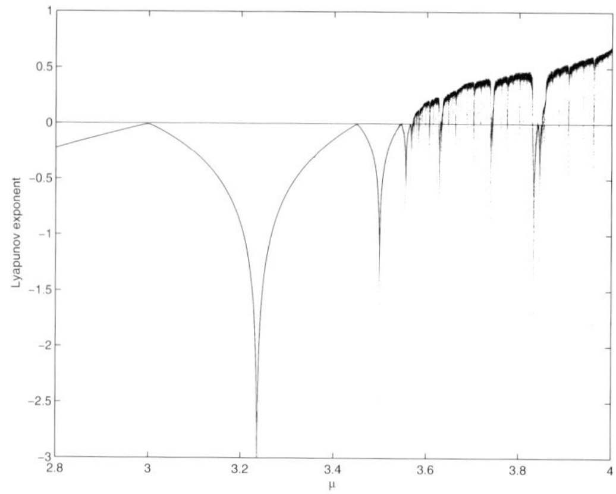

Fig. 3.7. Lyapunov exponent for the logistic map as a function of μ

diagram not as a curve but as a region – the *(periodic) windows.* The sequence of parameter values, $\{\mu_k\}$, converges to a number, μ_∞. At this point the orbit becomes aperiodic. Beyond this point, both chaotic orbits and odd-periodic limit cycles start to emerge. At $\mu = 4$ the motion is formally ergodic on the unit interval $(0, 1)$ [25]. However, the period-doubling bifurcation is not the only way that systems can become chaotic. Another frequently occurring route is called *intermittency.*

Intermittency states that the system behaves in such a way that the regular and stable periodic oscillations in relation to the observed variable are interrupted abruptly by "bursts" after a parameter is slightly varied. As the parameter is further varied, the bursts occur more frequently and the duration of the regular oscillations decreases.

The logistic map also readily displays an intermittent transition to chaos. Figure 3.8 shows the iteration sequences for μ in the neighborhood of $\mu_c = 1 + 2\sqrt{2} = 3.828\ldots$, a value for the onset of a large 3-cycle band in Fig. 3.2. As μ deviates from μ_c, the logistic map's iterates make the transition from a stable 3-cycle to chaos. For a value of μ nearby μ_c, where the logistic map exhibits intermittency as shown in the lower part of Fig. 3.8, we plot every third point in an iteration of the logistic map in Fig. 3.9 [26]. The stretches, in the order of 100 iterations long, where every third point comes back to the same value, show the laminar phases in this map. The chaotic bursts are the regions that connect these regular phases. It is clear in Fig. 3.9 that not all laminar regions contain the same number of iterations.

3.6 Center Manifold Theory

There are some successful techniques available for simplifying, without any significant loss of important information, the representation of the solution

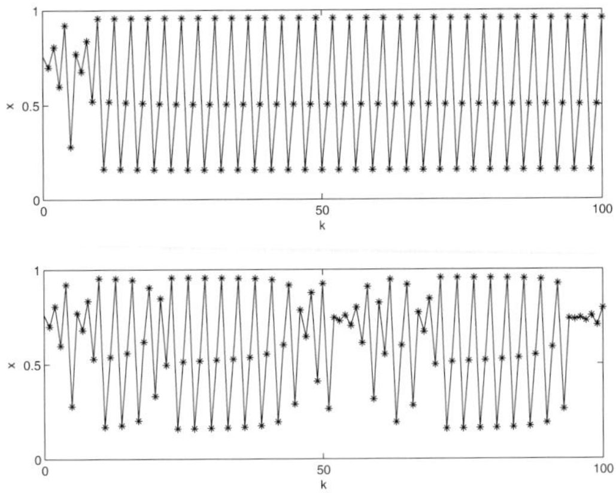

Fig. 3.8. Transition to intermittency in the logistic map. The upper sequences is for $\mu = 3.8304$ and the lower one is for $\mu = 3.8264$

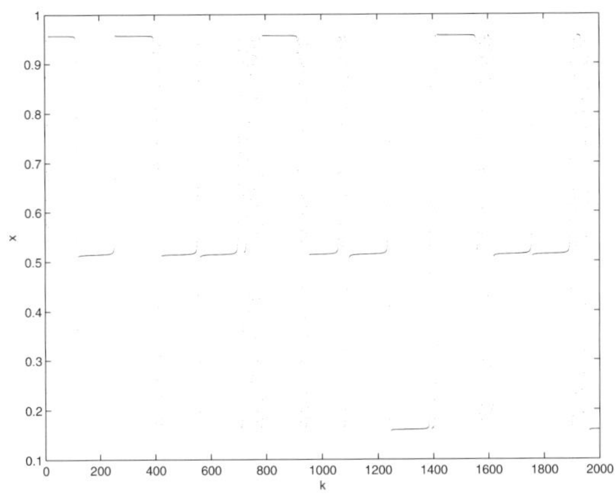

Fig. 3.9. Laminar phases in the logistic map where every third iterate is plotted for $\mu = 3.828422$

trajectories of a nonlinear dynamical system in a neighborhood of a nonhyperbolic equilibrium. The *center manifold theory* of differential geometry is one of such useful tools.

To introduce this technique, which generally works for higher-dimensional systems, we denote (μ_0, x^*) as an equilibrium (need not be hyperbolic) of the nonlinear dynamical system

$$\dot{x} = f(u, x). \tag{3.20}$$

If the linearization $Df(\mu, x^*)$ has no eigenvalue with zero real part, x^* is said to be a hyperbolic equilibrium, whereas if there exists an eigenvalue with zero real part, the equilibrium point is nonhyperbolic.

Let E_s, E_u and E_c be the corresponding *generalized eigenspaces* of the system's Jacobian, defined by the real part of the eigenvalue of the Jacobian, $\lambda = \lambda(\mu_0)$:

$$R(\lambda) \begin{cases} < 0 \ \text{ defines } E_s, \\ = 0 \ \text{ defines } E_c, \\ > 0 \ \text{ defines } E_u. \end{cases} \tag{3.21}$$

Then, there exist a *stable manifold*, M_s, an *unstable manifold*, M_u, and a *center manifold*, M_c, that are tangential to E_s, E_u and E_c, respectively, at the equilibrium (μ_0, x^*) [71, 72]. Here, the important role played by the center manifold is that the asymptotic behavior of the the nonlinear system's overall dynamics is preserved by those trajectories on the center manifold, locally in the neighborhood of the equilibrium.

The reduction of the system dynamics to those reproduced by the center manifold is the main subject of the theory. To characterize the trajectories of the reduced dynamics on the center manifold, one usually reformulates the system in a simpler form. Without loss of generality, we assume that the unstable manifold is empty, so as to simplify the notation. In this case, we have

$$\begin{cases} \dot{x_c} = A_c(\mu)x_c + f_c(\mu, x_c, x_s), \\ \dot{x_s} = A_s(\mu)x_s + f_s(\mu, x_c, x_s), \end{cases} \tag{3.22}$$

where $x_c \in \Re^{n_c}$ and $x_s \in \Re^{n_s}$, with $n_c + n_s = n$. In this form, which is simpler than the system (3.20), the constant matrix A_c has n_c eignevalues with zero real parts, and A_s has n_s eigenvalues with negative real parts. Moreover, the nonlinear functions f_c and f_s are C^2-smooth, vanishing simultaneously with their first derivatives at the equilibrium x^*. In system (3.22), x_c corresponds to the center manifold and x_s the stable manifold. By the *existence theorem* [71], there is a center manifold on which

$$x_s = h(x_c). \tag{3.23}$$

This can be obtained (often numerically, according to the *approximation theorem* [71]) from the second equation of system (3.22). By substituting it into the first equation of (3.22), we obtain

$$\dot{x_c} = A_c(\mu)x_c + f_c(\mu, x_c, h(x_c)), \tag{3.24}$$

which is sometimes called the *center manifold equation* of the nonlinear system (3.20). The *equivalence theorem* [71] states that the asymptotic behavior of equation (3.24), as $t \to \infty$, is topologically the same as that of system (3.20), in a neighborhood of the equilibrium x^*. It turns out that the reduced dynamics, at least in terms of dimension, of the center manifold equation can be much more easily understood and calculated, particularly for bifurcation analysis. For this reason, equation (3.24) is also called the *bifurcation equation* for the nonlinear system (3.20).

Essentially, the center manifold theory consists of the aforementioned three theorems: the existence, approximation, and equivalence theorems.

3.7 Normal Forms

Poincaré's theory of *normal forms* provides an alternative approach to local dynamics analysis, especially bifurcation analysis [72]. This technique, similar to the center manifold method, reduces a given nonlinear system to the simplest possible form that preserves the dynamics in a neighborhood of an equilibrium. The normal form technique can also be combined with the center manifold theory to reduce the local dynamics of the system.

For simplicity, we let the given system be in the form of

$$\dot{x} = Ax + f(x), \tag{3.25}$$

where A is a constant matrix in the Jordan canonical form and f a smooth nonlinear function. Let the eigenvalues of A be $\{\lambda_1, \ldots, \lambda_n\}$. These eigenvalues are said to be *resonant* if there is at least one index $i : 1 \leq i \leq n$ such that $\lambda_i = \sum_{j=1}^{n} m_j \lambda_j$ with nonnegative integer coefficients $\{m_j : j = 1, \ldots, n\}$ satisfying $\sum_{j=1}^{n} m_j \geq 2$. Without getting into details, we just state the following theorem, which characterize the normal form method [73].

Theorem 3.5 (Poincaré). *If the constant matrix A in system (3.25) has resonant eigenvalues, then the system can always be reduced to the linear system*

$$\dot{y} = Ay, \tag{3.26}$$

by a series of suitable transformations of variables that yields $x \rightarrow y$.

3.8 Control of Chaos

For many years, the main feature of chaos, i.e., the extreme sensitivity to initial conditions, made chaos undesirable, and most experimentalists consider such characteristics as something to be strongly avoided [74, 75]. In addition to this feature, chaotic systems have two other important ones. First, there are infinite many unstable periodic orbits embedded in the underlying chaotic attractor, and second, the dynamics in the chaotic attractor is ergodic, which implies that during its temporal evolution the system ergodically visits any small neighborhood of every point in each of the unstable periodic orbits embedded within the chaotic attractor.

Owing to these properties, a chaotic system can be seen as shadowing some periodic behavior at a given time, and erratically jumping from one to another periodic orbit. Thus, when a trajectory approaches ergodically a desired periodic orbit embedded in the chaotic attractor, one can apply small perturbations to stabilize such an orbit. Therefore, we can say that the extreme sensitivity of a chaotic system to changes in its initial conditions may be very desirable in practical experimental situations [75]. It suffices to note that, due to chaos, using the same chaotic system one is able to produce infinite many desired dynamical behaviors (either periodic or not periodic) only with properly chosen tiny perturbations. This property is not shared by a nonchaotic system, because the perturbations needed therein for producing a desired behavior must, in general, be of the same order of magnitude as the unperturbed dynamical variables.

Generally, chaos control approaches can be divided into two broad categories: feedback and nonfeedback (or say, open-loop) control approaches. Feedback control methods do not change the controlled systems and stabilize unstable periodic orbits embedded in chaotic attractors, while nonfeedback control methods slightly change the controlled system, mainly by a small tuning of control parameter, changing the system behavior from chaotic attractor to periodic orbit which is close to the initial attractor.

It is known that the nonfeedback approach is much less flexible, and requires more prior knowledge of motion. To apply such an approach, one does not have to follow the trajectory. The control can be activated at any time, and one can switch from one periodic orbit to another without returning into the chaotic behavior. This approach can be very useful in mechanical systems, where the feedback control systems are often very large (sometimes larger than the system controlled). The extremely simple, easily implementable, low-cost, and reliable nonfeedback approaches are widely applied in many physical experiments and industrial processes today, particularly for nonlinear dynamical systems, as a unified feedback control approach has not been fully established for general nonlinear dynamical systems. Roughly speaking, the nonfeedback approaches include the entrainment and migration control method [85, 90, 86, 87, 88, 89], control through external forcing [91, 92, 93, 74, 94, 95, 96], while the feedback approaches include the Ott-Grebogi-Yorke (OGY) method, engineering feedback control method, and Pyragas's time-delayed feedback control method [97, 98, 99]. In addition, chaos control approaches also include conventional linear and nonlinear control, adaptive control, neural networks-based control, fuzzy control (to be discussed in detail in this book), and another very important topic, synchronization of chaos. In the following, we just introduce a few typical chaos control methods.

3.8.1 Ott-Grebogi-Yorke Method

Ott, Grebogi and Yorke have developed a method, named OGY method thereafter, by which chaos can always be suppressed by shadowing one of the infinitely many unstable periodic orbits (or perhaps steady states) embedded in the chaotic attractor [78, 79, 81, 82, 83, 84]. By using the OGY method, one first needs to determine some of the unstable low-period periodic orbits that are embedded in the chaotic attractor, then examine the location and the stability of these orbits and choose one having the desired system performance. Finally, one applies small control to stabilize this desired periodic orbit.

There are some basic assumptions for using the OGY method [74].

i) The dynamics of the system can be described by an n-dimensional map of the form

$$\zeta_{n+1} = f(\zeta_n, p). \tag{3.27}$$

This map, in the case of continuous-time systems, can be constructed by, e.g., introducing a transversal surface of section for system trajectories (Poincaré map);

ii) p is some accessible system parameter, which can be changed within some small neighborhood of its nominal value p^*;

iii) For this value p^* there is a periodic orbit within the attractor around which we would like to stabilize the system;

iv) The position of this orbit changes smoothly as p varies, and there are small changes in the local system behavior for small variations of p.

Let ζ_F be a chosen fixed point of the system's map f existing for the parameter value p^*. In the close vicinity of this fixed point we can assume with good accuracy that the dynamics are linear and can approximately be expressed by

$$\zeta_{n+1} - \zeta_F = M(\zeta_n - \zeta_F). \tag{3.28}$$

The matrix M can be calculated using the measured chaotic time series and analyzing its behavior in the neighborhood of the fixed point. Further, the eigenvalues λ_s, λ_u and eigenvectors e_s, e_u of this matrix can be found. These eigenvectors determine the stable and unstable directions in the small neighborhood of the fixed point.

Denoting by f_s, f_u the contravariant eigenvectors ($f_s e_s = f_u e_u = 1$, $f_s e_u = f_u e_s = 0$), we can find the linear approximation valid for small $|p_n - p^*|$,

$$\zeta_{n+1} = p_n g + (\lambda_n e_n f_n + \lambda_s e_s f_s)(\zeta_n - p_n g), \tag{3.29}$$

where

$$g = \left.\frac{\partial \zeta_F(p)}{\partial p}\right|_{p+p^*}.$$

Since ζ_{n+1} should fall on the stable manifold of ζ_F, choose p_n such that $f_u \zeta_{n+1} = 0$:

$$p_n = \frac{\lambda_n \zeta_n f_u}{(\lambda_u - 1) g f_u}. \tag{3.30}$$

Diagrammatically, the OGY method is shown in Fig. 3.10. It has the following main properties.

i) No dynamical model is required. One can use either full information from the process or a delay coordinate embedding technique from single variable experimental time series;

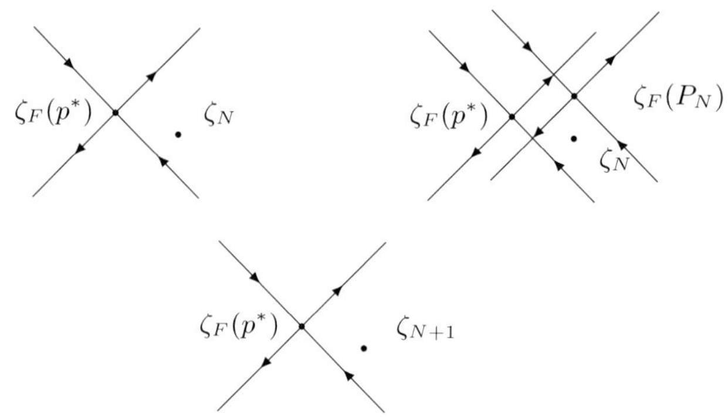

Fig. 3.10. The Ott-Grebogi-Yorke (OGY) method

ii) Any accessible variable system parameter can be used as the control parameter;
iii) In the absence of noise and error, the amplitude of the applied control signal must be large enough to achieve control;
iv) Unavoidable noise can destabilize the controlled orbit, resulting in occasional chaotic bursts;
v) Before settling into the neighborhood of the desired periodic orbit, the trajectory exhibits chaotic transients, the lengths of which depend on the actual starting point.

Here, the logistic map is used as an example to illustrate the effect of the OGY method.

Example 3.6. Recall the logistic map

$$x_{n+1} = f(\mu, x_n) = \mu x_n(1 - x_n), \tag{3.31}$$

where $x \in [0, 1]$, and μ is the control parameter.

It was already shown above that this map exhibits chaos via the period-doubling bifurcation route. For $0 < \mu < 1$, the asymptotic state of the map is $x = 0$; for $1 < \mu < 3$, the attractor is a nonzero fixed point $x = 1 - 1/\mu$; for $3 < \mu < 1+\sqrt{6}$, this fixed point is unstable and the attractor is a stable period-2 orbit. As μ is further varied, a sequence of period-doubling bifurcations occurs, and the period-doubling cascade accumulates at $\mu = \mu_\infty \approx 3.5699456$, after which chaos can arise.

Consider the case $\mu = 3.8$ for which the system is apparently chaotic. It was already discussed that there exist infinite many unstable periodic orbits embedded within a chaotic attractor. For instance, there are a fixed point $x_F \approx 0.7368$ and a period-2 orbit with components $x_1 \approx 0.3737$ and $x_2 \approx 0.8894$, where $x_1 = f(x_2)$ and $x_2 = f(x_1)$.

Now suppose that we want to avoid chaos at $\mu = 3.8$. Particularly, we hope to drive trajectories resulting from a randomly chosen initial condition x_0 to be as close as possible to the period-2 orbit, assuming that this period-2 orbit gives the desired system performance. Of course, the desired asymptotic state can be any of the infinite many unstable periodic orbits. Further suppose that the parameter μ can be finely tuned within a small range around $\mu_0 = 3.8$, i.e., μ can vary in the range $[\mu_0 - \delta, \mu_0 + \delta]$, where $\delta \ll 1$. Due to the ergodicity of the chaotic attractor, a trajectory starting from any initial value x_0 will fall, with probability one, into the neighborhood of the desired period-2 orbit some time later. The trajectory would then diverge quickly from the period-2 orbit in the absence of control. The OGY method is to tune the control parameter so that the trajectory stays in the neighborhood of the period-2 orbit as long as the control is present. In general, the small parameter perturbations are time-dependent.

Denote a target periodic orbit to be controlled by $x(i)$, $i = 1, \ldots, m$, where $x(i+1) = f(x(i))$ and $x(m+1) = x(1)$. Assume that at time n the trajectory falls into the neighborhood of component i of the period-m orbit. The logistic map can be linearized in the neighborhood of the component $(i+1)$ as

$$\begin{aligned} x_{n+1} - x(i+1) &= \frac{\partial f}{\partial x}(x_n - x(i)) + \frac{\partial f}{\partial \mu}\Delta\mu_n \\ &= \mu_0(1 - 2x(i))(x_n - x(i)) + x(i)(1 - x(i))\Delta\mu_n, \end{aligned} \tag{3.32}$$

where the partial derivatives are evaluated at $x = x(i)$ and $\mu = \mu_0$. We want x_{n+1} to stay in the neighborhood of $x(i+1)$. Thus, we let $x_{n+1} - x(i+1) = 0$, which gives

$$\Delta\mu_n = \mu_0 \frac{(2x(i) - 1)(x_n - x(i))}{x(i)(1 - x(i))}. \tag{3.33}$$

Eq. (3.33) holds only when the trajectory x_n enters a small neighborhood of the period-m orbit, i.e., when $|x_n - x(i)| < \varepsilon$ and, hence, the required parameter perturbation $\Delta\mu_n$ is small. Generally, the required maximum parameter perturbation δ is proportional to ε. Since ε can be arbitrarily small, δ can also be made arbitrarily small. It is remarked that the average transient time before a trajectory enters the neighborhood of the target periodic orbit depends on ε (or δ). When the trajectory is outside the neighborhood of the target periodic orbit, we do not apply any parameter perturbation, so the system evolves at its nominal parameter value μ_0. Hence, we set $\Delta\mu_n = 0$ when $\Delta\mu_n > \delta$. Note that the parameter perturbation $\Delta\mu_n$ depends on x_n and is time-dependent.

The OGY method is very flexible to stabilize different periodic orbits at different times. To switch from stabilizing a period-2 orbit to, say, stabilizing the fixed point in order to achieve a better system performance at a later time, we just need to turn off the parameter control with respect to the period-2 orbit. Thus, without control, the trajectory will diverge from the period-2 orbit exponentially, and resume to evolve with the parameter value μ_0. Later, the trajectory enters a small neighborhood of the fixed point. At this time, we turn on a new set of parameter perturbations in terms of the fixed point. The trajectory can then be stabilized around the fixed point.

It can be seen that a major advantage of the OGY method is that it can be applied to experimental systems in which *a priori* knowledge of the system is usually not available. A time series found by measuring one of the system's dynamical variables using the time delay embedded method [80], which transforms a scalar time series into a trajectory in phase space, is sufficient to determine the desired unstable periodic orbits to be controlled, and the relevant quantities required to compute the parameter perturbations. Another advantage of the OGY method is its flexibility in choosing the desired periodic orbit to be controlled. However, the disadvantage is also obvious, which is that small amounts of external noise may cause occasional large departures from the desired trajectory.

3.8.2 Feedback Control

Feedback control is the basic mechanism by which systems, whether mechanical, electrical, or biological, maintain their equilibrium or homeostasis. Feedback control may be defined as the use of difference signals, determined by comparing the actual values of system variables to their desired values, as a means of controlling a system. Classic feedback controllers have been designed for nonchaotic systems. Can the classic feedback control methods be applied to chaotic systems? The answer is positive. A fundamental reason lies in that chaotic systems with complex nonlinear dynamical behaviors and extreme sensitivity to initial conditions are deterministic by their very nature, and follow a set of deterministic rules that can be used in designing feedback controllers for different purposes [213].

To introduce the basic idea for feedback control of chaos, consider an n-dimensional controlled system,

$$\dot{x}(t) = f(x, u, t), \qquad x(t_0) = x_0, \tag{3.34}$$

where $x(t)$ is the system state vector, f a continuous or smooth nonlinear vector-valued function satisfying some necessary conditions, such as the well-posedness of the controlled system and uniqueness of the system trajectory under any admissible control $u(t)$, for any initial position x_0 in a region of interest, $t_0 \le t < \infty$.

Assume that the system is chaotic and has an unstable periodic orbit (*e.g.*, equilibrium) $\bar{x}$ of period $\tau > 0$ in the phase space, satisfying $\bar{x}(t+\tau) = \bar{x}(t)$ for all $t_0 \le t < \infty$. The task is to design a feedback controller, employing the target trajectory $\bar{x}$, of the following form,

$$u(t) = g(x, \bar{x}, t), \tag{3.35}$$

such that the following tracking control goal is achieved:

$$\lim_{t\to\infty} \|x(t) - \bar{x}(t)\| = 0. \tag{3.36}$$

Since the target periodic orbit, $\bar{x}$, is itself a solution of the original system ($u(t) = 0$), one has

$$\dot{\bar{x}} = f(\bar{x}, 0, t). \tag{3.37}$$

Substracting (3.37) from (3.34) yields the error dynamics

$$\dot{e}_x = f_e(e_x, \bar{x}, t), \tag{3.38}$$

where

$$\begin{cases} e_x(t) = x(t) - \bar{x}(t), \\ f_e(e_x, \bar{x}, t) = f(x, g(x, \bar{x}(t), t), t) - f(\bar{x}, 0, t). \end{cases}$$

It is noted that system (3.38) should only contain the dynamics of e_x but not x, considering that $x = e_x + \bar{x}$ and $\bar{x}$ is acceptable since it is merely a given time-function rather that a state variable in the error dynamics.

Thus, the design problem is to determine the controller, $u(t)$, such that

$$\lim_{t\to\infty} \|e_x(t)\| = 0. \tag{3.39}$$

It is clear from (3.38) and (3.39) that if zero is an equilibrium of the nonlinear system (3.38), then the original control problem is converted to the asymptotic stability problem for this equilibrium. In such a case, Lyapunov function methods can often be directly applied to obtain rigorous mathematical techniques for the controller design.

Throughout this book, the Lyapunov function methods are often utilized, therefore, we briefly introduce the Lyapunov methods here.

The Lyapunov methods are categorized as the first (or indirect) method of Lyapunov and the second (or direct) method of Lyapunov. The first method is applicable mainly to autonomous system and the second, to both autonomous and nonautonomous systems.

The first method of Lyapunov is also known as the *Jacobian* or *local linearization method.* It is to determine the asymptotic stability of an autonomous system, based on the fact that the stability of such a system, in a neighborhood of an equilibrium, is essentially the same at its linearized model operating at the same equilibrium point. While, the second method of Lyapunov originated from the concept of energy decay (energy dissipation) of a stable mechanical or electrical system.

A general nonautonomous system is represented as

$$\dot{x} = f(x, t), \tag{3.40}$$

where $f : \mathcal{S} \times [0, \infty) \to R^n$ is continuously differentiable in $\mathcal{S} \subset R^n$, a neighborhood of the origin, with a give initial state $x(0) = x_0 \in \mathcal{S}$. By a simple transform, it can be assumed, without loss of generality, that the origin, $x = 0$, is one of the system equilibria of interest.

The equilibrium, $x = 0$, of system (3.40) is called to be *asymptotically stable* if there exists a constant, $\delta = \delta(t_0) > 0$, such that

$$\|x(t_0)\| < \delta \Rightarrow \|x(t)\| \to 0 \qquad \text{as } t \to \infty. \tag{3.41}$$

This asymptotic stability is *uniform*, if the existing constant δ is independent of t_0 over $[0, \infty)$; it is *global*, if the convergence $\|x\| \to 0$ is independent of the starting point $x(t_0)$ over the entire domain $\mathcal{S} \subseteq R^n$ on which the system is defined.

An autonomous system is described as

$$\dot{x} = f(x), \qquad x(0) = x_0, \tag{3.42}$$

where $f : \mathcal{S} \to R^n$ is continuously differentiable in a neighborhood of the origin, i.e., $\mathcal{S} \subset R^n$.

In the following, the first method of Lyapunov for continuous-time autonomous systems and the second method of Lyapunov for continuous-time autonomous (and nonautonomous) systems are simply given without detailed explanation.

Theorem 3.7 (First method of Lyapunov for continuous autonomous systems). *In system* (3.42), *let*

$$J_0 = \left.\frac{\partial f}{\partial x}\right|_{x=0}$$

be the Jacobian of the system at the equilibrium $x = 0$. Then

i) $x = 0$ is asymptotically stable if all the eigenvalues of J_0 have a negative real part;

ii) $x = 0$ is unstable if at least one of the eigenvalues of J_0 has a positive real part.

Theorem 3.8 (Second method of Lyapunov for continuous nonautonomous systems). *Let $x = 0$ be an equilibrium of the nonautonomous system* (3.40), *and let*

$$\mathcal{K} \equiv \{g(t) | g(t) \text{ is continuous and nondecreasing on } [t_0, \infty) \text{ and } g(t_0) = 0\}.$$

The zero equilibrium of the system is globally, uniformly, and asymptotically stable, if there exist a scalar-valued function, $V(x, t)$, defined on $\mathcal{S} \times [t_0, \infty)$, and three functions $\alpha(\cdot), \beta(\cdot), \gamma(\cdot) \in \mathcal{K}$, such that

(i) $V(0, t_0) = 0$;
(ii) $V(x, t) > 0$ *for all* $x \neq 0$ *in* $\mathcal{S}$ *and all* $t \geq t_0$;
(iii) $\alpha(\|x\|) \leq V(x, t) \leq \beta(\|x\|)$ *for all* $t \geq t_0$;
(iv) $\dot{V}(x, t) \leq -\gamma(\|x\|) < 0$ *for all* $t \geq t_0$.

Theorem 3.9 (Second method of Lyapunov for continuous autonomous systems). *Let* $x = 0$ *be an equilibrium of the autonomous system (3.42). The zero equilibrium of the system is globally and asymptotically stable, if there exist a scalar-valued function,* $V(x)$, *defined on* $\mathcal{S}$, *such that*

(i) $V(0) = 0$;
(ii) $V(x) > 0$ *for all* $x \neq 0$ *in* $\mathcal{S}$;
(iii) $\dot{V}(x) < 0$ *for all* $x \neq 0$ *in* $\mathcal{S}$.

Stability criteria for discrete-time systems can be similarly established [213]. To this end, it is emphasized that the above stability conditions are only sufficient but not necessary for asymptotic stability. On the other hand, the choice of a Lyapunov function is not unique. It is possible for one choice of a Lyapunov function to yield a less conservative result than another one.

3.8.3 Pyragas Time-delayed Feedback Control

The time-delayed feedback control method was proposed by Lithuanian physicist K. Pyragas in 1992 [97, 98, 99] to stabilize an unstable τ-periodic orbit of a nonlinear control system,

$$\dot{x} = f(x, u), \tag{3.43}$$

by a simple feedback law,

$$u(t) = K(x(t) - x(t - \tau)), \tag{3.44}$$

where K is the transmission coefficient and τ is the time of delay. If τ is equal to the period of the existing periodic solution $\bar{x}(t)$ of (3.43) for $u = 0$, and also equal to the period of the solution $x(t)$ of the closed-loop system (3.43), (3.44) begins on the orbit $\Gamma = \{\bar{x}(t)\}$, then it remains in Γ for all $t \geq 0$. With this control, $x(t)$ can converge to Γ even if $x(0) \notin \Gamma$.

The feedback control law (3.44) is also used to stabilize the periodically exited process in system (3.43) with T-periodic solution. Then, τ must be taken equal to T.

The extreme sensitivity to the parameter, especially to the delay time τ, is a disadvantage of the control law (3.44). Obviously, if the system is T-periodic and the aim of control is to stabilize the forced T-periodic solution, then one must choose $\tau = T$.

This control method has been applied to stabilize the coherent modes of lasers [100, 101, 102], magnetoelastic system [103, 104], control of the heart conductivity model [105], step oscillations [106], traffic control [107, 108], or voltage transformers with pulse-width modulation [109], to name just a few among many others [76].

3.8.4 Entrainment and Migration Control

Since the entrainment and migration control method is not so popular, we just introduce the idea of this chaos control method in the following without going into the details.

The concept of the entrainment-goal method is based on the existence of certain convergence regions in the phase space of multi-attractor systems [85, 90, 86, 87, 88, 89]. In each of the convergence regions, all nearby orbits converge locally towards each other. An important observation is that, although many of the attractors have positive Lyapunov exponents, they nevertheless have such regions in their basins of attraction, where nearby orbits generally converge, implying that the Lyapunov exponents are merely an average measure of their dynamics.

The purpose of migration-goal control is to transfer system dynamics from one convergence region to another. There are many reasons for this kind of state transfer. For example, among the many coexisting attractors of a complex system, some may have very different types of dynamics, thus have many other complicated dynamical responses to the inputs. Multi-attractor phenomena can be found in fluid dynamics, thermal processes, nonlinear optics, adaptive control processes, neural networks, and biological systems.

4

Definition of Chaos in Metric Spaces of Fuzzy Sets

This chapter reviews the development of the definitions of chaos from the well-known Li-Yorke definition of chaos for difference equations in $\Re^1$ to those for difference equations in $\Re^n$ with either a snap-back repeller or saddle point as well as for maps in Banach spaces and complete metric spaces, among which Devaney's definition will be used in this book in the proof that chaos exists in anti-controlled fuzzy systems. Finally, a definition of chaos for maps in a space of fuzzy sets, namely, the metric space (ξ^n, D) of fuzzy sets on the base space $\Re^n$, is given, aiming to lay a theoretical foundation for further studies on the interactions between fuzzy logic and chaos theory. Some illustrative examples are presented.

4.1 Introduction

Although a unified, universally accepted, and rigorous mathematical definition of chaos is still not available in the scientific literature, various alternative, but closely related definitions of chaos have been proposed, among which those of Li-Yorke and Devaney seem to be the most popular ones. It was Li and Yorke who first introduced formally the term *chaos* into mathematics in 1975 [19], where they established a simple criterion for chaos in one-dimensional difference equations, i.e., the well-known "period three implies chaos" [129].

Consider a one-dimensional discrete dynamical system [19, 128]:

$$x_{k+1} = f(x_k), \quad k = 0, 1, 2, \ldots, \tag{4.1}$$

where $x_k \in J$ (an interval) and $f : J \to J$ is a continuous map. For $x \in J$, $f^0(x)$ denotes x while $f^{n+1}(x)$ denotes $f(f^n(x))$ for $n = 0, 1, 2, \ldots$. A point x^* is called a *period point with period* n (or an n*-period point*) if $x^* \in J$ and $x^* = f^n(x^*)$ but $x^* \neq f^k(x^*)$ for $1 \leq k < n$; if $n = 1$, then $x^* = f(x^*)$ is called a *fixed point*. A point x^* is said to be *periodic* or is called a *periodic point* if it is an n-periodic point for some $n \geq 1$. With this terminology, Li and

Zhong Li: *Fuzzy Chaotic Systems*, StudFuzz **199**, 53–72 (2006)
www.springerlink.com

Yorke introduced the first mathematical definition of chaos and established a very simple criterion, viz., "period three implies chaos" for its existence. This criterion, which plays a key role in predicting and analyzing one-dimensional chaotic dynamic systems, was described by Li and Yorke as follows:

Theorem 4.1. (Li-Yorke Theorem)
Let J be an interval and $f : J \to J$ be continuous. Assume that there is one point $a \in J$, for which the points $b = f(a)$, $c = f^2(a)$ and $d = f^3(a)$ satisfy

$$d \leq a < b < c \text{ (or } d \geq a > b > c).$$

Then:

(i) For every $k = 1, 2, \ldots$, there is a k-periodic point in J.
(ii) There is an uncountable set $S \subset J$, containing no periodic points, which satisfies the following conditions:
a) For every $p_s, q_s \in S$ with $p_s \neq q_s$,

$$\lim_{n\to\infty} \sup |f^n(p_s) - f^n(q_s)| > 0$$

and

$$\lim_{n\to\infty} \inf |f^n(p_s) - f^n(q_s)| = 0.$$

b) For every $p_s \in S$ and periodic points $q_{per} \in J$, with $p_s \neq q_{per}$,

$$\lim_{n\to\infty} \sup |f^n(p_s) - f^n(q_{per})| > 0.$$

The set S in part (a) of conclusion (ii) was called a *scrambled set* by Li and Yorke.

The first part of the Li-Yorke theorem is, in fact, a special case of Sharkovsky's theorem [114], which was proved by the Ukrainian mathematician A.N. Sharkovsky in 1964. It is, however, the second part of the Li-Yorke theorem that thoroughly unveils the nature and characteristics of chaos, specifically, the sensitive dependence on initial conditions and the resulting unpredictable nature of the long-term behavior of the dynamics.

In 1978, F.R. Marotto generalized the Li-Yorke theorem to higher-dimensional discrete dynamical systems [130]. He proved that if a difference equation in $\Re^n$ has a snap-back repeller, then it has a scrambled set similar to that defined in the Li-Yorke theorem and, thus, exhibits chaotic behavior.

Consider the following n-dimensional system:

$$x_{k+1} = f(x_k), \quad k = 0, 1, 2, \ldots, \tag{4.2}$$

where $x_k \in \Re^n$ and $f : \Re^n \to \Re^n$ is a continuous map, which is usually nonlinear. Denote by $B_r(x)$ the closed ball in $\Re^n$ of radius r centered at point x, and by $B_r^0(x)$ its interior. Also, let $\|x\|$ be the usual Euclidean norm of x in $\Re^n$. Then, assuming f to be differentiable in $B_r(x)$, Marotto claimed that the logical relationship $A \Rightarrow B$ ($\Rightarrow$ means "implying") holds, where :

A: *All eigenvalues of the Jacobian* $Df(x^*)$ *of system (4.2) at the fixed point* $x^* = f(x^*)$ *are greater that* 1 *in norm.*
B: *There exist some* $s > 1$ *and* $r > 0$ *such that* $\|f(x) - f(y)\| > s\|x - y\|$ *for all* $x, y \in B_r(x^*)$.

In other words, if A is satisfied, then B also holds, i.e., f is expanding in $B_r(x^*)$. Then, Marotto introduced the following concepts.

Definition 4.2. (Marotto Definitions)

1. *Expanding fixed point: Let* f *be differentiable in* $B_r(x^*)$. *The point* $x^* \in \Re^n$ *is an expanding fixed point of* f *in* $B_r(x^*)$ *if* $f(x^*) = x^*$ *and all eigenvalues of* $Df(x)$ *exceed* 1 *in norm for all* $x \in B_r(x^*)$.
2. *Snap-back repeller: Assume that* x^* *is an expanding fixed point of* f *in* $B_r(x^*)$ *for some* $r > 0$. *Then* x^* *is said to be a snap-back repeller of* f *if there exists a point* $x_0 \in B_r(x^*)$ *with* $x_0 \neq x^*$, $f^M(x_0) = x^*$ *and the determinant* $|Df^M(x_0)| \neq 0$ *for some positive integer* M.

Marotto showed that the presence of a snap-back repeller is a sufficient criterion for the existence of chaos [130].

Theorem 4.3. (Marotto Theorem)
If f *possesses a snap-back repeller, then system (4.2) is chaotic in the following generalized sense of Li-Yorke:*

(i) There is a positive integer N *such that for each integer* $p \geq N$ f *has a point of period* p;
(ii) There is a scrambled set of f, *i.e., an uncountable set* S *containing no periodic points of* f, *such that*
a) $f(S) \subset S$,
b) for every $x_s, y_s \in S$ *with* $x_s \neq y_s$,

$$\lim_{k \to \infty} \sup \|f^k(x_s) - f^k(y_s)\| > 0,$$

c) for every $x_s \in S$ *and any periodic point* y_{per} *of* f,

$$\lim_{k \to \infty} \sup \|f^k(x_s) - f^k(y_{per})\| > 0;$$

(iii) There is an uncountable subset S_0 *of* S *such that for every* $x_0, y_0 \in S_0$:

$$\lim_{k \to \infty} \inf \|f^k(x_0) - f^k(y_0)\| = 0.$$

It is apparent that the existence of a snap-back repeller for the one-dimensional map f is equivalent to the existence of a point of period-3 for the map f^n for some positive integer n, see [130].

Unfortunately, two counterexamples given in [128, 112] have shown that $A \Rightarrow B$ is not necessarily true. Since the Marotto theorem is based on the concept of "snap-back repeller", which was introduced from the assertion of $A \Rightarrow B$, there exists an error in the proof given by Marotto. Recently, an improved and corrected version of Marotto's theorem was given by Li and Chen [128], where the essential meanings of the two concepts of an expanding fixed point and a snap-back repeller of continuously differentiable maps in $\Re^n$ are clearly explained. For an earlier generalization of the Marroto theorem see [136], and for an extension to maps in metric spaces see [134] as well as below.

More generally, Devaney [113] calls a continuous map $f : X \to X$ in a metric space (X, d) *chaotic on* X, if

(i) *f is transitive on X, i.e., for any pair of nonempty open sets $U, V \subset X$, there exists an integer $k > 0$ such that $f^k(U) \cap V$ is nonempty;*
(ii) *the periodic points of f are dense in X;*
(iii) *f has sensitive dependence on initial conditions, i.e., if there exists a $\delta > 0$ such that for any $x \in X$ and for any neighborhood D of x, then there exists a $y \in D$ and an $k \geq 1$ such that $d(f^k(x), f^k(y)) > \delta$.*

It has been observed that conditions (i) and (ii) in this definition imply condition (iii) if X is not a finite set [110], and that condition (i) implies conditions (ii) and (iii) if X is an interval [22]. Hence, condition (iii) is in fact redundant in the above definition. Recently, Huang and Ye have shown that chaos in the sense of Devaney implies chaos in the sense of Li-Yorke [120].

For continuous-time nonlinear autonomous systems it is much more difficult to give a mathematically rigorous proof for the existence of chaos. Even one of the classic icons of modern nonlinear dynamics, the Lorenz attractor, now known for 40 years, was not proved rigorously to be chaotic until 1999. Warwick Tucker of the University of Uppsala showed, using normal form theory and careful computer simulations, in his doctoral dissertation [140, 141] that the Lorenz equations do indeed possess a robust chaotic attractor. A commonly agreed analytic criterion for proving the existence of chaos in continuous-time systems is based on the fundamental work of Shil'nikov, known as the Shil'nikov method or Shil'nikov criterion [137], whose role is in some sense equivalent to that of the Li-Yorke definition in the discrete setting. The Shil'nikov criterion guarantees that complex dynamics will occur near homoclinicity or heteroclinicity when an inequality (Shil'nikov inequality) is satisfied between the eigenvalues of the linearized flow around the saddle point(s), i.e., if the real eigenvalue is larger in modulus than the real part of the complex eigenvalue. Complex behavior always occurs when the saddle set is a limit cycle. More recently, Zhou *et al.* proposed a powerful tool, i.e., the undetermined coefficient method, to precisely express the homoclinic and heteroclinic orbits in the Chen system [210], which is topologically different and more complicated than the Lorenz system, in a series expansion form and proved uniform convergence [142]. Thus, one is convinced that the Chen system is indeed chaotic, with Smale horseshoes and the horseshoe type of chaos [72]. Further, this method was applied to prove the existence of chaos in the generalized Lorenz canonical form of dynamical systems [143, 144]. As compared with Tucker's method, this method is more straightforward and convenient.

Here, we only focus on discrete-time systems and discuss generalizations of Marotto's work, which are applicable to finite-dimensional difference equations with saddle points as well as to those with repellers, and to maps in Banach spaces and in complete metric spaces including maps from a metric space of fuzzy sets into itself.

4.2 Chaos of Difference Equations in $\Re^n$ with a Saddle Point

The Marotto theorem states that a difference equations in $\Re^n$ with a snap-back repeller behaves chaotically and is, thus, a generalization of the Li-Yorke theorem for difference equations in $\Re^1$. It differs in that the maps defining the difference equations are required to be continuously differentiable rather than only continuous. Thus, the inverse map theorem and the Brouwer fixed point theorem can be used to prove the existence of continuous inverse functions and periodic points. Marotto's result is, however, only applicable to difference equations with repellers, but not to those with saddle points. Therefore, it cannot be used for difference equations involving the horseshoe maps of Smale [138] or the twisted-horseshoe maps of Guckenheimer, Oster and Ipaktchi [118].

In this section, sufficient conditions for the chaotic behavior of difference equations in $\Re^n$ are given, which are applicable to difference equations with saddle points as well as to those with repellers. These conditions are valid for difference equations defined in terms of continuous maps, and in the special case of a difference equation with a snap-back repeller, they are more easily tested than those given by Marotto [130]. The proof is a modification of that used by Marotto, but there are two important differences. First, the maps in the difference equations are assumed to be continuous rather than continuously differentiable. The existence of continuous inverse maps follows from the fact that continuous one-to-one maps have continuous inverses on compact sets. Using this result rather than the inverse map theorem considerably simplifies the proof. Second, the Brouwer fixed point theorem is used on a homeomorph of an l-ball for some $1 \leq l \leq n$ rather than on a homeomorph of an n-ball as in Marotto's proof. Thus, this allows saddle points to be considered as well as repellers.

4.2.1 Sufficient Conditions for Chaos in $\Re^n$

In [123] a first order difference equation

$$x_{k+1} = f(x_k), \tag{4.3}$$

where $f : \Re^n \to \Re^n$ is a continuous map, was said to be *chaotic* if it is chaotic in the sense of the Marotto theorem, i.e., if there exist

(i) *a positive integer N such that (4.3) has a periodic point of period p for each $p \geq N$;*
(ii) *a scrambled set of (4.3) that is an uncountable set S containing no periodic points of (4.3) such that:*
  (a) *$f(S) \subset S$,*
  (b) *for every $x_s, y_s \in S$ with $x_s \neq y_s$*

$$\limsup_{x\to\infty} \|f^k(x_s) - f^k(y_s)\| > 0,$$

  (c) *for every $x_s \in S$ and any periodic point y_{per} of (4.3)*

$$\limsup_{x\to\infty} \|f^k(x_s) - f^k(y_{per})\| > 0;$$

(iii) *an uncountable subset S_0 of S such that for every $x_0, y_0 \in S_0$:*

$$\liminf_{x\to\infty} \|f^k(x_0) - f^k(y_0)\| = 0.$$

An l-ball is defined as a closed ball of finite radius in $\Re^l$ in terms of the Euclidean distance on $\Re^l$. Such a ball of radius r centered around a point $x \in \Re^l$ is denoted by $B^l_r(x)$. A map $f : \Re^n \to \Re^n$ is called *expanding* on a set $A \subset \Re^n$ if there exists a constant $\lambda > 1$ such that

$$\lambda\|x - y\| \leq \|f(x) - f(y)\| \tag{4.4}$$

for all $x, y \in A$. Note that such a map is one-to-one on A.

The following two lemmata will be used in the proof of the theorem below. The proof of the first one is straightforward and is thus omitted, while a proof of the second lemma can be found in [115].

Lemma 4.4. *Let $f : \Re^n \to \Re^n$ be a continuous map, which is one-to-one on a compact subset $K \subset \Re^n$. Then, there exists a continuous map $g : f(K) \to K$ such that $g(f(x)) = x$ for all $x \in K$.*

The map g in Lemma 4.4 is a continuous inverse of map f on the compact set K. It is denoted by f_K^{-1} in the sequel.

Lemma 4.5. *Let $f : \Re^n \to \Re^n$ be a continuous map and let $\{K_i\}_{i=0}^{\infty}$ be a sequence of compact sets in $\Re^n$ such that $K_{i+1} \subseteq f(K_i)$ for $i = 0, 1, 2, \ldots$. Then, there exists a nonempty compact set $K \subseteq K_0$ such that $f^i(x_0) \in K_i$ for all $x_0 \in K$ and all $i \geq 0$.*

The principle result in this section is the following generalization of the Marroto theorem due to Kloeden [123].

Theorem 4.6. (Kloeden) *Let $f : \Re^n \to \Re^n$ be a continuous map and suppose that there exist nonempty compact sets A and B, and integers $1 \leq l \leq n$ and $n_1, n_2 \geq 1$ such that:*

(a) A is homeomorphic to an l-ball;
(b) $A \subseteq f(A)$;
(c) f is expanding on A;
(d) $B \subseteq A$;
(e) $f^{n_1}(B) \cap A = \emptyset$;
(f) $A \subseteq f^{n_1+n_2}(B)$;
(g) $f^{n_1+n_2}$ is one-to-one on B.

Then difference equation (4.3) defined in terms of the map f is chaotic in the sense of the Marotto theorem.

Proof: The proof is similar to that used by Marotto in [130], except that Lemma 4.4 is used instead of the inverse map theorem, and that the Brouwer fixed point theorem is used on homeomorphisms of l-balls rather than n-balls. The Brouwer fixed point theorem says that a smooth map from an n-dimensional closed ball into itself must have a fixed point.

From the continuity of f and assumption (f) there exists a nonempty, compact subset $C \subseteq B$ such that $A = f^{n_1+n_2}(C)$. By (g), $f^{n_1+n_2}$ is one-to-one on C, and by Lemma 4.4, there exists a continuous function $g : A \to C$ such that $g(f^{n_1+n_2}(x)) = x$ for all $x \in C$. Note that $f^{n_1}(C) \cap A = \emptyset$ by (e).

Now f is one-to-one on A by (c), so by Lemma 4.4, f has a continuous inverse $f_A^{-1} : f(A) \to A$. By (b), $C \subset A \subseteq f(A)$, so $f_A^{-k}(C) \subset A$ holds for all $k \geq 0$.

For each $k \geq 0$ the map $f_A^{-k} \circ g : A \to A$ is a continuous map from a homeomorph of an l-ball into itself. So by the Brouwer fixed point theorem there exists a point $y_k \in A$ such that $f_A^{-k}(g(y_k)) = y_k$. In fact, $y_k \in f^{-k}(C)$ and so $f^{n_1+k}(y_k) = f^{n_1+k}(f_A^{-k}(g(y_k))) = f^{n_1}(g(y_k)) \in f^{n_1}(C)$ as $g(y_k) \in C$. Hence, $f^{n_1+n_k}(y_k) \notin A$ as $f^{n_1}(C) \cap A = \emptyset$. Also, $f^{n_1+n_2+k}(y_k) = f^{n_1+n_2}(g(y_k)) = y_k$.

Now for $k \geq n_1+n_2$ the point y_k is a periodic point of period $p = n_1+n_2+k$. To see this, note that p cannot be less than or equal to k, because $f^j(y_k) \in f_A^{-k+j}(C) \subset A$ for $1 \leq j \leq k$ and, then, the whole cycle would belong to A in contradiction to the fact that $f^{n_1+k}(y_k) \notin A$. Also, p cannot lie between k and n_1+n_2+k when $k \geq n_1+n_2$, because $f^{n_1+n_2+k}(y_k) = y_k$ and so p would have to divide n_1+n_2+k exactly, which is impossible when $k \geq n_1+n_2$. Hence, the difference equation (4.3) has a periodic point of period p for each $p \geq N = 2(n_1+n_2)$.

Write $D = f^{n_1}(C)$ and $h = f^N$. Then, $A \cap D = \emptyset$ and

$$h(D) = f^N(D) = f^{2n_1+n_2}(f^{n_2}(D)) = f^{2n_1+n_2}(A) \supseteq A \tag{4.5}$$

in view of (b) and the definition of C. Also

$$h(A) = f^N(A) \supseteq A \tag{4.6}$$

by (b) and

$$h(A) = f^N(A) \supseteq f^{2(n_1+n_2)}(f_A^{-n_1-2n_2}(C)) = f^{n_1}(C) = D \tag{4.7}$$

as $f_A^{-n_1-2n_2}(C) \subset A$. Moreover, as A and D are nonempty, disjoint compact sets it follows that

$$\inf\{\|x-y\|; x \in A, y \in D\} > 0. \tag{4.8}$$

The existence of a scrambled set S then follows exactly as in Marotto's proof [130] or in Li and Yorke [19]. It will be briefly outlined here for completeness.

Let $\mathcal{E}$ be the set of sequences $\xi = \{E_k\}_{k=1}^{\infty}$ where E_k is either A or D, and $E_{k+1} = E_{k+2} = A$ if $E_k = D$. Let $r(\xi, k)$ be the number of sets E_j equal to D for $1 \leq j \leq k$, and for each $\eta \in (0,1)$ choose $\xi^\eta = \{E_k^\eta\}_{k=1}^{\infty}$ to be a sequence in $\mathcal{E}$ satisfying

$$\lim_{k\to\infty} \frac{r(\xi^\eta, k^2)}{k} = \eta.$$

Let $\mathcal{F} = \{\xi^\eta \, ; \, \eta \in (0,1)\} \subset \mathcal{E}$. Then, $\mathcal{F}$ is uncountable. Also from (4.5)–(4.7) $h(E_k^\eta) \supseteq E_{k+1}^\eta$, and so by Lemma 4.5 for each $\xi^\eta \in \mathcal{F}$ there is a point $x_\eta \in A \cup D$ with $h^k(x_\eta) \in E_k^\eta$ for all $k \geq 1$. Let $S_h = \{h^k(x_\eta) \, ; \, k \geq 0 \text{ and } \xi^\eta \in \mathcal{F}\}$. Then, $h(S_h) \subset S_h$, S_h contains no periodic points of h, and there exists an infinite number of k's such that $h^k(x) \in A$ and $h^k(y) \in D$ for any $x, y \in S_h$ with $x \neq y$. Hence, from (4.8) for any $x, y \in S_h$ with $x \neq y$

$$L_1 = \limsup_{k\to\infty} \|h^k(x) - h^k(y)\| > 0.$$

Thus, letting $S = \{f^k(x); x \in S_h \text{ and } k \geq 0\}$ it follows that $f(S) \subset S$, S contains no periodic points of f, and for any $x, y \in S$ with $x \neq y$

$$\limsup_{k\to\infty} \|f^k(x) - f^k(y)\| \geq L_1 > 0.$$

This proves that the set S has properties (ii)a and (ii)b of a scrambled set. The remaining property (ii)c can be proven similarly. See Li and Yorke for further details [19].

Now it remains to establish the existence of an uncountable subset S_0 of the scrambled set S with the properties listed in part (iii) of the definition of chaotic behavior. In contrast to Marotto's proof, this is the first place where assumption (c), that f is expanding on A, is required. Until now, all that has been required is that f is one-to-one on A. From this, (b) and Lemma 4.4 follow the existence of a continuous inverse $f_A^{-1} : A \to A$. Hence, by the Brouwer fixed point theorem, there exists a point $a \in A$ such that $f_A^{-1}(a) = a$, or equivalently $f(a) = a$.

As f is expanding on A, it follows that f_A^{-1} is contracting A, i.e.,

$$\|f_A^{-1}(x) - f_A^{-1}(y)\| \leq \lambda^{-1}\|x-y\|$$

for all $x, y \in A$ where $\lambda > 1$ is the coefficient of expansion of f on A. Hence, for any $k \geq 1$ and all $x, y \in A$

$$\|f_A^{-k}(x) - f_A^{-k}(y)\| \leq \lambda^{-k}\|x - y\|,$$

and in particular for any $x \in C \subset A$ and for $y = a$

$$\|f_A^{-k}(x) - a\| \leq \lambda^{-k}\|x - a\|, \tag{4.9}$$

so $f_A^{-k}(x) \to a$ as $k \to \infty$ for all $x \in C$. Consequently, for any $\varepsilon > 0$ there exists an integer $j = j(x, \varepsilon)$ such that $f_A^{-j}(x) \in A \cap B^n(a; \varepsilon)$. Then, by continuity there exists a $\delta = \delta(x, \varepsilon) > 0$ such that $f_A^{-1}(A \cap \text{int}B^n(x; \delta)) \subset A \cap B^n(a; \varepsilon)$. Now, the collection $\varsigma = \{\text{int } B^n(x; \delta); x \in C\}$ constitutes an open cover of the compact set C, so there exists a finite sub-collection $\varsigma_0 = \{\text{int } B^n(x_i; \delta_i); i = 1, 2, \ldots, L\}$ which also covers C. Let $T = T(\varepsilon) = \max\{j(x_i; \varepsilon); i = 1, 2, \ldots, L\}$. Then, $f_A^{-T}(x) \in B^n(a; \varepsilon) \cap A$ for all $x \in C$, and so by (4.9) $f_A^{-k}(C) \subset B^n(a; \varepsilon) \cap A$ for all $k \geq T(\varepsilon)$.

Let $H_k = h_A^{-k}(C)$ for all $k \geq 0$, where h_A^{-1} is a continuous inverse of $h = f^N$ on A. Then, for any $\varepsilon > 0$ there exists a $J = J(\varepsilon)$ such that $\|x - a\| < \varepsilon/2$ for all $x \in H_k$ and all $k > J$.

The remainder of the proof parallels that in Marotto [130] and in Li and Yorke [19]. The sequences $\xi^n = \{E_k^n\}_{k=1}^{\infty} \in \mathcal{E}$ will be further restricted as follows: if $E_k^n = D$ then $k = m^2$ for some integer m, and if $E_k^\eta = D$ for both $k = m^2$ and $k = (m+1)^2$ then $E_k^\eta = H_{2m-j}$ for $k = m^2 + j$ for $j = 1, 2, \ldots, 2m$. Finally, for the remaining k's, $E_k^\eta = A$. As these sequences still satisfy $h(E_k^\eta) \supset E_{k+1}^\eta$, by Lemma 4.5 there exists a point x_η with $h^k(x_\eta) \in E_k^\eta$ for all $k \geq 0$. Let $S_0 = \{x_\eta : \eta \in (\frac{4}{5}, 1)\}$. Then S_0 is uncountable, $S_0 \subset S_h \subset S$ and for any $s, t \in (\frac{4}{5}, 1)$ there exist infinitely many m's such that $h^k(x_s) \in E_k^s = H_{2m-1}$ and $h^k(x_t) \in E_k^t = H_{2ml-1}$ where $k = m^2 + 1$. But from the above, given any $\varepsilon > 0$, $\|x - a\| < \varepsilon/2$ for all $x \in H_{2m-1}$ provided m is sufficiently large. Hence, for any $\varepsilon > 0$ there exists an integer m such that $\|h^k(x_s) - h^k(x_t)\| < \varepsilon$ where $k = m^2 + 1$. As $\varepsilon > 0$ is arbitrary, it follows that

$$L_2 = \liminf_{k \to \infty} \|h^k(x_s) - h^k(x_t)\| = 0.$$

Thus, for any $x, y \in S_0$

$$\liminf_{k \to \infty} \|hk(x_s) - h^k(x_t)\| \leq L_2 = 0.$$

This completes the proof of Theorem 4.6.

4.2.2 Some Examples

Two examples are given here to illustrate the application of Theorem 4.6. The first example is a one-dimensional difference equation with a snap-back repeller involving the tent or baker's map. It forms one of the components of the second example, the two-dimensional twisted-horseshoe difference equation of Guckenheimer, Oster and Ipacktchi [118], which has a saddle point.

Example 4.7. Consider the difference equation on the unit interval $I = [0,1]$, which is defined in terms of the baker's map

$$f(x) = \begin{cases} 2x & \text{for } 0 \le x \le \frac{1}{2}, \\ 2-2x & \text{for } \frac{1}{2} < x \le 1. \end{cases}$$

This map f maps I into itself and has two fixed points 0 and $\frac{2}{3}$, both of which are easily seen to be snap-back repellers.

The conditions of Theorem 4.6 are satisfied by $A = [\frac{9}{16}, \frac{7}{8}]$, $B = [\frac{3}{4}, \frac{7}{8}]$, $n = l = 1$ and $n_1 = n_2 = 1$. To see this, note that

$$f(A) = \left[\frac{1}{4}, \frac{7}{8}\right], \qquad f(b) = \left[\frac{1}{4}, \frac{1}{2}\right], \qquad f^2(B) = \left[\frac{1}{2}, 1\right],$$

so

$$f(A) \supset A, \qquad f(B) \cap A = \emptyset, \qquad f^2(B) \supset A.$$

Also, f is expanding on A, because for $x, y \in A$

$$|f(x) - f(y)| = |(2-2x) - (2-2y)| = 2|x-y|,$$

and f^2 is one-to-one on B, because for all $x \in B$

$$f^2(x) = 2(2-2x) = 4 - 4x.$$

Hence, this difference equation exhibits chaotic behavior.

Example 4.8. Consider the difference equation on the unit square I^2 in $\Re^2$, which is defined in terms of the continuous map $f = (f_1, f_2)$, where

$$f_1(x,y) = \begin{cases} 2x & \text{for } 0 \le x \le \frac{1}{2}, \\ 2-2x & \text{for } \frac{1}{2} < x \le 1, \end{cases} \qquad f_2(x,y) = \frac{x}{2} + \frac{y}{10} + \frac{1}{4}.$$

This map describes a twisted horseshoe on I^2, and has been investigated in detail by Guckenheimer, Oster and Ipaktchi [118]. It has a fixed point $(\bar{x}, \bar{y}) = (\frac{2}{3}, \frac{35}{54})$, which is a saddle point with eigenvalues -2 and $\frac{1}{10}$. Consequently, Marotto's snap-back repeller theorem cannot be used here, but Theorem 4.6 can.

Let L_1 be the line $90x + 378y = 305$ and L_2 the line $90x - 378y = -125$. Then, $(\bar{x}, \bar{y}) \in L_1$. Also let

$$A = \left\{(x,y) \in L_1 \ ; \ \frac{9}{16} \le x \le \frac{7}{8}\right\}, \qquad B = \left\{(x,y) \in L_1 \ ; \ \frac{3}{4} \le x \le \frac{7}{8}\right\}.$$

Then,

$$f(A) = \left\{(x,y) \in L_1 \ ; \ \frac{1}{4} \le x \le \frac{7}{8}\right\}, \quad f(B) = \left\{(x,y) \in L_1 \ ; \ \frac{1}{4} \le x \le \frac{1}{2}\right\},$$

$$f^2(B) = \left\{ (x, y) \in L_2 \; ; \; \frac{1}{2} \leq x \leq 1 \right\},$$

and $f^3(B) \supset L_1 \cap I^2$.

Hence, $f(A) \supset A$, $f(B) \cap A = \emptyset$ and $f^3(B) \supset A$, so conditions (b), (d), (e) and (f) of Theorem 4.6 are satisfied with $n_1 = 1$ and $n_2 = 2$. Also, A is homeomorphic to a 1-ball and f is expanding on A, because for $(x, y) \in A$

$$f_1(x, y) = 2 - 2x, \qquad f_2(x, y) = \frac{35}{18} - 2y,$$

so for any two points $(x', y'), (x'', y'') \in A$

$$\|f(x', y') - f(x'', y'')\| = 2\|(x', y') - (x'', y'')\|.$$

Finally, for all $(x, y) \in B$

$$f_1^3(x, y) = 2 \cdot 2 \cdot (2 - 2x)) = 8 - 8x$$

and

$$f_2^3(x, y) = \frac{381}{200} x + \frac{1}{1000} y - \frac{249}{400},$$

which gives the nonsingular Jacobian matrix

$$\begin{bmatrix} -8 & 0 \\ \frac{381}{200} & \frac{1}{1000} \end{bmatrix}.$$

Hence, f^3 is one-to-one on B.

Thus, all the conditions of Theorem 4.6 are satisfied, so this twisted-horseshoe difference equation behaves chaotically.

4.3 Chaotic Maps in Banach Spaces

The proof of Theorem 4.6 above can easily be modified by using the Schauder fixed point theorem, in which case X can be a Banach space, rather than the finite-dimensional Euclidian space $\Re^n$ [135]. The Schauder fixed point theorem states that a compact map f from a closed bounded convex set K in a Banach space X into itself has a fixed point. In this setting Theorem 4.6 becomes

Theorem 4.9. (Kloeden) *[124]*
Let $f : X \to X$ be a continuous map of a Banach space X into itself, and suppose that there exist non-empty compact subsets A and B of X, and integers $n_1, n_2 \geq 1$ such that

(i) *A is homeomorphic to a convex subset of X,*
(ii) *$A \subseteq f(A)$,*

(iii) f is expanding on A, i.e., there exists a constant $\lambda > 1$ such that

$$\lambda\|x - y\| \leq \|f(x) - f(y)\|$$

for all $x, y \in A$,

(iv) $B \subset A$,

(v) $f^{n_1}(B) \cap A = \emptyset$,

(vi) $A \subseteq f^{n_1+n_2}(B)$,

(vii) $f^{n_1+n_2}$ is one-to-one on B.

Then, the map f is chaotic in the generalized sense of Li and Yorke given in Theorem 4.6.

The proof is essentially a repetition of that given above for $X = \Re^n$, but requires the Schauder fixed point theorem rather than the Brouwer fixed point theorem. For difference equations on $\Re^1$, conditions (iii) and (vii) of the theorem are superfluous as the intermediate value theorem can be used instead of the Schauder fixed theorem to establish the existence of cyclic points. Without these conditions the theorem then contains the sufficient conditions for chaotic behavior of Barna [111] and Sharkovsky [132] as special cases.

This theorem applies to the baker's map and to the twisted horseshoe map with the same sets A and B as in the previous section, noting that intervals and connected segments of straight lines are convex sets.

However, the above theorem is not applicable to diffeomorphisms such as the Hénon map and the Smale horseshoe map and, hence, in general not to the Poincaré maps for ordinary differential equations.

4.4 Chaos of Discrete Systems in Complete Metric Spaces

Even more generally, some criteria for chaos of difference equations in general complete metric spaces will be given in this section. In contrast to the Euclidian spaces and Banach spaces, these metric spaces may not have a linear structure which allows one to, say, take the difference of two points. Recall that the n-dimensional Euclidian space $\Re^n$ is complete, and any bounded and closed subset therein is compact. Furthermore, a compact subset of a general metric space is complete as a subspace. Therefore, difference equations defined in terms of continuous maps in compact subsets of metric spaces and the corresponding criteria of chaos will be discussed. Thus, the existing relevant results of chaos in $\Re^n$ and Banach spaces are extended and improved [134]. Here, we just list the main results of [134] without giving proofs. Readers interested in the details are referred to [134].

The criteria of chaos obtained in this section are related to Cantor sets in metric spaces and a symbolic dynamical system, which has rich dynamical structures.

Definition 4.10. *Let X be a topological space and Λ be a subset in X. Then, Λ is called a Cantor set if it is compact, totally disconnected, and perfect. A set in X is totally disconnected if each of its connected components is a single point; a set is perfect if it is closed and every point in it is an accumulation point or a limit point of other points in the set.*

Consider the space of sequences

$$\Sigma_2^+ := \{s = (s_0, s_1, s_2, \ldots) : s_j = 0 \text{ or } 1\}$$

and define a distance between two points $s = (s_0, s_1, s_2, \ldots)$ and $t = (t_0, t_1, t_2, \ldots)$ by

$$\rho(s,t) = \sum_{i=0}^{\infty} 2^i |s_i - t_i|.$$

For any $s, t \in \Sigma_2^+$, $\rho(s,t) \leq 2^{-n}$ if $s_i = t_i$ for $0 \leq i \leq n$. Conversely, if $\rho(s,t) < 2^{-n}$, then $s_i = t_i$ for $0 \leq i \leq n$.

Lemma 4.11. *(Σ_2^+, ρ) is a complete, compact, totally disconnected and perfect metric space.*

Definition 4.12. *The shift map $\sigma : \Sigma_2^+ \to \Sigma_2^+$ defined by $\sigma(s_0, s_1, s_2, \ldots) = (s_1, s_2, \ldots)$ is continuous. The dynamical system governed by σ is called a symbolic dynamical system on Σ_2^+.*

The shift map σ has the following properties:

1. $Card\ Per_n(\sigma) = 2^n$,
2. $Per(\sigma)$ is dense in Σ_2^+,
3. there exists a dense orbit of σ in Σ_2^+,

where $Card\ Per_n(\sigma)$ denotes the number of periodic points of period n for σ.

Theorem 4.13. *Let (X,d) be a complete metric space and V_0, V_1 be nonempty, closed, and bounded subsets of X with $d(V_0, V_1) > 0$. If a continuous map $f : V_0 \cup V_1 \to X$ satisfies*

(1) $f(V_j) \supset V_0 \cup V_1$ for $j = 0, 1$;
(2) f is expanding in V_0 and V_1, respectively, i.e., there exists a constant $\lambda_0 > 1$ such that

$$d(f(x), f(y)) \geq \lambda_0 d(x,y)\ \forall x, y \in V_0 \text{ and } \forall x, y \in V_1;$$

(3) there exists a constant $\mu_0 > 0$ such that

$$d(f(x), f(y)) \leq \mu_0 d(x,y)\ \forall x, y \in V_0 \text{ and } \forall x, y \in V_1;$$

then there exist a Cantor set $\Lambda \subset V_0 \cup V_1$ such that $f : \Lambda \to \Lambda$ is topologically conjugate to the symbolic dynamical system $\sigma : \Sigma_2^+ \to \Sigma_2^+$. Consequently, f is chaotic on Λ in the sense of Devaney.

Recall from the fundamental theory of topology that a compact subset of a metric space is closed, bounded, and complete as a subspace; a closed subset of a compact space is compact; and the distance between two disjoint compact subsets of a metric space is positive. Therefore, if V_0 and V_1 are compact subsets of a metric space (X, d), assumption (3) in Theorem 4.13 can be dropped.

The following is the corresponding result for chaos of difference equations defined in terms of continuous maps in two compact subsets of a metric space.

Theorem 4.14. *Let (X, d) be a metric space and V_0, V_1 be two disjoint compact subset of X. If the continuous map $f : V_0 \cup V_1 \to X$ satisfies*

(1) $f(V_j) \supset V_0 \cup V_1$ for $j = 0, 1$;
(2) there exists a constant $\lambda_0 > 1$ such that

$$d(f(x), f(y)) \geq \lambda_0 d(x, y) \ \forall x, y \in V_0 \text{ and } \forall x, y \in V_1,$$

then there exists a Cantor set $\Lambda \in V_0 \cup V_1$ such that $f : \Lambda \to \Lambda$ is topologically conjugate to the symbolic dynamical system $\sigma : \Sigma_2^+ \to \Sigma_2^+$. Consequently, f is chaotic on Λ in the sense of Devaney.

It should be noticed that by Theorems 4.13 and 4.14 the appearance of chaos of f is only relevant to the properties of f on V_0 and V_1, but has no relationship with the properties of f at any other points. The following example is used to illustrate the application of the theorems.

Example 4.15. Consider the discrete dynamical system

$$x_{n+1} = \mu x_n(1 - x_n)$$

governed by the Logistic map $f(x) = \mu x(1-x)$, where $\mu > 0$ is the parameter.

This map has exactly two fixed points: $x_1^* = 0$ and $x_2^* = 1 - \mu^{-1}$. It is clear that f is continuously differentiable on $\Re$ and if $\mu > 2 + \sqrt{5}$ then

$$|f'(x)| > 1 \quad \text{for } x \in [0, x_1] \cup [x_2, 1],$$

where $x_1 = 2^{-1} - \sqrt{4^{-1} - \mu^{-1}} > 0$, $x_2 = 2^{-1} + \sqrt{4^{-1} - \mu^{-1}} < x_2^*$. This implies that

$$|f(x) - f(y)| \geq \lambda_0 |x - y| \ \forall x, y \in [0, x_1] \text{ and } \forall x, y \in [x_2, 1],$$

where $\lambda_0 = \sqrt{\mu^2 - 4\mu} > 1$. On the other hand, we have

$$f([0, x_1]) = [0, 1] \supset [0, x_1] \cup [x_2, 1],$$

$$f([x_2, 1]) = [0, 1] \supset [0, x_1] \cup [x_2, 1],$$

Clearly, $[0, x_1]$ and $[x_2, 1]$ are compact, so that all assumptions in Theorem 4.14 are satisfied, and for $\mu > 2 + \sqrt{5}$ there exists a Cantor set

$\Lambda \in [0, x_1] \cup [x_2, 1]$ such that $f : \Lambda \to \Lambda$ is topologically conjugate to the symbolic dynamical system $\sigma : \Sigma_2^+ \to \Sigma_2^+$.

Now, consider the following map:

$$g(x) = \begin{cases} \mu x(1-x), \; x \in [0, x_1], \\ \quad h(x), \quad\; x \in (x_1, x_2), \\ \mu x(1-x), \; x \in [x_2, 1], \end{cases}$$

where $h(x)$ can be any function on (x_1, x_2). It is stressed that g may not even be continuous on $[x_1, x_2]$. By the above discussion on the logistic map and by Theorem 4.14, one can conclude that for $\mu > 2 + \sqrt{5}$ there exists a Cantor set $\Lambda' \subset [0, x_1] \cup [x_2, 1]$ such that $g : \Lambda' \to \Lambda'$ is topologically conjugate to the symbolic dynamical system $\sigma : \Sigma_2^+ \to \Sigma_2^+$ and, consequently, g is chaotic on Λ'. We notice that the Cantor set Λ' may be taken to be the set Λ.

Furthermore, by means of snap-back repeller arguments, two criteria of chaos for difference equations defined in terms of continuous maps in complete metric spaces and compact subsets of metric spaces will be established in the following.

Theorem 4.16. *Let (X, d) be a complete metric space and $f : X \to X$ be a map. Assume that*

(1) f has a regular nondegenerate snap-back repeller $z \in X$, i.e., there exist positive constants r_1 and $\lambda_1 > 1$ such that $f(B_{r_1}(z))$ is open and

$$d(f(x), f(y)) \geq \lambda_1 d(x, y) \;\forall x, y \in \bar{B}_{r_1}(z),$$

and there exist a point $x_0 B_{r_1}(z)$, $x_0 \neq z$, a positive integer m, and positive constants δ_1 and γ such that $f^m(x_0) = z$, $B_{\delta_1}(x_0) \subset B_{r_1}(z)$, z is an interior point of $f^m(B_{\delta_1}(x_0))$, and

$$d(f^m(x), f^m(y)) \geq \gamma d(x, y) \;\forall x, y \in \bar{B}_{\delta_1}(x_0); \tag{4.10}$$

(2) there exists a positive constant μ_1 such that

$$d(f(x), f(y)) \leq \mu_1 d(x, y) \;\forall x, y \in \bar{B}_{r_1}(z); \tag{4.11}$$

(3) there exists a positive constant μ_2 such that

$$d(f^m(x), f^m(y)) \leq \mu_2 d(x, y) \;\forall x, y \in \bar{B}_{\delta_1}(x_0). \tag{4.12}$$

In addition, assume that f is continuous on $\bar{B}_{r_1}(z)$ and that f^m is continuous on $\bar{B}_{\delta_1}(x_0)$. Then, for each neighborhood U of z, there exist a positive integer $n > m$ and a Cantor set $\Lambda \subset U$ such that $f^n : \Lambda \to \Lambda$ is topologically conjugate to the symbolic dynamical system $\sigma : \Sigma_2^+ \to \Sigma_2^+$. Consequently, f^n is chaotic on Λ in the sense of Devaney.

By Theorem 4.14, the following result for metric spaces with a certain compactness property similar to that of finite-dimensional Euclidian spaces can be established.

Theorem 4.17. *Let (X, d) be a metric space in which each bounded and closed subset is compact. Assume that $f : X \to X$ has a regular nondegenerate snap-back repeller z, associated with x_0, m, and r as specified in Marotto's definitions, f is continuous on $\bar{B}_r(z)$, and f^m is continuous in a neighborhood of x_0. Then, for any neighborhood U of z there exist a positive integer n and a Cantor set $\Lambda \subset U$ such that $f^n : \Lambda \to \Lambda$ is topologically conjugate to the symbolic dynamical system $\sigma : \Sigma_2^+ \to \Sigma_2^+$. Consequently, f^n is chaotic on Λ in the sense of Devaney.*

4.5 Chaos of Difference Equations in Metric Spaces of Fuzzy Sets

In this section, the Li-Yorke and Marotto definitions are generalized to be applicable to maps from a space of fuzzy sets into itself, namely, the metric space $(\mathcal{E}^n, D)$ of fuzzy sets on the base space $\Re^n$.

4.5.1 Chaotic Maps on Fuzzy Sets

The following definitions and results are taken from Kaleva [121], see also the monograph [117].

The set $\mathcal{E}^n$ consists of all functions, called fuzzy sets here, $u : \Re^n \to [0, 1]$ for which

(i) u is normal, i.e., there exists an $x_0 \in \Re^n$ such that $u(x_0) = 1$,
(ii) u is fuzzy-convex, i.e., for any $x, y \in \Re^n$ and $0 \le \lambda \le 1$

$$u(\lambda x + (1 - \lambda)y) \ge \min\{u(x), u(y)\},$$

(iii) u is uppersemicontinuous,
(iv) the closure of $\{x \in \Re^n; u(x) > 0\}$, denoted by $[u]^0$, is compact.

Let $u \in \mathcal{E}^n$. Then, for each $0 < \alpha \le 1$ the α-level set $[u]^\alpha$ of u, defined by

$$[u]^\alpha = \{x \in \Re^n; u(x) \ge \alpha\},$$

is a non-empty compact convex subset of $\Re^n$, as is the support $[u]^0$ of u. Let d be the Hausdorff metric for nonempty compact subsets of $\Re^n$. Then,

$$D(u, v) = \sup_{0 \le \alpha \le 1} d([u]^\alpha, [v]^\alpha),$$

where $u, v \in \mathcal{E}^n$, is a metric on $\mathcal{E}^n$. Moreover, $(\mathcal{E}^n, D)$ is a complete metric space.

Let $u, v \in \mathcal{E}^n$ and let c be a positive number. Then, addition $u + v$ and (positive) scalar multiplication cu in $\mathcal{E}^n$ are defined in terms of the α-level sets by

$$[u+v]^\alpha = [u]^\alpha + [v]^\alpha,$$

and

$$[cu]^\alpha = c[u]^\alpha,$$

for each $0 \leq \alpha \leq 1$, where

$$A + B = \{x + y; x \in A, y \in B\} \text{ and } cA = \{cx; x \in A\}$$

for nonempty subsets A and B of $\Re^n$. This defines a linear structure (but without subtraction) on $\mathcal{E}^n$, such that

$$D(u+w, v+w) = D(u,v)$$

and

$$D(cu, cv) = cD(u,v),$$

for all $u, v, w \in \mathcal{E}^n$ and $c > 0$. There is, however, no norm on $\mathcal{E}^n$ equivalent to the metric D which makes $\mathcal{E}^n$ into a normed linear space with the above natural linear structure. Nevertheless, by an embedding theorem of Rådström, $\mathcal{E}^n$ can be embedded isometrically and isomorphically as a convex cone in some Banach space. Consequently, many well-known results for Banach spaces can be adapted to the metric space $(\mathcal{E}^n, D)$ of fuzzy sets. An example is the following fixed point theorem of Kaleva [121].

Theorem 4.18. (Kaleva)
Let $f : \mathcal{E}^n \to \mathcal{E}^n$ be continuous and let $\mathcal{X}$ be a nonempty compact convex subset of $\mathcal{E}^n$ such that $f(\mathcal{X}) \subseteq \mathcal{X}$. Then, f has a fixed point $\bar{u} = f(\bar{u}) \in \mathcal{X}$.

Consider now an iterative scheme of fuzzy sets

$$u_{k+1} = f(u_k), \quad k = 1, 2, \ldots, \tag{4.13}$$

where f is a continuous map from the space of fuzzy sets $\mathcal{E}^n$ into itself. Such an iterative scheme or map f will be called to be chaotic if conditions analogous to those of Theorem 4.1 are satisfied. Using the Kaleva fixed point theorem, the following analogue of Theorem 4.6 can be shown to hold. It provides sufficient conditions for a map on fuzzy sets to be chaotic. (Alternatively, one could apply the results of the previous section here).

Theorem 4.19. (Kloeden) *[125]*
Let $f : \mathcal{E}^n \to \mathcal{E}^n$ be continuous and suppose that there exist nonempty compact subsets $\mathcal{A}$ and $\mathcal{B}$ of $\mathcal{E}^n$, and integers $n_1, n_2 \geq 1$ such that

(i) $\mathcal{A}$ is homeomorphic to a convex subset of $\mathcal{E}^n$,
(ii) $\mathcal{A} \subseteq f(\mathcal{A})$,

(iii) f is expanding on $\mathcal{A}$, that there exists a constant $\lambda > 1$ such that

$$\lambda D(u,v) \leq D(f(u), f(v))$$

for all $u, v \in \mathcal{A}$,
(iv) $\mathcal{B} \subset \mathcal{A}$,
(v) $f^{n_1}(\mathcal{B}) \cap \mathcal{A} = \emptyset$,
(vi) $\mathcal{A} \subseteq f^{n_1+n_2}(\mathcal{B})$,
(vii) $f^{n_1+n_2}$ is one-to-one on $\mathcal{B}$.

Then, the map f is chaotic.

The proof of Theorem 4.19 essentially mimics the proof of Theorem 4.6, using the Kaleva fixed point theorem instead of the Brouwer fixed point theorem. Thus, it is omitted. A useful characterization of compact subsets of $\mathcal{E}^n$ has been presented by Diamond and Kloeden in [116] and [117].

4.5.2 Example of a Chaotic Map on Fuzzy Sets

To illustrate the application of Theorem 4.19, a simple example of a chaotic map $f : \mathcal{E}^1 \to \mathcal{E}^1$, which satisfies the hypotheses of the theorem, will be constructed. For this purpose observe that for each $u \in \mathcal{E}^1$ there exist functions (their dependence on u is not explicitly stated) $a, b : [0,1] \to \Re$ such that the α-level sets of u are the intervals $[u]^\alpha = [a(\alpha), b(\alpha)]$. Moreover, for any $0 \leq \alpha \leq \alpha' \leq 1$ the following inequalities hold

$$a(0) \leq a(\alpha) \leq a(\alpha') \leq a(1) \leq b(1) \leq b(\alpha') \leq b(\alpha) \leq b(0).$$

Consider the following subsets of $\mathcal{E}^1$:

(a) $\mathcal{E}_0^1 = \{u \in \mathcal{E}^1 \; ; \; a(0) = 0\}$,
(b) $\Im_0^1 = \{u \in \mathcal{E}_0^1 \; ; \; a(\alpha) = \frac{1}{2}\alpha(b(0) - L)$ and $b(\alpha) = b(0) - \frac{1}{2}\alpha(b(0) - L)$ for some $0 \leq L \leq b(0)\}$,
(c) $\Delta_0^1 = \{u \in \Im_0^1 \; ; \; L = 0\}$.

For any $u \in \mathcal{E}_0^1$ the support $[u]^0$ is a nonnegative interval anchored on $x = 0$. The endograph of any $u \in \Im_0^1$ is a symmetric trapezium centered on $x = \frac{1}{2}b(0)$, with base length $b(0)$ and top length L. For any $u \in \Delta_0^1$ the endograph is an isosceles triangle.

Now define the following maps on fuzzy sets:

(1) $f_1 : \mathcal{E}^1 \to \mathcal{E}_0^1$ by $[f_1(u)]^\alpha = [a(\alpha) - a(0), b(\alpha) - b(0)]$,
(2) $f_2 : \mathcal{E}_0^1 \to \Im_0^1$ by $[f_2(u)]^\alpha = [\alpha M, b(0) - \alpha M]$, where $M = \frac{1}{2}b(0) - \frac{1}{8}(b(1) - a(1))$;
(3) $f_3 : \Im_0^1 \to \Im_0^1$ by $[f_3(u)]^\alpha = g(b(0))[u]^\alpha = [g(b(0))a(\alpha), g(b(0))b(\alpha)]$,

where $g : \Re^+ \to \Re^+$ is the function

$$g(x) = \begin{cases} 2 & \text{if } 0 \leq x \leq \frac{1}{2}, \\ -2 + 2/x & \text{if } \frac{1}{2} \leq x \leq 1, \\ 0 & \text{if } 1 \leq x. \end{cases}$$

Also define $h : \Re^+ \to \Re^+$ by

$$h(x) = xg(x) = \begin{cases} 2x & \text{if } 0 \leq x \leq \frac{1}{2}, \\ 2 - 2x & \text{if } \frac{1}{2} \leq x \leq 1, \\ 0 & \text{if } 1 \leq x. \end{cases}$$

Finally, define $f : \mathcal{E}^1 \to \mathcal{E}^1$ by $f = f_3 \circ f_2 \circ f_1$. This map f is clearly continuous with respect to the D-metric and maps Δ_0^1 into itself. Now any $u \in \Delta_0^1$ is determined uniquely by its value of $b(0)$ henceforth written b, and will be denoted by u_b. Then, $f(u_b) = u_{h(b)}$.

Theorem 4.19 applies for the map f with $n_1 = n_2 = 1$ and with the compact convex subsets of $\mathcal{E}^1$

$$\mathcal{A} = \left\{ u_b \in \Delta_0^1 \; ; \; \frac{9}{16} \leq b \leq \frac{7}{8} \right\}, \qquad \mathcal{B} = \left\{ u_b \in \Delta_0^1 \; ; \; \frac{3}{4} \leq b \leq \frac{7}{8} \right\},$$

with

$$f(\mathcal{A}) = \left\{ u_b \in \Delta_0^1 \; ; \; \frac{1}{4} \leq b \leq \frac{7}{8} \right\}, \qquad f(\mathcal{B}) = \left\{ u_b \in \Delta_0^1 \; \frac{1}{4} \leq b \leq \frac{1}{2} \right\}$$

and

$$f^2(\mathcal{B}) = \left\{ u_b \in \Delta_0^1 \; \frac{1}{2} \leq b \leq 1 \right\},$$

so

$$\mathcal{A} \subseteq f(\mathcal{A}), \quad f(\mathcal{B}) \cap \mathcal{A} = \emptyset, \quad \mathcal{A} \subseteq f^2(\mathcal{B}).$$

Moreover, f is expanding on $\mathcal{A}$, because h is expanding on $[\frac{9}{16}, \frac{7}{8}]$ with

$$|h(x) - h(y)| = |(2 - 2x) - (2 - 2y)| = 2|x - y|,$$

there, so

$$D(f(u_x), f(u_y)) = 2D(u_x, u_y)$$

for any $u_x, u_y \in \mathcal{A}$. Finally, h^2 is one-to-one on $[\frac{3}{4}, \frac{7}{8}]$, because

$$h^2(x) = 2(2 - 2x) = 4 - 4x,$$

there, which implies that f^2 is one-to-one on $\mathcal{B}$.

The map f is thus chaotic. Its chaotic action is most apparent in the compact subset $\{u_b \in \Delta_0^1; 0 \leq b \leq 1\}$ of $\mathcal{E}^1$. It is not hard to show that for any $u \in \mathcal{E}^1$, the successive iterates $f^k(u)$ asymptote towards this set. Their endographs become more and more triangular in shape, unless $b(0) - a(0) > 1$, in which case they collapse onto the singleton fuzzy set χ_0.

It is finally remarked that the definition of chaos for difference equations in metric spaces of fuzzy sets is still difficult to be applied to a general fuzzy system. Often in practice, however, to establish the existence of chaos in a particular dynamical system still depends mainly on numerical calculations to estimate quantities such as the maximum Lyapunov exponent and topological entropy. A unified, well accepted, easy-to-test, and rigorous mathematical definition of chaos is still in the process of being revealed, and this work is far from complete.

5

Fuzzy Modeling of Chaotic Systems – I (Mamdani Model)

In this chapter we introduce an approach to model chaotic dynamics in a linguistic manner based on the Mamdani fuzzy model. This approach allows to design robust chaotic generators by means of few fuzzy sets and using a small number of fuzzy rules. The generated chaotic signals can be of assigned characteristics (e.g., Lyapunov exponents). As examples, fuzzy descriptions of well-known discrete chaotic maps, such as the logistic map, a double-scroll attractor and the 2-dimensional Hénon map, are given to illustrate the effectiveness of the proposed approach.

5.1 Introduction

Fuzzy logic allows to model processes in a linguistic manner. As described in Chapter 2, the basic configuration of a fuzzy logic system is composed of a fuzzyfier, a fuzzy rule-base, a fuzzy inference engine and a defuzzyfier, where the fuzzy rule-base consists of a collection of fuzzy IF-THEN rules, and the fuzzy inference engine uses these fuzzy IF-THEN rules to determine a map from fuzzy inputs to fuzzy outputs based on fuzzy composition rules.

The fuzzy IF-THEN rules are of the following form:

$$\begin{aligned} R^i\text{:}\ & \text{IF } x_1 \text{ is } \Gamma_1^i \text{ and } \cdots \text{ and } x_n \text{ is } \Gamma_n^i \\ & \text{THEN y is } G^i, \end{aligned} \tag{5.1}$$

where Γ_j^i and G^i are fuzzy sets, $x = (x_1, \ldots, x_n)^T \in U$ and $y \in V$ are input and output linguistic variables, respectively, and $i = 1, 2, \ldots, q$, in which q is the number of the rules. The fuzzy IF-THEN rules provide a convenient framework to incorporate human experts' knowledge. Each fuzzy IF-THEN rule of (5.1) defines a fuzzy set $\Gamma_1^i \times \cdots \times \Gamma_n^i \to G^i$ in the product space $U \times V$. The most commonly used fuzzy compositional rule in the fuzzy inference engine is the so-called sup-star composition. It is a general framework in which

Zhong Li: *Fuzzy Chaotic Systems*, StudFuzz **199**, 73–89 (2006)
www.springerlink.com

linguistic information from human experts is quantified and fuzzy inference is used to make systematic use of linguistic information [59, 129].

To derive a Mamdani fuzzy model of a chaotic system, it is required to well describe the chaotic system. The definition of chaotic behaviors involves the three fundamental concepts of transitivity, density of periodic orbits, and sensitivity to initial conditions [66]. Furthermore, from a qualitative point of view, chaos can be defined by monitoring the time evolution of trajectories emanating from nearby points on the attractor. In a chaotic system, points which are close to each other repel themselves so that the flow stretches. Then, a folding action must take place for the chaotic behavior to combine with the boundedness of the attractor.

The stretching and folding features of the flow are responsible for the sensitivity to initial conditions, and characterize the chaotic behavior. If the initial state is known, even with the smallest uncertainty, the flow magnifies it until the uncertainty interval involves the whole region of the state space where the attractor is embedded. In this sense a chaotic system is unpredictable.

To develop a fuzzy model of the evolution of a chaotic signal x, two variables need to be considered as inputs, i.e., the center value $x(k)$, which is the nominal value of the state x at the instant k, and the uncertainty $d(k)$ on the center value. In terms of fuzzy description, this means that the model contains four linguistic variables, i.e., $x(k)$, $x(k+1)$, $d(k)$ and $d(k+1)$. The whole set of rules has to determine the values $x(k+1)$ and $d(k+1)$ from the values $x(k)$ and $d(k)$.

Therefore, to suitably represent a chaotic motion, first the rules of the fuzzy model must produce an oscillatory behavior in the desired interval. Moreover, the uncertainty must enlarge until the boundary region is reached, when the folding process takes place. The diverging trajectories are then pushed toward the inner region of the attractor. The interaction between these features leads to a chaotic motion.

Famous chaotic systems, such as the Chua circuit [145], the Duffing oscillator [146] and the Rössler system, can be represented as third order nonlinear autonomous systems. However, chaotic dynamics can also be generated by simple discrete maps, like the logistic map:

$$x(k+1) = ax(k)(1 - x(k)),$$

and the Hénon map:

$$\begin{cases} x(k+1) = y(k) + 1 - ax^2(k) \\ y(k+1) = bx(k) \end{cases}.$$

The one-dimensional map $x(k+1) = f(x(k))$ is of special interest, because there exists a simple expression for Lyapunov exponents:

$$\lambda = \lim_{n \to \infty} \frac{1}{n} \sum_{k=1}^{n} \ln |f'(x(k))|.$$

Here, for simplicity, two typical chaotic systems, i.e., double scroll and single scroll systems, are taken as examples to show the modeling process, in which the fuzzy model with the Mamdani implication, the center-of-sums defuzzification method and the product as t-norm will be employed [147].

5.2 Double-scroll Chaotic System

In the fuzzy model of a double scroll system [148, 37], the linguistic variables $x(k)$ and $x(k+1)$ can take four linguistic values: large left (LL), small left (SL), small right (SR) and large right (LR). The linguistic variables $d(k)$, $d(k+1)$ can take five linguistic values: zero (Z), small (S), medium (M), large (L) and very large (VL). The fuzzy sets associated to these linguistic values are shown in Fig. 5.1 [129].

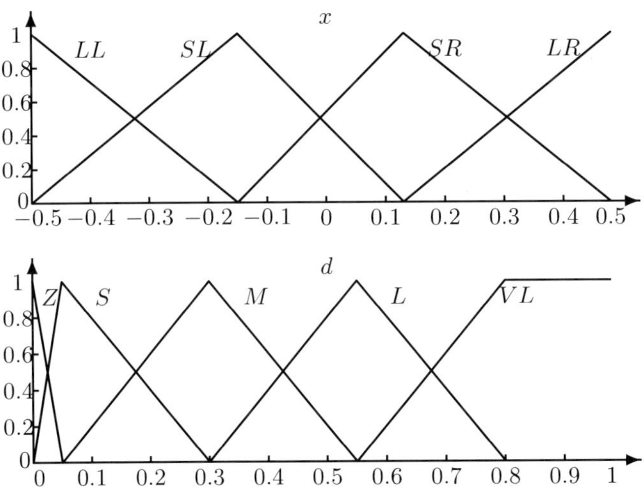

Fig. 5.1. Fuzzy sets for x (upper) and d (lower)

A double scroll system is characterized by two attractors, one in the positive axis and one in the negative axis, that let the state oscillate around each one of these attractors. Thus, considering the positive attractor, it happens that if the value of $x(k)$ is LR, then the value of $x(k+1)$ is SR and *vice versa*. The rules are like these in the following:

$$R_1: \quad \text{IF } x(k) \text{ is } LR \text{ THEN } x(k+1) \text{ is } SR;$$
$$R_2: \quad \text{IF } x(k) \text{ is } SR \text{ THEN } x(k+1) \text{ is } LR.$$

Obviously, a similar behavior is observed when $x(k)$ is negative (i.e., when $x(k)$ is LL or SL).

Once the region is determined in which the attractor is embedded, the uncertainty d must enlarge until the boundary region is reached. Thus, one has

$$\begin{aligned} R_3: &\quad \text{IF } d(k) \text{ is } S \text{ THEN } x(k+1) \text{ is } M; \\ R_4: &\quad \text{IF } d(k) \text{ is } L \text{ THEN } x(k+1) \text{ is } VL. \end{aligned}$$

Moreover, in order to allow x to oscillate around each one of the attractors for a certain time, a rule must be imposed to keep the enlargement of x limited:

$$R_5: \quad \text{IF } d(k) \text{ is } M \text{ THEN } x(k+1) \text{ is } M.$$

Once the uncertainty has reached the boundary region, the folding process takes place. This process consists of two different actions. One action is the shrinking of the uncertainty d in order to avoid behaviors leading to stability.

$$R_6: \quad \text{IF } x(k) \text{ is } LR \text{ and } d(k) \text{ is } VL \text{ THEN } d(k+1) \text{ is } S.$$

The other action consists in the changing of the lobe.

$$R_7: \quad \text{IF } x(k) \text{ is } LR \text{ and } d(k) \text{ is } VL \text{ THEN } x(k+1) \text{ is } SL.$$

Note that this rule is fired only when x is LR or LL, but not when x is SR or LR. Thus, oscillations begin around one of the two attractors just after x has passed from the positive axis to the negative one (or *vice versa*), but bounce backs are prevented. For this purpose, if $x(k)$ is small, the diverging trajectories are pushed toward the inner region of the attractor.

$$R_8: \quad \text{IF } x(k) \text{ is } SR \text{ and } d(k) \text{ is } VL \text{ THEN } x(k+1) \text{ is } LR.$$

The complete set of rules for the double scroll system is summed up in Table 5.1.

Table 5.1. Fuzzy rules generating a double scroll chaotic system

$x(k)/d(k)$	Z	S	M	L	VL
LL	SL/Z	SL/M	SL/M	SL/VL	SR/L
SL	LL/Z	LL/M	LL/M	LL/VL	LL/S
SR	LR/Z	LR/M	LR/M	LR/VL	LR/S
LR	SR/Z	SR/M	SR/M	SR/VL	SL/S

Figure 5.2 shows the evolution of the time series $x(k)$ generated by the fuzzy system for $x(0) = 0.01$ and $d(0) = 0.01$. As expected, the state x oscillates around one of the attractors, then it jumps again, and so on. Thus, the behavior of the state reproduces that of Chua's circuit.

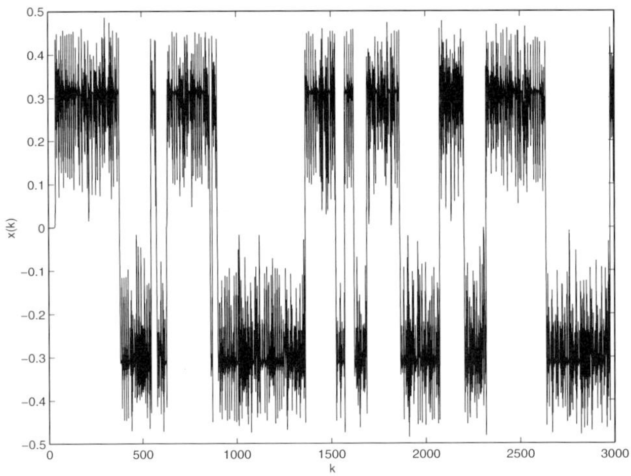

Fig. 5.2. The chaotic time series generated by the fuzzy system

Figure 5.3 shows the trajectory, on the phase plane $x(k) - x(k+1)$, of the nonlinear map implemented with Mamdani fuzzy modeling. As we can see, the system has two distinguished zones of chaos. Further, by connecting the fore-and-aft points in Fig. 5.3, we obtain Fig. 5.4 to more clearly show the chaotic motion of the fuzzy system.

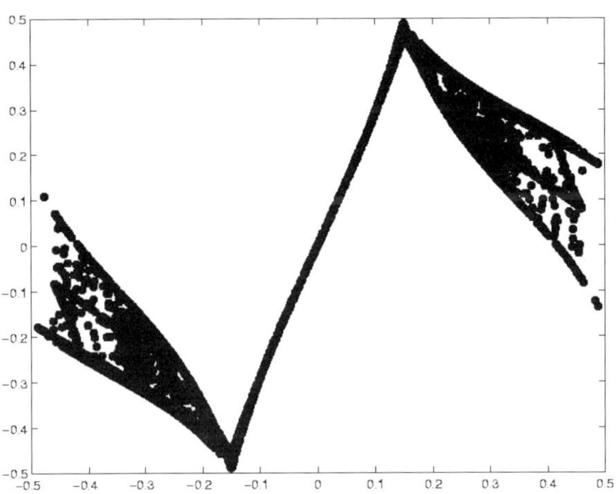

Fig. 5.3. The center value $x(k+1)$ on $x(k)$

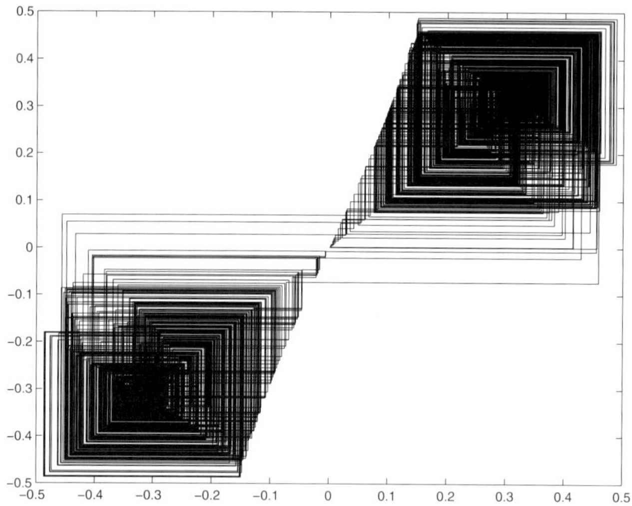

Fig. 5.4. The map generated by the fuzzy model of the two-lobe system

5.3 Single-scroll Chaotic System: Logistic Map

The most well-known one-dimensional map, which exhibits a single scroll chaotic behavior, is the logistic map $x(k+1) = \mu x(k)(1 - x(k))$ as shown in Fig. 5.5. Assuming $\mu = 4$, this map has two unstable fixed points, $x_1^* = 0$ and $x_2^* = 3/4$ (see Section 3.1), which influence its behavior.

To construct the fuzzy model of such chaotic dynamics, all the linguistic variables of the system, $(x(k), x(k+1), d(k), d(k+1))$, can take five linguistic values: zero (Z), small (S), medium (M), large (L) and very large (VL). The fuzzy sets associated to these linguistic values are shown in Fig. 5.6. They are constructed in such a way that the equilibrium point $x_2^* = 3/4$ is between the fuzzy set M and the fuzzy set L.

In this single scroll system, x tends to move out from the trivial equilibrium point $x_1^* = 0$ until x begins to oscillate around the nontrivial equilibrium point $x_2^* = 3/4$. The increasing amplitude of the oscillations forces the trajectory to enter again the neighborhood of $x_1^* = 0$, where, due to its instability, the above process repeats.

In other words, when x is smaller than the nontrivial equilibrium point $x_2^* = 3/4$, it tends to increase and, when x is very large, it tends to decrease.

$$\begin{aligned} R_1 &: \quad \text{IF } x(k) \text{ is } S \text{ THEN } x(k+1) \text{ is } M; \\ R_2 &: \quad \text{IF } x(k) \text{ is } VL \text{ THEN } x(k+1) \text{ is } Z. \end{aligned}$$

On the contrary, when x is close enough to $x_2^* = 3/4$, it tends to oscillate around the equilibrium point. Thus, it happens that, if the value of $x(k)$ is

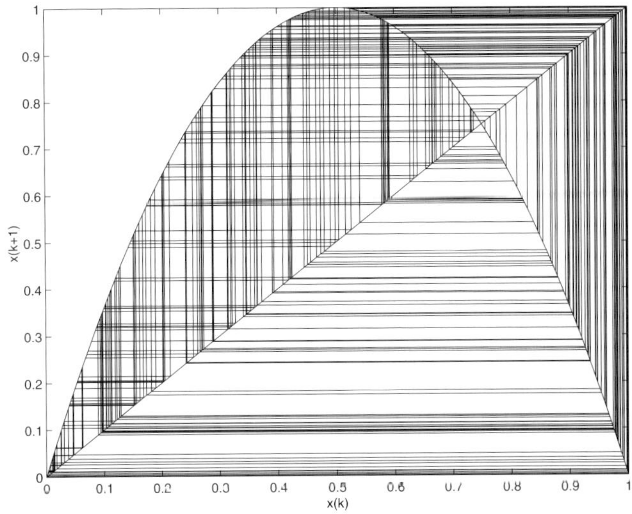

Fig. 5.5. Logistic map: $\mu = 4$ (chaos)

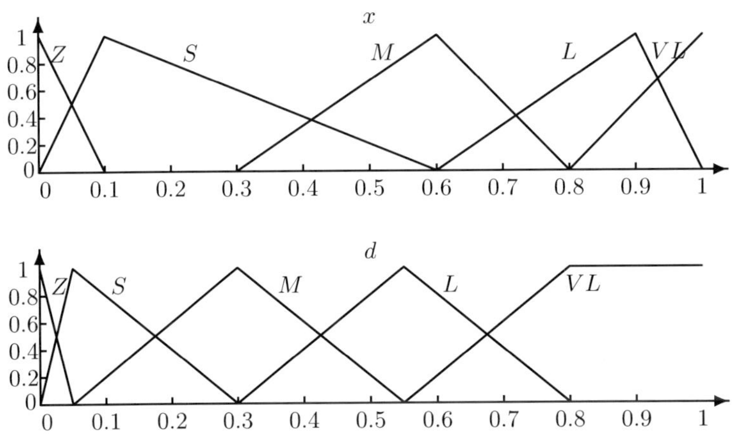

Fig. 5.6. The fuzzy sets for x (upper) and d (lower): logistic map

medium then the value of $x(k+1)$ is large, and if the value of $x(k)$ is large then the value of $x(k+1)$ is medium.

$$R_3: \quad \text{IF } x(k) \text{ is } M \text{ THEN } x(k+1) \text{ is } L;$$
$$R_4: \quad \text{IF } x(k) \text{ is } L \text{ THEN } x(k+1) \text{ is } M.$$

Moreover, due to the asymmetric shape of these fuzzy sets (L and M) it is necessary to obtain a motion like that of Fig. 5.5.

After the uncertainty has reached the boundary region, the folding process takes place, which consists of two different actions. One is the shrinking of the uncertainty d in order to avoid behaviors leading to instability:

$$R_5: \quad \text{IF } x(k) \text{ is } M \text{ and } d(k) \text{ is } VL \text{ THEN } d(k+1) \text{ is } S.$$

The other action is the escape of x from the neighborhood of the nontrivial equilibrium point to prevent it from the onset of a limit cycle:

$$R_6: \quad \text{IF } x(k) \text{ is } M \text{ and } d(k) \text{ is } VL \text{ THEN } x(k+1) \text{ is } VL.$$

In such a way, a complete set of fuzzy rules can be concluded as compiled in Table 5.2.

Table 5.2. Fuzzy rules implementing a single scroll chaotic system

$x(k)/d(k)$	Z	S	M	L	VL
Z	Z/Z	Z/M	Z/M	S/VL	L/L
S	M/Z	M/M	M/M	M/VL	L/S
M	L/Z	L/M	L/M	L/VL	VL/S
L	M/Z	M/M	M/M	M/VL	Z/S
VL	Z/Z	Z/M	Z/M	Z/VL	Z/L

Figure 5.7 represents the evolution of the time series $x(k)$ generated by the fuzzy system for $x(0) = 0.01$ and $d(0) = 0.01$.

From the trajectory on the phase space $x(k) - x(k+1)$ in Fig. 5.8 we can observe the range in which x is bounded and the nontrivial equilibrium

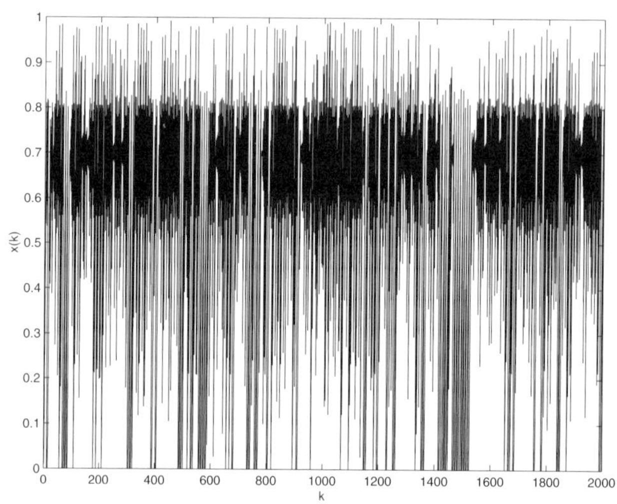

Fig. 5.7. The chaotic time series generated by the fuzzy system: logistic map

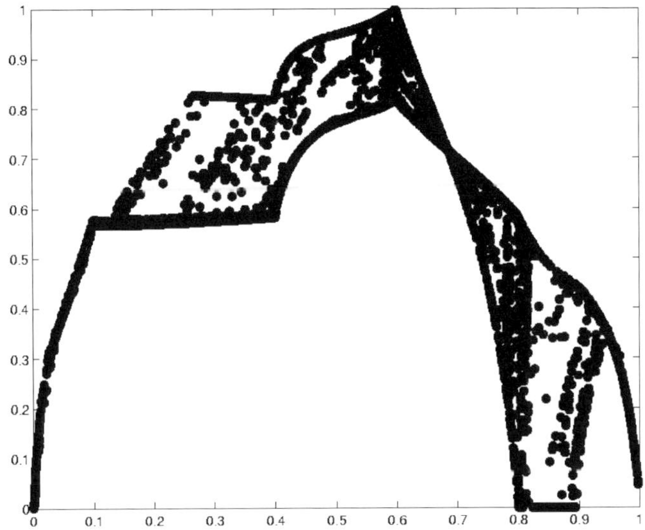

Fig. 5.8. The center value $x(k+1)$ on $x(k)$: logistic map

point around which the fuzzy map evolves. Besides, as shown in Fig. 5.9, the trajectory of the fuzzy system in the phase space can be drawn by connecting the fore-and-aft points. Comparing Fig. 5.5 with Fig. 5.9, the similarity between them is obvious. Moreover, it is worth noticing that the approximation is better when the fuzzy sets are dense, and worse when they are sparse (see Fig. 5.6). Thus, if we want to have a better approximation in a fixed region, we only need to increase the number of fuzzy sets describing x.

5.4 Lyapunov Exponents

For the one-dimensional map $x(k+1) = f(x(k))$, a simple expression for the Lyapunov exponents reads (referring to Section 3.4)

$$\lambda = \lim_{n\to\infty} \frac{1}{n} \sum_{k=1}^{n} \ln |f'(x(k))|.$$

More generally [66], this expression can be rewritten only by means of the values of uncertainty,

$$\lambda = \lim_{n\to\infty} \frac{1}{n} \sum_{k=1}^{n} \ln \left| \frac{E_{k+1}}{E_k} \right|,$$

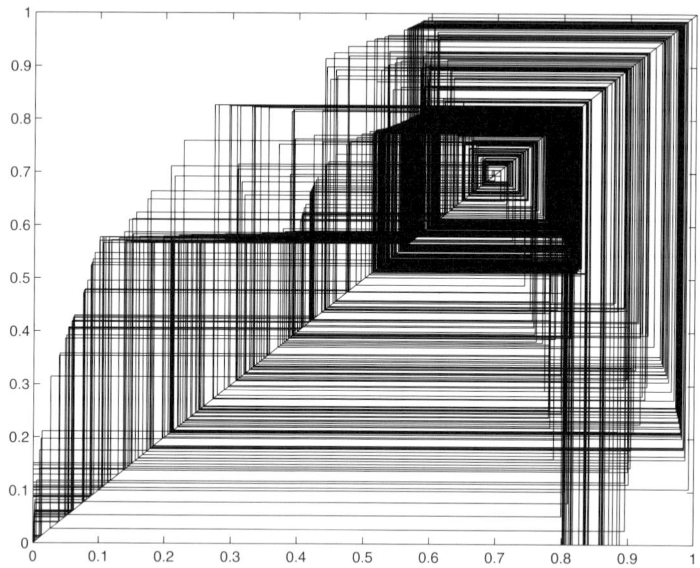

Fig. 5.9. The logistic map generated by the fuzzy model

where $E_{k+1} = f(x(k) + E_k) - f(x(k))$ only when the exact expression of the map f is known. It can also be written as

$$\left|\frac{E_{k+1}}{E_k}\right| = e^{\bar{\lambda}},$$

where $\bar{\lambda}$ can be considered as the desired value of the exponent. Considering the fuzzy models, whose uncertainties evolve following the rules previously explained, and can be written as:

$$\left|\frac{C_M}{C_S}\right| = e^{\bar{\lambda}}, \left|\frac{C_{VL}}{C_L}\right| = e^{\bar{\lambda}},$$

where C_* denotes the centers of the membership functions of the uncertainty $d(k)$. This fact means that it is possible to design a chaotic time series and derive its Lyapunov exponent only by changing the distance between the centers of the fuzzy sets. Figure 5.10 shows that there is a quite good accordance among computed values, obtained using the method described in [37], and desired values of Lyapunov exponents, at least for values between 0 and 0.25. Even if this range of values seems to be small, the Lyapunov exponent of a very large number of chaotic systems is contained in it [149].

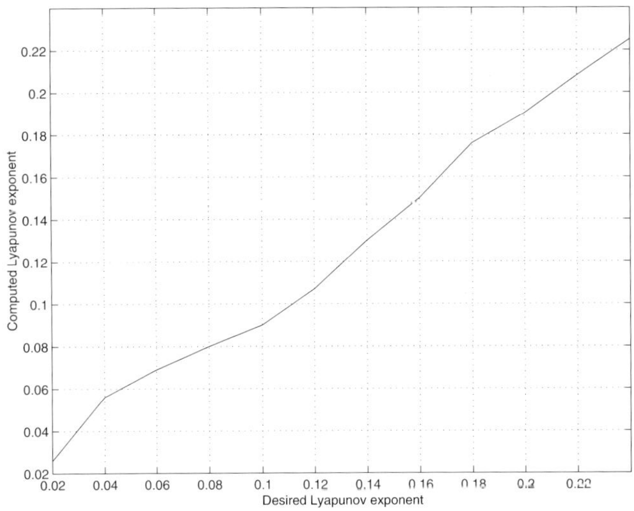

Fig. 5.10. The desired Lyapunov exponents against calculated ones

Remark 5.1. :

i) If some chaotic attractors are well known and can be well described with linguistic terms, fuzzy models can be constructed, although a routine of fuzzy modeling is not yet available;
ii) Fuzzy systems are capable of uniformly approximating a nonlinear function on U to any degree of accuracy if U is compact [59, 62];
iii) The finer the universe of discourse of the linguistic variables is partitioned, the more accurate the fuzzy model and the larger the rule-base will be. However, this results in long calculation time;
iv) From fuzzy modeling of chaotic systems shown above, it is straightforward that fuzzy systems can be chaotic.

5.5 Two-dimensional Map

The Hénon map can be considered in some sense as a two-dimensional extension of the logistic map, which is described as

$$\begin{cases} x(k+1) = y(k) + 1 - ax^2(k) \\ y(k+1) = bx(k). \end{cases} \tag{5.2}$$

For some values of parameters a and b, its Poincaré map denotes a fractal-like limit set of a typical chaotic system (see Fig. 5.11). The time evolution of the state variable x is depicted in Fig. 5.12. The same behavior can be

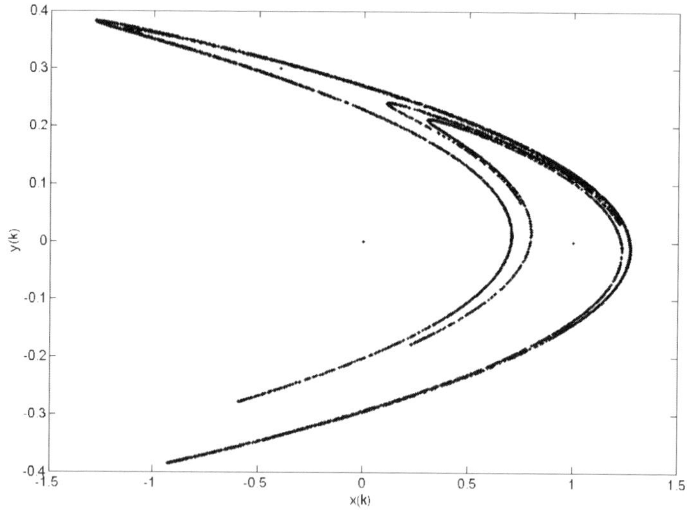

Fig. 5.11. The Poincaré map of the Hénon map with $a = 1.4$ and $b = 0.3$

observed for y, which is proportional to a one-step delayed sequence of x, according to Eq. (5.2).

If uncertainties d_x and d_y for each one of the two variables are considered using the above described approach, this may lead to a quite complex definition of the qualitative fuzzy model, compared with the analytical description of the system. In order to avoid these complications, here, only d_x is considered, which influences y through the second equation of (5.2) in a linear way. Due to the proportionality between $y(k+1)$ and $x(k)$, the membership functions of x and y can be chosen to be identical. However, the light influence of $y(k)$ on $x(k+1)$ with the parameters $a = 1.4$ and $b = 0.3$ suggests to use fewer fuzzy sets for $y(k)$, which acts only if its absolute value is high. It is easy to derive that there is a stable equilibrium point $(x^*, y^*) = (0.63, 0.19)$ asymmetric with respect to the range of x (and of y).

The fuzzy sets associated to the linguistic variables are shown in Fig. 5.13. They have been constructed in such a way that the nontrivial equilibrium point is between the fuzzy set M and the fuzzy set L. Only three fuzzy sets are adopted for y as they have the main influence on the evolution of x according to (5.2), ignoring the fuzzy sets of medium values and close to zero. Fuzzy sets of the uncertainty d_x have the same shapes as those in previous section, but with different ranges.

A qualitative linguistic description of the evolution of x is given as follows: x tends to move towards the nontrivial equilibrium point, until it begins to oscillate around it. This happens until the influence of the unstable equilibrium point perturbs its trajectory moving it out from the neighborhood of the nontrivial equilibrium point. To this aim, when the influence of $y(k)$ on $x(k+1)$

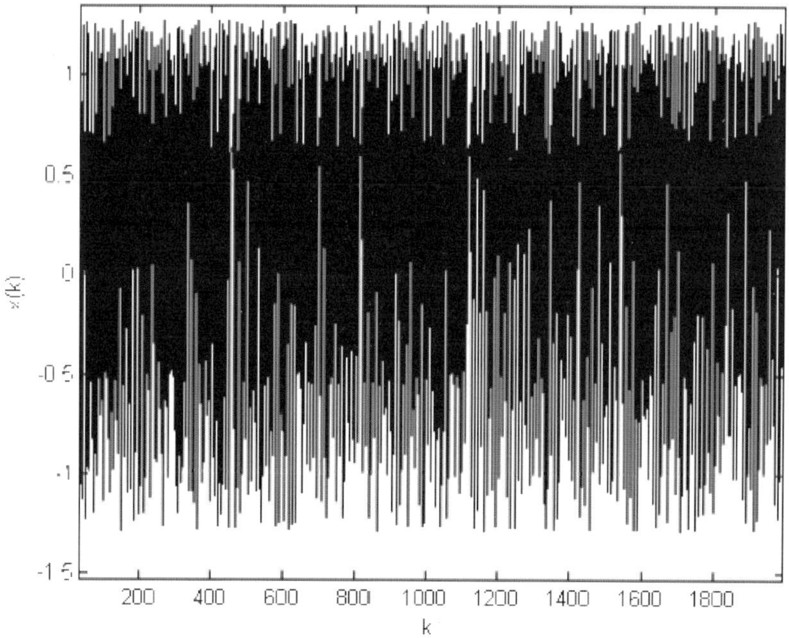

Fig. 5.12. Chaotic time series generated by the Hénon map with $a = 1.4$ and $b = 0.3$

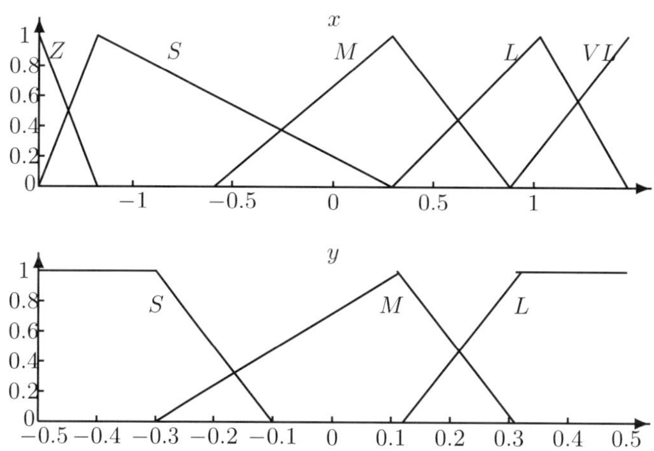

Fig. 5.13. The fuzzy sets for x (upper) and y (lower): Hénon map

is light (approximately, when $y(k)$ is M), it is possible to apply the same rule as in the previous section (see Table 5.3). The influence becomes relevant when $y(k)$ is L or S, even if limited to the transitions L-VL and S-Z of x. Therefore, when $y(k)$ is S, the new rules that have to be added are the following:

$$HS_1: \text{ IF } x(k) \text{ is } Z \text{ and } d(k) \text{ is } L \text{ and } y(k) \text{ is } S \text{ THEN } x(k+1) \text{ is } Z \text{ (instead of } S);$$

$$HS_2: \text{ IF } x(k) \text{ is } M \text{ and } d(k) \text{ is } VL \text{ and } y(k) \text{ is } S \text{ THEN } x(k+1) \text{ is } L \text{ (instead of } VL).$$

Table 5.3. The set of rules for the evaluation of $x(k+1)$ and $d(k+1)$

$x(k)/d(k)$	Z	S	M	L	VL
Z	Z/Z	Z/M	Z/M	S/VL	L/L
S	M/Z	M/M	M/M	M/VL	L/S
M	L/Z	L/M	L/M	L/VL	VL/S
L	M/Z	M/M	M/M	M/VL	Z/S
VL	Z/Z	Z/M	Z/M	Z/VL	Z/L

The complete set of fuzzy rules is given in Table 5.4. The differences with respect to Table 5.3 are underlined. The action of y decreases $x(k+1)$ from S to Z or from VL to L. On the other hand, when $y(k)$ is L, only increasing transitions (from L to VL or from S to Z) take place. Therefore, the new rules to be added are the following:

$$HS_3: \text{ IF } x(k) \text{ is } Z \text{ and } d(k) \text{ is } Z \text{ and } y(k) \text{ is } L \text{ THEN } x(k+1) \text{ is } S \text{ (instead of } Z);$$

$$HS_4: \text{ IF } x(k) \text{ is } Z \text{ and } d(k) \text{ is } S \text{ and } y(k) \text{ is } L \text{ THEN } x(k+1) \text{ is } S \text{ (instead of } Z);$$

$$HS_5: \text{ IF } x(k) \text{ is } Z \text{ and } d(k) \text{ is } M \text{ and } y(k) \text{ is } L \text{ THEN } x(k+1) \text{ is } S \text{ (instead of } Z);$$

$$HS_6: \text{ IF } x(k) \text{ is } Z \text{ and } d(k) \text{ is } VL \text{ and } y(k) \text{ is } L \text{ THEN } x(k+1) \text{ is } VL \text{ (instead of } L);$$

$$HS_7: \text{ IF } x(k) \text{ is } M \text{ and } y(k) \text{ is } L \text{ THEN } x(k+1) \text{ is } VL \text{ (instead of } L);$$

$$HS_8: \text{ IF } x(k) \text{ is } L \text{ and } d(k) \text{ is } VL \text{ and } y(k) \text{ is } L \text{ THEN } x(k+1) \text{ is } S \text{ (instead of } Z):$$

The complete set of fuzzy rules is derived in Table 5.5, and the changes made with respect to Table 5.3 are underlined.

In order to complete the whole set of rules for this fuzzy system, the dynamics of $y(k)$ have to be considered. Due to the second equation of (5.2), these simple rules can be added:

Table 5.4. The sets of rules for the evaluation of $x(k+1)$ and $d(k+1)$

$x(k)/d(k)$	Z	S	M	L	VL
Z	Z/Z	Z/M	Z/M	Z/VL	L/L
S	M/Z	M/M	M/M	M/VL	L/S
M	L/Z	L/M	L/M	L/VL	L/S
L	M/Z	M/M	M/M	M/VL	Z/S
VL	Z/Z	Z/M	Z/M	Z/VL	Z/L

Table 5.5. The set of rules for the evaluation of $x(k+1)$ and $d(k+1)$

$x(k)/d(k)$	Z	S	M	L	VL
Z	S/Z	S/M	S/M	S/VL	VL/L
S	M/Z	M/M	M/M	M/VL	L/S
M	VL/Z	VL/M	VL/M	VL/VL	VL/S
L	M/Z	M/M	M/M	M/VL	S/S
VL	Z/Z	Z/M	Z/M	Z/VL	Z/L

$$\begin{aligned}
HS_9 &: \quad \text{IF } x(k) \text{ is } Z \text{ THEN } y(k+1) \text{ is } S;\\
HS_{10} &: \quad \text{IF } x(k) \text{ is } S \text{ THEN } y(k+1) \text{ is } S;\\
HS_{11} &: \quad \text{IF } x(k) \text{ is } M \text{ THEN } y(k+1) \text{ is } M;\\
HS_{12} &: \quad \text{IF } x(k) \text{ is } L \text{ THEN } y(k+1) \text{ is } L;\\
HS_{13} &: \quad \text{IF } x(k) \text{ is } VL \text{ THEN } y(k+1) \text{ is } L.
\end{aligned}$$

These statements are summarized as a set of fuzzy rules in Table 5.6.

Table 5.6. The set of rules for the evaluation of $y(k+1)$

$x(k)/d(k)$	S	M	L
Z	S	S	S
S	S	S	S
M	M	M	M
L	L	L	L
VL	L	L	L

The so-designed fuzzy system is thus completed and its dynamic evolution can be derived. Taking the initial condition $[x(0), y(0), d(0)] = [0.01, 0.01, 0.01]$, the time evolution of $x(k)$ is shown in Fig 5.14, and the phase trajectory in phase space $x - y$ is depicted in Fig. 5.15.

We conclude with some remarks about the qualitative approach for fuzzy modeling of chaotic dynamics.

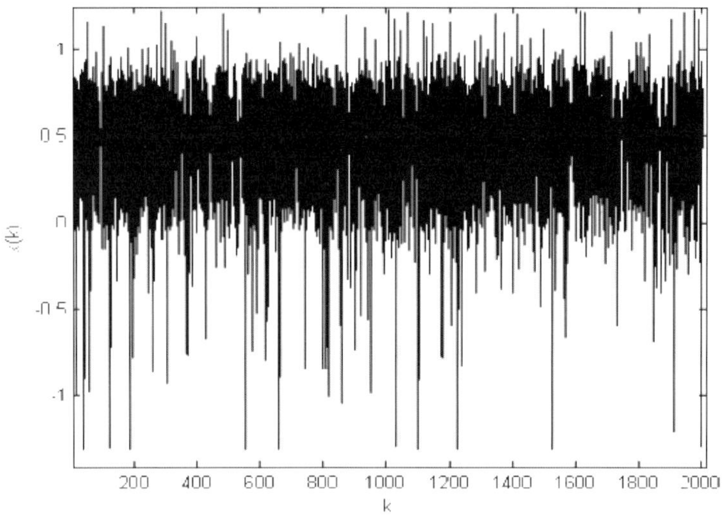

Fig. 5.14. Time evolution of state variable $x(k)$ generated from the fuzzy system

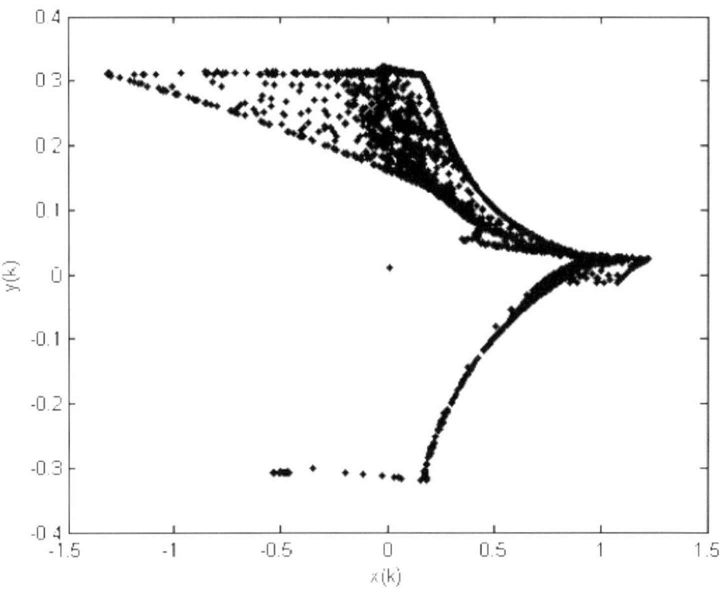

Fig. 5.15. State-space plot of variables $x(k)$ and $y(k)$ generated from the fuzzy system

Remark 5.2. :

i) Simple fuzzy systems are able to generate complex dynamics;
ii) The precision in the approximation of the time series depends on the number and the shape of the fuzzy sets for the variable x, only;
iii) The chaoticity of the system depends on the shape of the fuzzy sets for the uncertainty d, only;
iv) The analysis of a chaotic system via a linguistic description allows for a better understanding of the system itself;
v) Accurate generators of chaos with desired characteristics can be built using fuzzy models;
vi) In some cases, high-dimensional chaotic maps do not need a large number of rules to be modeled.

6

Fuzzy Modeling of Chaotic Systems – II (TS Model)

In this chapter, another typical fuzzy modeling approach, based on the TS fuzzy model, for chaotic dynamics is introduced. The TS fuzzy model has a rigorous mathematical expression and, thus, eases the stability analysis and controller design. A couple of examples will be given to show the procedure of how to construct TS fuzzy models of chaotic systems.

6.1 Introduction

It is already known that the main drawback of fuzzy logic systems is the lack of a systematic modeling and control design methodology. Particularly, stability analysis of fuzzy systems is not easy, and parameter tuning is generally a time-consuming procedure, due to the nonlinear and multi-parametric nature of fuzzy control systems. To resolve these problems, a linear system can be adopted as the consequent part of a fuzzy rule, which leads to the so-called TS model [10].

If a TS fuzzy model does not exactly model a given nonlinear system, the designed controller may not be able to guarantee the control performance and the stability of the closed-loop controlled system. Therefore, in order for a TS fuzzy model to exactly express the nonlinear system, often the linearization method is utilized. The TS fuzzy model gives an input-output relationship identical to that of the nonlinear system only at the operating points of the state space at which the linearization is taken, and only if adequate membership functions are chosen. This method is manual and a case-by-case scheme. To develop a systematic procedure, alternative automatic TS fuzzy modeling methods have recently been developed using soft computing methodologies such as the genetic algorithm (GA) [150] and the neural network theory [151]. An exact TS fuzzy modeling method for nonlinear systems is discussed in [152, 153]. Here, the word "exact" means that the defuzzified output of the constructed TS fuzzy model is mathematically identical to that of the original nonlinear system.

Zhong Li: *Fuzzy Chaotic Systems*, StudFuzz **199**, 91–119 (2006)
www.springerlink.com

In this chapter, both the continuous-time and discrete-time TS fuzzy models will first be reviewed, and a discretization theorem will be derived. Then, the process of TS fuzzy modeling of chaotic systems will be introduced together with a couple of illustrative examples. Finally, the bifurcation phenomena in a simple TS fuzzy system will be investigated and discussed.

6.2 TS Fuzzy Models

Consider a class of continuous-time nonlinear control systems in the following form:

$$\dot{x}(t) = f(x(t)) + g(x(t))u(t)\,, \tag{6.1}$$

where $x(t) \in \Re^n$ is the state vector, $u(t) \in \Re^m$ is the control input vector, and $f(x(t)) \in \Re^n$ and $g(x(t)) \in \Re^n$ are nonlinear vector functions.

A TS fuzzy model is described by a set of fuzzy implications, which characterize local relations of the system in the state space. The main feature of a TS model is to express the local dynamics of each fuzzy rule (implication) by a linear state-space system model, and the overall fuzzy system is then modeled by fuzzy "blending" of these local linear system models through some suitable membership functions. More precisely, the ith rule of the continuous-time TS fuzzy model is formulated in the following form:

Continuous-time TS fuzzy model:

Plant Rule i:

IF $x_1(t)$ is Γ_1^i and $\cdots$ and $x_n(t)$ is Γ_n^i

THEN $\dot{x}(t) = A_i x(t) + B_i u(t)$, $i = 1, 2, \ldots, q\,,$ (6.2)

where Γ_j^i is a fuzzy set, $x(t) \in \Re^n$ is the state vector, $u(t) \in \Re^m$ is the control input vector, $A_i \in \Re^{n\times n}$ and $B_i \in \Re^{n\times m}$ are system matrix and input matrix, respectively, and q is the number of rules of this TS fuzzy model.

The defuzzified output of this TS fuzzy system (6.2) is represented as follows:

$$\dot{x}(t) = \sum_{i=1}^{q} \mu_i(x(t))(A_i x(t) + B_i u(t))\,, \tag{6.3}$$

where

$$\omega_i(x(t)) = \prod_{j=1}^{n} \Gamma_j^i(x_j(t)), \text{ and } \mu_i(x(t)) = \frac{\omega_i(x(t))}{\sum_{i=1}^{q} \omega_i(x(t))}\,,$$

in which $\Gamma_j^i(x_j(t))$ is the grade of membership of $x_j(t)$ in Γ_j^i. Some basic properties of $\omega_i(t)$ are:

$$\omega_i(x(t)) \geq 0, \text{ and } \sum_{i=1}^{q} \omega_i(x(t)) > 0, \; i = 1, 2, \ldots, q\,.$$

It is clear that

$$\mu_i(x(t)) \geq 0, \text{ and } \sum_{i=1}^{q} \mu_i(x(t)) = 1, \ i = 1, 2, \ldots, q.$$

Similarly to the continuous-time case, consider a class of discrete-time nonlinear control systems in the form

$$x(t+1) = f(x(t)) + g(x(t))u(t), \tag{6.4}$$

where t and $t+1$ denote the indices of the time steps. The discrete-time TS fuzzy model is constructed as follows:

$$\begin{aligned}
&\text{Discrete-time fuzzy model:}\\
&\text{Plant Rule } i\text{:}\\
&\quad \text{IF } x_1(t) \text{ is } \Gamma_1^i \text{ and } \cdots \text{ and } x_n(t) \text{ is } \Gamma_n^i\\
&\quad \text{THEN } x(t+1) = G_i x(t) + H_i u(t), i = 1, 2, \ldots, q.
\end{aligned} \tag{6.5}$$

There are a few methods to discretize a linear time-invariant (LTI) continuous-time system. Unfortunately, these discretization methods cannot directly be applied to the discretization of the continuous-time TS fuzzy model, since the defuzzified system is not LTI but is linear time-varying. It is very difficult to obtain the state transition matrix needed for discretization. However, the following theorem has recently been established [155], which lays a rigorous mathematical foundation for the discretization of a continuous-time TS fuzzy model.

Theorem 6.1. *A continuous-time TS fuzzy model of the form*

$$\begin{aligned}
&\textit{Continuous-time plant rule } i:\\
&\quad \textit{IF } x_1(t) \textit{ is } \Gamma_1^i \textit{ and } \cdots \textit{ and } x_n(t) \textit{ is } \Gamma_n^i\\
&\quad \textit{THEN } \dot{x}(t) = A_i x(t) + B_i u(t)
\end{aligned}$$

can be converted to the following model:

$$\begin{aligned}
&\textit{Discrete-time plant rule } i\\
&\quad \textit{IF } x_1(t) \textit{ is } \Gamma_1^i \textit{ and } \cdots \textit{ and } x_n(t) \textit{ is } \Gamma_n^i\\
&\quad \textit{THEN } x(t+1) = G_i x(t) + H_i u(t),
\end{aligned}$$

where

$$G_i = exp(A_i T_s) = I + A_i T_s + A_i^2 \frac{T_s^2}{2!} + \cdots \tag{6.6}$$

$$H_i = \int_0^{T_s} exp(A_i \tau) B_i d\tau = (G_i - I) A_i^{-1} B_i, \tag{6.7}$$

and T_s is the sampling time.

Proof. The exact solution of (6.3) at $t = kT_s + T_s$, where $T_s > 0$ is the sampling period, is given by

$$\begin{aligned} x(kT_s + T_s) = {} & \Phi\left(\sum_{i=1}^{q} h_i(kT_s + T_s), \sum_{i=1}^{q} h_i(kT_s)\right) x(kT_s) \\ & + \int_{kT_s}^{kT_s+T_s} \Phi\left(\sum_{i=1}^{q} h_i(kT_s + T_s), \sum_{i=1}^{q} h_i(\tau)\right) \\ & \times \left[\sum_{i=1}^{q} (h_i B_i)\right] u(\tau) d\tau, \end{aligned} \tag{6.8}$$

where

$$\Phi\left(\sum_{i=1}^{q} h_i(kT_s + T_s), \sum_{i=1}^{q} h_i(kT_s)\right) = \Psi\left(\sum_{i=1}^{q} h_i(kT_s + T_s)\right) \Psi\left(\sum_{i=1}^{q} h_i(kT_s)\right)$$

is the state transition matrix of (6.3), and Ψ is the fundamental matrix of the uncontrolled TS fuzzy system (with $u(t) = 0$), which is nonsingular for all t.

For a sufficiently small T_s, the input $u(t)$ can approximately be regarded as piecewise constant over the integration interval; namely, $u(t) = u(kT_s)$ for $kT_s \leq t < kT_s + T_s$. Then, (6.8) can be rewritten as

$$x(kT_s + T_s) = \bar{G} x(kT_s) + \bar{H} u(kT_s), \tag{6.9}$$

where

$$\begin{aligned} \bar{G} &= \Phi\left(\sum_{i=1}^{q} h_i(kT_s + T_s), \sum_{i=1}^{q} h_i(kT_s)\right), \\ \bar{H} &= \int_{kT_s}^{kT_s+T_s} \Phi\left(\sum_{i=1}^{q} h_i(kT_s + T_s), \sum_{i=1}^{q} h_i(\tau)\right) \times \left[\sum_{i=1}^{q} (h_i B_i)\right] d\tau. \end{aligned}$$

The exact derivation of the state transition matrix $\Phi(\cdot, \cdot)$ is very difficult, if not impossible, since the continuous-time TS fuzzy model (6.3) is time-varying. To solve this problem, we select a set of discrete-time points, kT_s, such that $\sum_{i=1}^{q} h_i(t) A_i$ and $\sum_{i=1}^{q} h_i(t) B_i$ can be approximated by constant matrices $\sum_{i=1}^{q} h_i(kT_s) A_i$ and $\sum_{i=1}^{q} h_i(kT_s) B_i$, respectively, over each interval $[kT_s, kT_s + T_s]$. Then, a set of difference equations can be used to describe the discrete-time TS fuzzy model at each kT_s [154].

In the time interval $kT_s \leq t < kT_s + T_s$, $\bar{G}$ and $\bar{H}$ have the following representations:

$$
\begin{aligned}
\bar{G}(kT_s) &= \exp\left(\left(\sum_{i=1}^{q} h_i(kT_s)A_i\right) T_s\right), \\
\bar{H}(kT_s) &= \int_{kT_s}^{kT_s+T_s} \exp\left(\left(\sum_{i=1}^{q} h_i(kT_s)A_i\right) \times (kT_s + T_s - \tau)\right) \left(\sum_{i=1}^{q} h_i(kT_s)B_i\right) d\tau \\
&= \left(\exp\left(\left(\sum_{i=1}^{q} h_i(kT_s)A_i\right) T_s\right) - I\right) \left(\sum_{i=1}^{q} h_i(kT_s)A_i\right)^{-1} \\
&\quad \cdot \left(\sum_{i=1}^{q} h_i(kT_s)B_i\right).
\end{aligned}
\tag{6.10}
$$

For a sufficiently small sampling period $T_s > 0$, and by using a power series expansion, one has

$$
\begin{aligned}
\bar{G}(kT_s) &= \exp\left(\left(\sum_{i=1}^{q} h_i(kT_s)A_i\right) T_s\right) \\
&= I + \sum_{i=1}^{q} h_i(kT_s)A_iT_s + O(T_s^2) \\
&\approx \sum_{i=1}^{q} h_i(kT_s)(I + A_iT_s) \\
&\approx \sum_{i=1}^{q} h_i(kT_s)\exp(A_iT_s) \\
&= \sum_{i=1}^{q} h_i(kT_s)G_i,
\end{aligned}
$$

and by using the short-hand notation $(\exp(x) - I)x^{-1} = I + (1/2!)x + (1/3!)x^2 + \ldots$, one also has

$$
\begin{aligned}
\bar{H}(kT_s) &= \int_{kT_s}^{kT_s+T_s} \exp\left(\left(\sum_{i=1}^{q} h_i(kT_s)A_i\right) \times (kT_s + T_s - \tau)\right) \left(\sum_{i=1}^{q} h_i(kT_s)B_i\right) d\tau \\
&= \left(\exp\left(\left(\sum_{i=1}^{q} h_i(kT_s)A_i\right) T_s\right)\right)_I \\
&\quad \times \left(\sum_{i=1}^{q} h_i(kT_s)A_i\right)^{-1} \left(\sum_{i=1}^{q} h_i(kT_s)B_i\right) \\
&= \left[\left(\sum_{i=1}^{q} h_i(kT_s)A_i\right) T_s + O(T_s^2)\right]
\end{aligned}
$$

$$\times \left(\sum_{i=1}^{q} h_i(kT_s)A_i\right)^{-1} \left(\sum_{i=1}^{q} h_i(kT_s)B_i\right)$$
$$\approx \sum_{i=1}^{q} h_i(kT_s)B_iT_s$$
$$= \sum_{i=1}^{q} h_i(kT_s)H_i,$$

where

$$G_i = \exp(A_iT_s) = I + A_iT_s + A_i^2\frac{T_s^2}{2!} + \ldots \approx I + A_iT_s$$

and

$$H_i = B_iT_s \approx \int_0^{T_s} \exp(A_i\tau)d\tau = (G_i - I)A_i^{-1}B_i.$$

Thus, the proof is completed.

6.3 Preliminary Theorem

For TS fuzzy modeling of nonlinear systems, the following theorem can be used to convert the nonlinear terms in the nonlinear systems to weighted linear sums of some linear functions [155].

Theorem 6.2. Consider the following nonlinear term:

$$f_n = x_1x_2\cdots x_n\,, \tag{6.11}$$

where $x_i \in \left[M_1^i\,, M_2^i\right]$. Formula (6.11) can exactly be represented by a linear weighted sum of the form

$$f_n = \left(\sum_{i_2,i_3,\ldots,i_n=1}^{2} \mu_{i_2i_3\cdots i_n}\cdot g_{i_2i_3\cdots i_n}\right)x_1, \tag{6.12}$$

where

$$g_{i_2i_3\cdots i_n} = \prod_{j=2}^{n} M_{i_j}^j\,, \quad \mu_{i_2i_3\cdots i_n} = \prod_{j=2}^{n} \Gamma_{i_j}^j\,,$$

in which $\Gamma_{i_j}^j$ is positive semi-definite for all $x_j \in [M_1\,, M_2]$, defined as follows:

$$\Gamma_1^j = \frac{-x_j + M_2^j}{M_2^j - M_1^j}\,, \quad \Gamma_2^j = \frac{x_j - M_1^j}{M_2^j - M_1^j}\,.$$

Proof. Theorem 6.2 can be proved by using inductive reasoning.

If $n = 1$, then Eq. (6.12) is obviously true.

When $n = 2$, the nonlinear equation is $f_2 = x_1 x_2$, which can be represented as the weighted sum of linear functions of x_1 as follows:

$$f_2 = \left(\sum_{i_2=1}^{2} \mu_{i_2} g_{i_2} \right) x_1 = x_2 x_1 ,$$

where

$$g_1 = M_1^2 , \quad g_2 = M_2^2 ,$$
$$\mu_1 = \frac{-x_2 + M_2^2}{M_2^2 - M_1^2} , \qquad \mu_2 = \frac{x_2 - M_1^2}{M_2^2 - M_1^2} .$$

Assuming that Eq. (6.12) holds when $n = k$, then the nonlinear function $f_{k+1} = x_1 x_2 \cdots x_{k+1}$ can be represented by a weighted linear sum of linear functions of x_1 in the following form:

$$\begin{aligned}
f_{k+1} &= \left(\left(\sum_{i_2, i_3 \cdots, i_k}^{2} \mu_{i_2 i_3 \cdots i_k} g_{i_2 i_3 \cdots i_k} \right) \left(\Gamma_1^{k+1} M_1^{k+1} + \Gamma_2^{k+1} M_2^{k+1} \right) \right) x_1 \\
&= \left(\sum_{i_2, i_3, \cdots, i_k}^{2} \mu_{i_2 i_3 \cdots i_k 1} \cdot g_{i_2 i_3 \cdots i_k 1} + \mu_{i_2 i_3 \cdots i_k 2} \cdot g_{i_2 i_3 \cdots i_k 2} \right) x_1 \\
&= \left(\sum_{i_2, i_3, \ldots, i_{k+1}=1}^{2} \mu_{i_2 i_3 \cdots i_{k+1}} \cdot g_{i_2 i_3 \cdots i_{k+1}} \right) x_1 .
\end{aligned}$$

Hence, Eq. (6.12) holds for all n.

6.4 TS Fuzzy Modeling of Chaotic Systems: Examples

In the following, we give some examples to show the process of TS fuzzy modeling for some chaotic systems.

6.4.1 Continuous-time Chaotic Lorenz System

The Lorenz system is a representative continuous-time chaotic system, which is derived from the simplified model of convection rolls in the atmosphere [156]. The Lorenz system shows very typical chaotic phenomena with some simple nonlinearity.

The Lorenz equations are as follows:

$$\frac{d}{dt}\begin{bmatrix} x \\ y \\ z \end{bmatrix} = \begin{bmatrix} -\sigma x + \sigma y \\ rx - y - xz \\ xy - bz \end{bmatrix}, \tag{6.13}$$

where $\sigma, r, b > 0$ are parameters (actually, σ is the Prandtl number, r is the Rayleigh number, and b is a scaling constant). The nominal values of (σ, r, b) are $\left(10, 28, \frac{8}{3}\right)$ for chaos to emerge. Figure 6.1 shows the phase trajectory of (6.13) starting from $(x(0), y(0), z(0)) = (10, -10, -10)$. The system (6.13) has two nonlinear quadratic terms, xy and xz. Therefore, this system can be divided into a linear system with a nonlinear term as follows:

$$\frac{d}{dt}\begin{bmatrix} x \\ y \\ z \end{bmatrix} = \begin{bmatrix} -\sigma & \sigma & 0 \\ r & -1 & 0 \\ 0 & 0 & -b \end{bmatrix}\begin{bmatrix} x \\ y \\ z \end{bmatrix} + \begin{bmatrix} 0 \\ -xz \\ xy \end{bmatrix}. \tag{6.14}$$

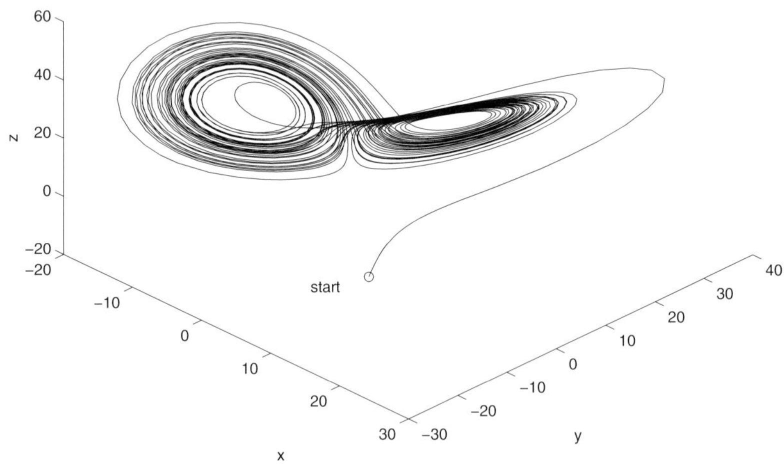

Fig. 6.1. Trajectory of the chaotic Lorenz system

By Theorem 6.2, the nonlinear terms xz and xy in (6.14) can be expressed as weighted linear sums of some linear functions.

Corollary 6.3. Assume $x \in [M_1, \quad M_2]$. The nonlinear term

$$f(x, y) = xy$$

can be represented by a linear weighted sum of linear functions of the form

$$f(x,y) = \left(\sum_{i_2=1}^{2} \mu_{i_2} g_{i_2}(x,y)\right) y\,,$$

where

$$g_1(x,y) = M_1\,, \quad g_2(x,y) = M_2,$$

and

$$\mu_1 = \Gamma_1^2\,, \quad \mu_2 = \Gamma_2^2\,,$$
$$\Gamma_1^2 = \frac{-x + M_2}{M_2 - M_1}\,, \quad \Gamma_2^2 = \frac{x - M_1}{M_2 - M_1}\,.$$

Now, we can construct an exact TS fuzzy model of system (6.13). Using Corollary 6.3, system (6.14) can be expressed as follows:

Plant Rules:

Rule 1: IF $x(t)$ is about M_1 THEN $\frac{d}{dt}\begin{bmatrix} x(t) \\ y(t) \\ z(t) \end{bmatrix} = A_1 \begin{bmatrix} x(t) \\ y(t) \\ z(t) \end{bmatrix}$

Rule 2: IF $x(t)$ is about M_2 THEN $\frac{d}{dt}\begin{bmatrix} x(t) \\ y(t) \\ z(t) \end{bmatrix} = A_2 \begin{bmatrix} x(t) \\ y(t) \\ z(t) \end{bmatrix}$

where

$$A_1 = \begin{bmatrix} -\sigma & \sigma & 0 \\ r & -1 & -M_1 \\ 0 & M_1 & -b \end{bmatrix}, \quad A_2 = \begin{bmatrix} -\sigma & \sigma & 0 \\ r & -1 & -M_2 \\ 0 & M_2 & -b \end{bmatrix},$$

and the membership functions are

$$\Gamma_1 = \frac{-x + M_2}{M_2 - M_1}\,, \quad \Gamma_2 = \frac{x - M_1}{M_2 - M_1}\,,$$

where $\Gamma_i, i = 1, 2$, are positive semi-definite for all $x \in [M_1, \quad M_2]$. The trajectory of the TS fuzzy model of the Lorenz system is shown in Fig. 6.2. It is noticed that Fig. 6.1 is identical to Fig. 6.2, because these two models have exactly the same outputs.

It is emphasized that the TS fuzzy model of the Lorenz system is not an approximation of the original system, but is a perfect fuzzy model, since the defuzzified output of the TS fuzzy model is identical to that of the original chaotic Lorenz system.

6.4.2 Continuous-time Flexible-joint Robot Arm

Here, the TS fuzzy model of a flexible-joint robot arm is derived, and its structure is shown in Fig. 6.3.

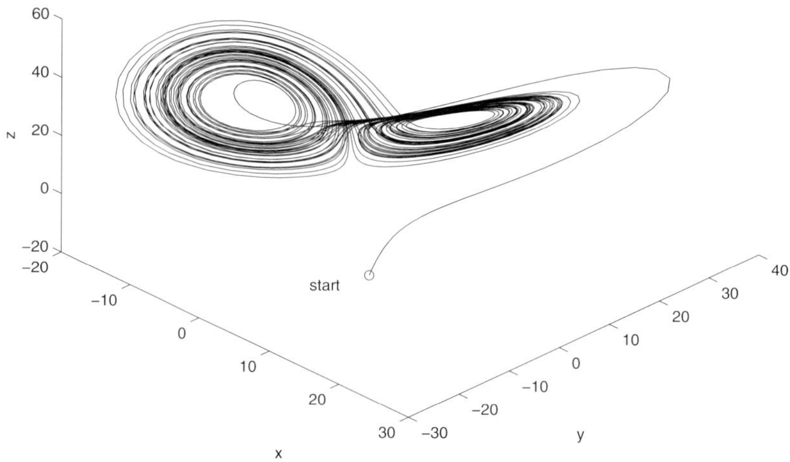

Fig. 6.2. Trajectory of the TS fuzzy model of the chaotic Lorenz system

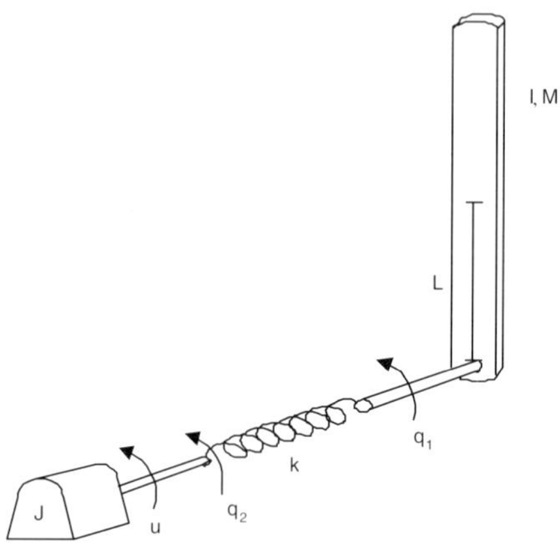

Fig. 6.3. Structure of the flexible-joint robot arm

The flexible joint of this robot arm model is modeled as a linear spring of stiffness k. Let $\theta_1(t)$ be the link-angular variable with the vertical axis as its reference, and $\theta_2(t)$ be the actuator-shaft angle, of which reference is not important. The link is assumed to be rigid with rotational inertia I about the axis of rotation. Assume also. that the rotor inertia of the actuator shaft is J. Let $u(t)$ be the torque, considered as the control input to the arm system, generated by the actuator. Ignoring damping for simplicity, then the Euler-Lagrange equations of motions are

$$\begin{cases} I\ddot{\theta}_1(t) + Mgl\sin(\theta_1(t)) + k(\theta_1(t) - \theta_2(t)) = 0, \\ J\ddot{\theta}_2(t) - k(\theta_1(t) - \theta_2(t)) = u(t), \end{cases} \tag{6.15}$$

where M is the total mass of the arm, l the distance to the joint from the mass-center of the axis of rotation, and g the gravity constant [157]. The following actual system parameter values are borrowed from [158]: $I = 0.03\ (kgm^2), J = 0.004\ (kgm^2), M = 0.2687\ (kg), g = 9.8\ (m/sec^2), l = 1\ (m), k = 31.00\ (Nm/rad)$.

Choosing the state vector $x(t)$ as $[\theta_1(t), \dot{\theta}(t)_1, \theta_2(t), \dot{\theta}_2(t)]^T$ yields the following representation in form of state-space equations.

$$\begin{bmatrix} \dot{x}_1(t) \\ \dot{x}_2(t) \\ \dot{x}_3(t) \\ \dot{x}_4(t) \end{bmatrix} = \begin{bmatrix} x_2(t) \\ -\frac{Mgl}{I}\sin(x_1(t)) - \frac{k}{I}(x_1(t) - x_3(t)) \\ x_4(t) \\ \frac{k}{J}(x_1(t) - x_3(t)) \end{bmatrix} + \begin{bmatrix} 0 \\ 0 \\ 0 \\ \frac{1}{J} \end{bmatrix} u(t). \quad (6.16)$$

System (6.16) has a nonlinear term, $\sin(x_1(t))$. If this nonlinear term can be represented as a weighted linear sums of some linear functions, then the TS fuzzy model of (6.16) can be constructed. Similar to Corollary 6.3, we introduce the following corollary.

Corollary 6.4. Assume $x(t) \in [M_1\,, M_2]$. The nonlinear term

$$f(x(t)) = \sin(x(t))$$

can be represented by a linear weighted sum of linear functions of the form

$$f(x(t)) = \left(\sum_{i=1}^{2} \mu_i g_i(x(t)) \right) x(t)\,,$$

where

$$g_1(x(t)) = 1\,, \quad g_2(x(t)) = \alpha\,,$$

and

$$\mu_1 = \Gamma_1\,, \quad \mu_2 = \Gamma_2\,,$$
$$\begin{cases} \Gamma_1 = \dfrac{\sin(x(t)) - \alpha x(t)}{(1-\alpha)x(t)}\,, \quad \Gamma_2 = \dfrac{x(t) - \sin(x(t))}{(1-\alpha)x(t)}\,, & \text{for } x(t) \neq 0 \\ \Gamma_1 = 1\,, \quad \Gamma_2 = 0\,, \quad \text{for } x(t) = 0\,, \end{cases}$$

and $\alpha = \sin^{-1}(\max(M_1\,, M_2))$.

Using Corollary 6.4, the TS fuzzy model of (6.16) can be represented as follows:

Plant Rules:
Rule 1: IF $x(t)$ is about M_1 THEN $\dot{x}(t) = A_1 x(t) + B_1 u(t)$
Rule 2: IF $x(t)$ is about M_2 THEN $\dot{x}(t) = A_2 x(t) + B_2 u(t)$

where

$$A_1 = \begin{bmatrix} 0 & 1 & 0 & 0 \\ -\frac{Mgl}{I} & -\frac{k}{I} & 0 & \frac{k}{I} & 0 \\ 0 & 0 & 0 & 1 \\ \frac{k}{J} & 0 & -\frac{k}{J} & 0 \end{bmatrix}, \quad A_2 = \begin{bmatrix} 0 & 1 & 0 & 0 \\ -\frac{\alpha Mgl}{I} & -\frac{k}{I} & 0 & \frac{k}{I} & 0 \\ 0 & 0 & 0 & 1 \\ \frac{k}{J} & 0 & -\frac{k}{J} & 0 \end{bmatrix},$$

$$B_1 = B_2 = \begin{bmatrix} 0 \\ 0 \\ 0 \\ \frac{k}{J} \end{bmatrix},$$

and the membership functions are

$$\begin{cases} \Gamma_1 = \dfrac{\sin(x(t)) - \alpha x(t)}{(1-\alpha)x(t)}, \quad \Gamma_2 = \dfrac{x(t) - \sin(x(t))}{(1-\alpha)x(t)}, \quad \text{for } x(t) \neq 0, \\ \Gamma_1 = 1, \quad \Gamma_2 = 0, \quad \text{for } x(t) = 0, \end{cases}$$

where $\alpha = \sin^{-1}(\max(M_1, M_2))$, and $\Gamma_i, i = 1, 2$, are positive semi-definite for all $x_1(t) \in [M_1, M_2]$.

It is obvious that the TS fuzzy model of the flexible-joint robot arm does not have any modeling uncertainties, since the defuzzified output of the TS fuzzy model is exactly same as that of the original flexible-joint robot arm system (6.16).

6.4.3 Duffing-like Chaotic Oscillator

Consider the following Duffing-like chaotic oscillator [159]:

$$\ddot{y}(t) - ay(t) + by(t)|y(t)| = -\epsilon(\zeta\dot{y}(t) - c\sin(\omega t)) \tag{6.17}$$

where $a = 1.1, b = 1$, and $c = 21$ are some positive constants, and $\zeta = 3$ and $\epsilon = 0.1$ are small positive constants for chaos to emerge. The trajectory of this system shown in Fig. 6.4 is chaotic and irregular. Let the system state be $x_c(t) = \left[y(t), \dot{y}(t)\right]^T$, and rewrite (6.17) as

$$\frac{d}{dt}x_c(t) = \begin{bmatrix} x_{c2}(t) \\ ax_{c1}(t) - bx_{c1}(t)|x_{c1}(t)| - \epsilon\zeta x_{c2}(t) \end{bmatrix} + \begin{bmatrix} 0 \\ \epsilon c\sin(\omega t) \end{bmatrix} \tag{6.18}$$

First, in order to construct the TS fuzzy system, the nonlinear term $x_{c1}(t)|x_{c1}(t)|$ should be expressed as a convex sum of the state as follows:

$$x_{c1}(t)|x_{c1}(t)| = \Gamma_1^1(x_{c1}(t)) \cdot 0 + \Gamma_1^2(x_{c1}(t)) \cdot Mx_{c1}(t) \tag{6.19}$$

$$\Gamma_1^1(x_{c1}(t)) + \Gamma_1^2(x_{c1}(t)) = 1 \tag{6.20}$$

where $x_{c1}(t) \in (-M, M)$. Reasonably, M is chosen as 2.5. Solving (6.19) and (6.20) yields

$$\begin{cases} \Gamma_1^1(x_{c1}(t)) = 1 - \frac{|x_{c1}(t)|}{M} \\ \Gamma_1^2(x_{c1}(t)) = \frac{|x_{c1}(t)|}{M} \end{cases} \tag{6.21}$$

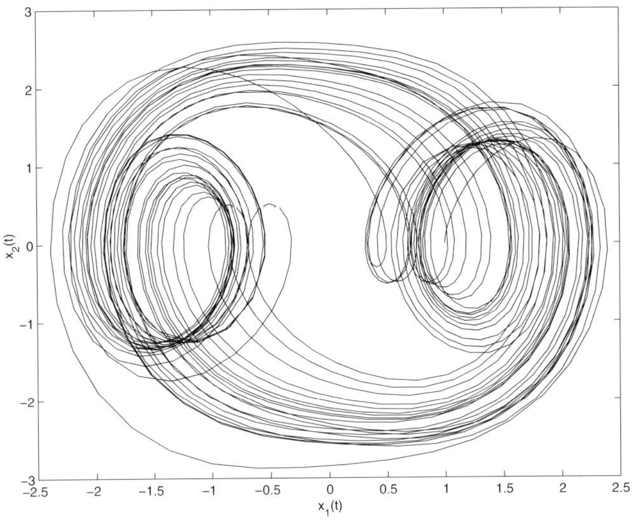

Fig. 6.4. The trajectory of the Duffing-like chaotic oscillator

Next, since only the autonomous dynamical system is considered, it is desirable that the sinusoidal function is included into the state vector. Hence, the state of the system can be redefined as

$$x_c(t) = \left[y(t)\ \dot{y}(t)\ \sin(\omega t)\ \cos(\omega t)\right]^T$$

Now, the analytic TS fuzzy system of (6.17) is given by

$$R^1 : \text{IF } x_{c1}(t) \text{ is about } \Gamma_1^1, \text{ THEN } \dot{x}_c(t) = A_1 x_c(t)$$
$$R^2 : \text{IF } x_{c1}(t) \text{ is about } \Gamma_1^2, \text{ THEN } \dot{x}_c(t) = A_2 x_c(t)$$

where

$$A_1 = \begin{bmatrix} 0 & 1 & 0 & 0 \\ a & -\epsilon\zeta & \epsilon c & 0 \\ 0 & 0 & 0 & \omega \\ 0 & 0 & -\omega & 0 \end{bmatrix}, \quad A_2 = \begin{bmatrix} 0 & 1 & 0 & 0 \\ a - bM & -\epsilon\zeta & \epsilon c & 0 \\ 0 & 0 & 0 & \omega \\ 0 & 0 & -\omega & 0 \end{bmatrix}$$

Since the mathematical equations of the original Duffing-like chaotic oscillator and its TS fuzzy system are the same, one can easily expect that their trajectories are identical.

6.4.4 Other Examples

Rössler's Equation with an Input Term

Rössler's equation with an input term is described as

$$\begin{cases} \dot{x}_1(t) = -x_2(t) - x_3(t), \\ \dot{x}_2(t) = x_1(t) + ax_2(t), \\ \dot{x}_3(t) = bx_1(t) - (c - x_1(t))x_3(t) + u(t), \end{cases}$$

where a, b and c are constants. Assume that $x_1(t) \in [c-d, \quad c+d]$ and $d > 0$. Then, we obtain the following fuzzy model, which exactly represents the nonlinear equation under $x_1(t) \in [c-d, \quad c+d]$:

$$\begin{aligned} &\text{Rule 1 : IF } x_1(t) \text{ is } M_1 \text{ THEN } \dot{x}(t) = A_1x(t) + Bu(t), \\ &\text{Rule 2 : IF } x_1(t) \text{ is } M_2 \text{ THEN } \dot{x}(t) = A_2x(t) + Bu(t), \end{aligned}$$

where $x(t) = [x_1(t) \quad x_2(t) \quad x_3(t)]^T$,

$$A_1 = \begin{bmatrix} 0 & -1 & -1 \\ 1 & a & 0 \\ b & 0 & -d \end{bmatrix}, \quad A_2 = \begin{bmatrix} 0 & -1 & -1 \\ 1 & a & 0 \\ b & 0 & d \end{bmatrix},$$

$$B = \begin{bmatrix} 0 \\ 0 \\ 1 \end{bmatrix},$$

$$M_1(x_1(t)) = \frac{1}{2}\left(1 + \frac{c - x_1(t)}{d}\right), \text{ and } M_2(x_1(t)) = \frac{1}{2}\left(1 - \frac{c - x_1(t)}{d}\right).$$

Duffing Forced-Oscillation

Duffing's forced-oscillation is formulated as

$$\begin{cases} \dot{x}_1(t) = x_2(t) \\ \dot{x}_2(t) = -x_1^3(t) - 0.1x_2(t) + 12\cos(t) + u(t) \end{cases}$$

Assume that $x_1(t) \in [-d, \quad d]$ and $d > 0$. Then, we can have the following fuzzy model as well:

$$\begin{aligned} &\text{Rule 1 : IF } x_1(t) \text{ is } M_1 \text{ THEN } \dot{x}(t) = A_1x(t) + Bu^*(t), \\ &\text{Rule 2 : IF } x_1(t) \text{ is } M_2 \text{ THEN } \dot{x}(t) = A_2x(t) + Bu^*(t), \end{aligned}$$

where $x(t) = [x_1(t) \quad x_2(t)]^T$, $u^*(t) = u(t) + 12\cos(t)$,

$$A_1 = \begin{bmatrix} 0 & 1 \\ 0 & -0.1 \end{bmatrix}, \quad A_2 = \begin{bmatrix} 0 & 1 \\ -d^2 & -0.1 \end{bmatrix},$$

$$B = \begin{bmatrix} 0 \\ 1 \end{bmatrix},$$

$$M_1(x_1(t)) = 1 - \frac{x_1^2(t)}{d^2}, \text{ and } M_2(x_1(t)) = \frac{x_1^2(t)}{d^2}.$$

Hénon Map

The Hénon map has the form

$$x_1(t+1) = -x_1^2(t) + 0.3x_2(t) + 1.4 + u(t),$$
$$x_2(t+1) = x_1(t),$$

Assume that $x_1(t) \in [-d, \quad d]$ and $d > 0$. The following equivalent fuzzy model can be constructed as:

$$\text{Rule 1 : IF } x_1(t) \text{ is } M_1 \text{ THEN } x(t+1) = A_1 x(t) + Bu^*(t),$$
$$\text{Rule 2 : IF } x_1(t) \text{ is } M_2 \text{ THEN } x(t+1) = A_2 x(t) + Bu^*(t),$$

where $x(t) = [x_1(t), \quad x_2(t)]^T$, $u^*(t) = u(t) + 1.4$,

$$A_1 = \begin{bmatrix} d & 0.3 \\ 1 & 0 \end{bmatrix}, \quad A_2 = \begin{bmatrix} -d & 0.3 \\ 1 & 0 \end{bmatrix},$$
$$B = \begin{bmatrix} 1 \\ 0 \end{bmatrix},$$
$$M_1(x_1(t)) = \frac{1}{2}\left(1 - \frac{x_1(t)}{d}\right), \text{ and } \quad M_2(x_1(t)) = \frac{1}{2}\left(1 + \frac{x_1(t)}{d}\right).$$

6.5 Discretization of Continuous-time TS Fuzzy Models

The discretized TS fuzzy model of the continuous-time chaotic Lorenz system is shown as an example in the following:

Plant Rules:

$$\text{Rule 1: IF } x(t) \text{ is about } M_1 \text{ THEN } \begin{bmatrix} x(t+1) \\ y(t+1) \\ z(t+1) \end{bmatrix} = G_1 \begin{bmatrix} x(t) \\ y(t) \\ z(t) \end{bmatrix}$$
$$\text{Rule 2: IF } x(t) \text{ is about } M_2 \text{ THEN } \begin{bmatrix} x(t+1) \\ y(t+1) \\ z(t+1) \end{bmatrix} = G_2 \begin{bmatrix} x(t) \\ y(t) \\ z(t) \end{bmatrix},$$

where

$$G_1 = \begin{bmatrix} 1-\sigma T_s & \sigma T_s & 0 \\ rT_s & 1-T_s & -M_1 T_s \\ 0 & M_1 T_s & 1-bT_s \end{bmatrix}, \quad G_2 = \begin{bmatrix} 1-\sigma T_s & \sigma T_s & 0 \\ rT_s & 1-T_s & -M_2 T_s \\ 0 & M_2 T_s & 1-bT_s \end{bmatrix}.$$

Figure 6.5 shows the phase trajectory of the discrete-time version of the continuous-time TS fuzzy Lorenz model, with $T_s = 0.002$ sec. The overall shape of the phase trajectory is almost the same as that in Fig. 6.1.

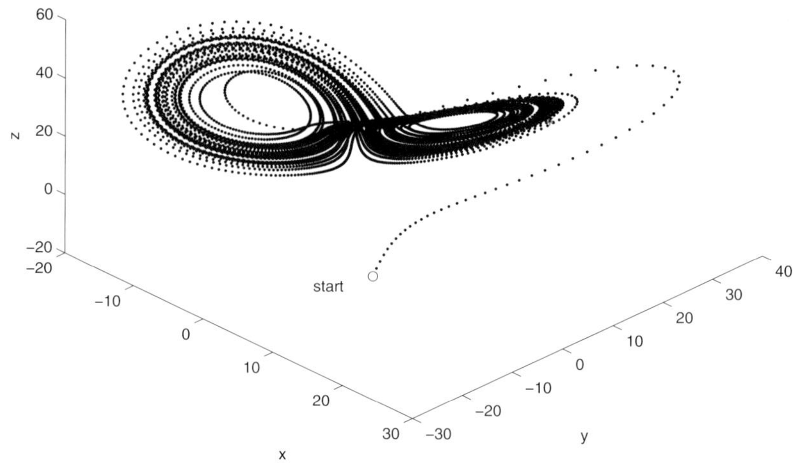

Fig. 6.5. The trajectory of the discretized TS fuzzy model of the chaotic Lorenz system

Next, the continuous time TS fuzzy model of the flexible-joint robot arm is discretized. Based on Theorem 6.1, the discretized version is:

Plant Rules:

Rule 1: IF $x(t)$ is about M_1 THEN $x(t+1) = G_1 x(t) + H_1 u(t)$

Rule 2: IF $x(t)$ is about M_2 THEN $x(t+1) = G_2 x(t) + H_2 u(t)$

where

$$G_1 = \begin{bmatrix} 0 & 1 & 0 & 0 \\ -\frac{MglT_s}{I} - \frac{kT_s}{I} & 0 & \frac{kT_s}{I} & 0 \\ 0 & 0 & 0 & 1 \\ \frac{kT_s}{J} & 0 & -\frac{kT_s}{J} & 0 \end{bmatrix}, \quad G_2 = \begin{bmatrix} 0 & 1 & 0 & 0 \\ -\frac{\alpha MglT_s}{I} - \frac{kT_s}{I} & 0 & \frac{kT_s}{I} & 0 \\ 0 & 0 & 0 & 1 \\ \frac{kT_s}{J} & 0 & -\frac{kT_s}{J} & 0 \end{bmatrix},$$

and

$$H_1 = H_2 = \begin{bmatrix} 0 \\ 0 \\ 0 \\ \frac{kT_s}{J} \end{bmatrix}.$$

where $T_s = 0.0002$ is the sampling time.

6.6 Bifurcation Phenomena in TS Fuzzy Systems

It is known from the fuzzy modeling of chaotic systems discussed above that TS fuzzy systems can be chaotic. Indeed, fuzzy Systems are nonlinear systems and, consequently, they can exhibit multiple equilibria, periodic orbits and even chaotic attractors. In this section, the bifurcation phenomena, which are very closely related to chaos, in some simple TS fuzzy systems will be qualitatively investigated.

Roughly speaking, a system is said to undergo a bifurcation phenomenon, when a qualitative change in the system response is observed as system parameters are varied. These phenomena can lead to changes in the number of stationary solutions (equilibrium points), to the appearance of oscillations, or even to more complex behaviors such as chaos. Thus, from a qualitative point of view, possible bifurcations are related to issues of robustness, since the system only displays behaviors that are structurally stable far from the bifurcation points [160, 161, 162, 163, 164].

It should be remarked that in terms of bifurcation theory it is possible to determine not only the parameter values for which the qualitative behavior of a system changes, but also the kind of behavior that the system will exhibit after such changes. It will be shown that some of the qualitative characteristics of these complex behaviors can be captured through bifurcation analysis [165, 166].

6.6.1 Bifurcation in TS Fuzzy Systems

Consider an affine TS fuzzy system:

$$\begin{aligned} \text{Rule } i: \text{IF } x_1 \text{ is } \Gamma_1^i \text{ and } x_2 \text{ is } \Gamma_2^i, \ \ldots, \text{ and } x_n \text{ is } \Gamma_n^i \\ \text{THEN } \dot{x} = A_i x + C_i, \quad i = 1, 2, \ldots, q, \end{aligned} \tag{6.22}$$

where Γ_j^i are fuzzy sets, $x = (x_1, x_2, \ldots, x_n)^T$, and A_i, C_i are constant matrices of adequate dimensions. The defuzzified output of this TS fuzzy system (6.22) is represented as follows:

$$\dot{x} = \sum_{i=1}^{q} \mu_i(x)(A_i x + C_i) \tag{6.23}$$

where

$$\mu_i(x) = \frac{\prod_{j=1}^{n} \Gamma_j^i(x_j)}{\sum_{i=1}^{q} \prod_{j=1}^{n} \Gamma_j^i(x_j)} \tag{6.24}$$

are nonlinear functions of the state vector $x \in \Re^n$, representing the weight or contribution of each rule R_i to the dynamics of the system.

Equation (6.23) can be regarded as a parametric system $\dot{x} = f(x, p)$, where $p \in \mathcal{P}$ stands for the parameters of the system, because in a fuzzy dynamic system the parameter space $\mathcal{P}$ can be formed by the parameters appearing in the membership functions and the consequents.

Bifurcation phenomena can be classified into two main classes, namely, local and global ones. Roughly speaking, *local bifurcations* are due to changes in the dynamics of a small region of the phase space, typically, a neighborhood of an equilibrium point. When all the eigenvalues of the system's linearization at one equilibrium point have real parts unequal to zero, the equilibrium is called *hyperbolic*. It is known that hyperbolic equilibria whose eigenvalues have negative real parts are asymptotically stable, while if there is some eigenvalue with positive real part, then the equilibrium is unstable. Thus, starting, for instance, from a stable equilibrium, a crossing of certain eigenvalues through the imaginary axis will lead to a local bifurcation. It was already mentioned in Sections 3.2 and 3.3 that the most basic bifurcations are those corresponding to a zero eigenvalue (saddle-node, pitchfork and transcritical bifurcations) or to a pair of pure imaginary eigenvalues (Hopf bifurcation) [162, 164].

Furthermore, there can be *global bifurcations*. In this case, a bifurcation is produced in such a way that it involves phenomena not reducible to locality. That occurs, for instance, when the interaction of a limit cycle with a saddle point is produced. In such a case, an attractor can suddenly appear or disappear. Apart from this situation, which is called a *homoclinic or saddle connection*, and its analogue involving two equilibria (*heteroclinic connection*) [162, 164], other global bifurcations are, for instance, the saddle-node bifurcation of periodic orbits (two periodic orbits of different stability character collide to disappear or vice versa) and the Hopf bifurcation from infinity, where a periodic orbit of great amplitude comes from (goes to) infinity [169]. Global bifurcations cannot be detected by local analysis and normally need numerical or approximate global methods. In addtion, the harmonic balance method can be used [170, 171, 172].

The relevance of qualitative analysis and bifurcations in TS fuzzy systems has already been pointed out ([173], and the references therein). However, theoretical analysis of the complicated structure is still lacking from the mathematical point of view.

Due to the piecewise linear membership functions with local support, the state space of TS systems can be divided into operating regions X_i as shown in Fig. 6.6, where only one rule and the corresponding affine dynamics is active, and interpolation regions in between them, where several dynamics are present. Nevertheless, the problem remains nontrivial to analyze (bifurcation theory normally assumes a high degree of smoothness for the system).

For simplicity, only the simplest TS systems are considered, i.e., $x \in \Re^2$ and only x_1 with two fuzzy sets are included in the antecedents, which yields two rules ($q = 2$ in (6.24)). Suppose that the TS fuzzy systems have the following structure,

$$\begin{aligned} &\text{IF } x_1 \text{ is } N \text{ THEN } \dot{x} = A_1 x + C_1, \\ &\text{IF } x_1 \text{ is } P \text{ THEN } \dot{x} = A_2 x + C_2, \end{aligned} \tag{6.25}$$

where the linguistic terms N and P are represented by the normalized trapezoidal membership functions shown in Fig. 6.7, which results in the following dynamical system,

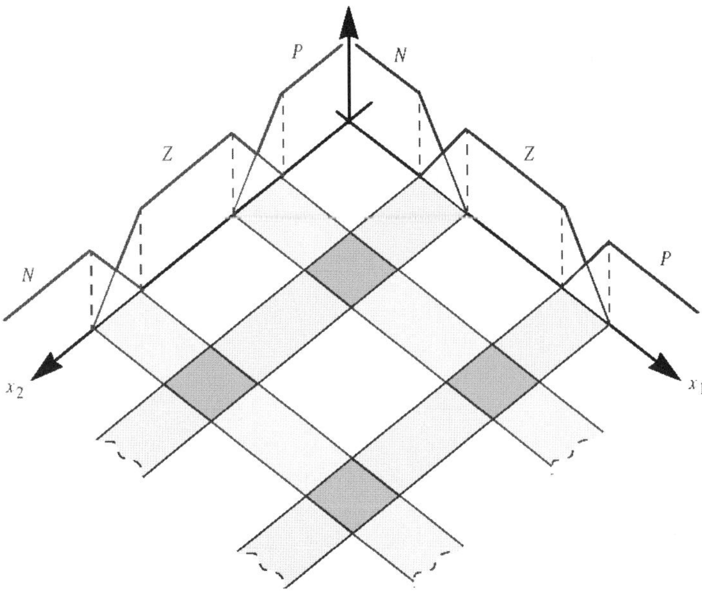

Fig. 6.6. State space partition induced by piecewise linear membership functions

$$\dot{x} = \mu_1(x)(A_1x + C_1) + \mu_2(x)(A_2x + C_2), \tag{6.26}$$

where

$$\mu_1(x) = N(x_1) = \begin{cases} 1 & \text{for} \quad x_1 < -1, \\ \frac{1}{2} - \frac{1}{2}x_1 & \text{for} \quad |x_1| \leq 1, \\ 0 & \text{for} \quad x_1 > 1, \end{cases}$$
$$\mu_2(x) = P(x_1) = \begin{cases} 0 & \text{for} \quad x_1 < -1, \\ \frac{1}{2}x_1 + \frac{1}{2} & \text{for} \quad |x_1| \leq 1, \\ 1 & \text{for} \quad x_1 > 1. \end{cases} \tag{6.27}$$

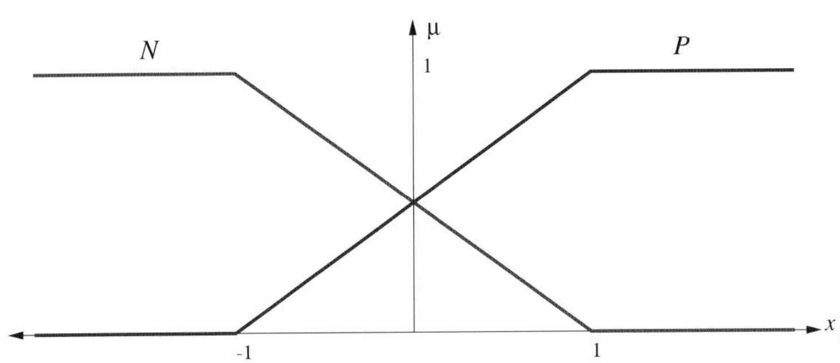

Fig. 6.7. Membership functions

Note that $N(x_1) + P(x_1) = 1$, for all x_1, and that two operating regions and only one interpolating region will appear.

Eqs. (6.25)–(6.27) define a dynamical system in $\Re^2$, which is governed by a piecewise smooth vector field (it is only of class C^0 for $x_1 = 1$ and $x_1 = -1$).

Therefore, the dynamic system (6.25)–(6.27) is equivalent to

$$\dot{x} = \begin{cases} A_1 x + C_1, & \text{for } x_1 < -1, \\ A_2 x + C_2, & \text{for } x_1 > 1, \end{cases} \tag{6.28}$$

and gives rise to a quadratic system in the interpolation region, i.e., for $|x_1| < 1$.

In this setting, the bifurcation analysis can take advantage of the state space partition induced by the membership functions. Since in every operating region the system becomes purely affine, it is easy to derive the equilibrium points and the Jacobian, i.e., $A_i, i = 1, 2$. It should be noticed that even the *virtual* equilibria, i.e., the solutions of $A_i x + C_i = 0, \quad i = 1, 2$, which are not in the corresponding region, do govern the dynamics in the region. In the interpolating region, the analysis of its intrinsic dynamics is more complex, as it is a quadratic system. The problem arises when the regional dynamics are merged into a unique state space. For instance, the merging of only two affine regions is enough to produce limit cycles, see [174], which is impossible for each separate region. The interaction of trajectories in different regions can produce interesting global phenomena, in spite of the linear nature of the two systems involved [176].

To facilitate the analysis of periodic orbits, the Poincaré section method will be employed. This technique reduces the problem to the study of discrete dynamics on a space of dimension $n - 1$, where n is the dimension of the original system.

If γ is a periodic orbit of $\dot{x} = f(x)$ with period $T > 0$, and Φ is the flow of the system, then $\Phi_{t+T}(x) = \Phi_t(x)$ for $x \in \gamma$ and arbitrary $t \in \Re$. Moreover, given a local transversal section of $\mathcal{S}$ at $x \in \gamma$, it can be shown that there is an open set $\mathcal{U}$ of x in $\mathcal{S}$ and a unique differentiable function $\rho : \mathcal{U} \to \Re$, so that

$$\rho(x) = T, \quad \Phi_{\rho(y)}(y) \in \mathcal{S}, \forall y \in \mathcal{U}. \tag{6.29}$$

where $\rho(y)$ is the time required by the orbit staring at point $y \in \mathcal{U}$ to reach the section $\mathcal{S}$ at a point $P(y)$ (see Fig. 6.8a).

Thus, there exists a map $P : \mathcal{U} \to \mathcal{S}$, defined by

$$P(y) = \Phi_{\rho(y)}(y), \quad y \in \mathcal{U}. \tag{6.30}$$

where P is the Poincaré map or first return map associated to the flow in a neighborhood of the closed periodic orbit γ. Therefore, $P(x) = x$, with $x \in \gamma$, is a fixed point of the Poincaré map.

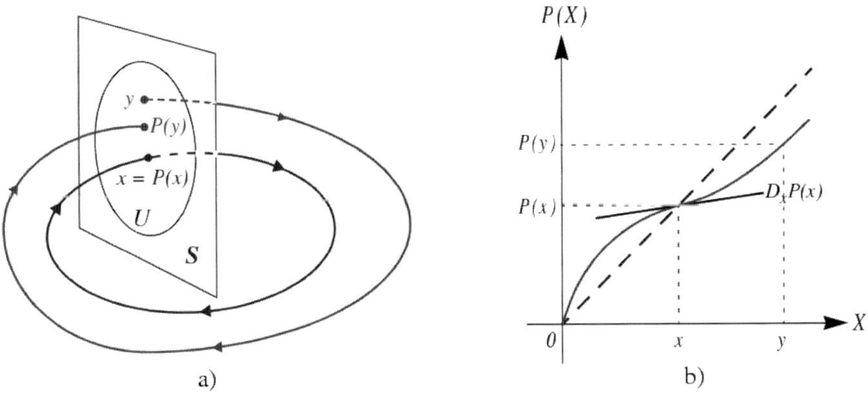

Fig. 6.8. (**a**) Poincaré section in a three-dimensional state space; (**b**) resulting Poincaré map in a bi-dimensional space

The linear approximation of P in x is $D_xP(x)$, and it can be shown [177] that the eigenvalues of $D_xP(x)$ are independent on the point $x \in \gamma$ chosen for tracing the section and on the section S itself. Therefore, it is possible to analyze the stability of a periodic orbit from the eigenvalues of $D_xP(x)$, which in case of dimension 2 ($n = 2$) are given by the slope of the curve $P(x)$ at x (see Fig. 6.8b).

In the following, we take a simple TS fuzzy system with linear consequents as an example to carry out the bifurcation analysis.

6.6.2 TS Fuzzy Systems with Linear Consequents

Consider the TS fuzzy system

$$\begin{aligned} &\text{IF } x_1 \text{ is } N \text{ THEN } \dot{\mathbf{x}} = \begin{bmatrix} 1 & -\frac{1}{2} \\ 1 & 0 \end{bmatrix} \mathbf{x}, \\ &\text{IF } x_1 \text{ is } P \text{ THEN } \dot{\mathbf{x}} = \begin{bmatrix} -\beta & -\frac{1}{2} \\ \beta & 0 \end{bmatrix} \mathbf{x}, \end{aligned} \tag{6.31}$$

where β is the bifurcation parameter. The linguistic terms N and P are represented by normalized trapezoidal membership functions as shown in Fig. 6.7.

The phase space can be partitioned into two operating regions and one interpolating region according to the membership functions, so that

$$
\begin{aligned}
\dot{\mathbf{x}} &= \begin{bmatrix} 1 & -\frac{1}{2} \\ 1 & 0 \end{bmatrix} \mathbf{x}, \quad \text{for } x_1 < -1, \\
\dot{\mathbf{x}} &= \tfrac{1-x_1}{2} \begin{bmatrix} 1 & -\frac{1}{2} \\ 1 & 0 \end{bmatrix} \mathbf{x} + \tfrac{1+x_1}{2} \begin{bmatrix} -\beta & -\frac{1}{2} \\ \beta & 0 \end{bmatrix} \mathbf{x} \\
&= \begin{bmatrix} \frac{1-\beta}{2} x_1 - \frac{1+\beta}{2} {x_1}^2 - \frac{1}{2} x_2 \\ \frac{1+\beta}{2} x_1 - \frac{1-\beta}{2} {x_1}^2 \end{bmatrix}, \text{ for } |x_1| \le 1, \\
\dot{\mathbf{x}} &= \begin{bmatrix} -\beta & -\frac{1}{2} \\ \beta & 0 \end{bmatrix} \mathbf{x}, \quad \text{for } x_1 > 1.
\end{aligned}
\tag{6.32}
$$

The global dynamics in the phase space is the composition of the three regional dynamics. Therefore, all the trajectories can be determined by gluing together (not only continuously but also with a continuous derivative) trajectories in each region. It can be seen that the dynamics in the left operating region ($x_1 < -1$) is a linear one, independent of the bifurcation parameter. For this region the origin is the only equilibrium governing the dynamics (in fact, one unstable focus), but note that it is out of the region and, so, it constitutes a virtual equilibrium. Trajectories enter this region from the middle region at $x_1 = -1$ for $x_2 > -2$, and they always return to the middle region at $x_1 = -1$ for $x_2 < -2$.

Now, considering the right operating region, the corresponding dynamics, which are also linear, depend on the value of β. Again, for $\beta \neq 0$, the only equilibrium governing the dynamics is the origin (a virtual equilibrium). By linear analysis it is possible to make the following assertions:

- For $\beta < 0$, the virtual equilibrium at the origin is a saddle.
- For $\beta = 0$, there appears a continuum of equilibria at the x_1 axis. For $x_1 \geq 1$, these points are actual equilibrium points and the dynamics are rather degenerate, which originates from the lack of differentiability, as all trajectories are horizontal straight lines (entering the region for $x_2 < 0$ and leaving it for $x_2 > 0$).
- For $0 < \beta < 2$, the dynamics are of a stable focus type. Trajectories enter the region at $x_1 = 1$ for $x_2 < -2\beta$ and always return to the middle region at $x_1 = 1$ for $x_2 > -2\beta$.
- For $\beta = 2$, the dynamics are governed by a stable node. Trajectories enter the right region at $x_1 = 1$ for $x_2 < -4$ and always return to the middle region at $x_1 = 1$ for $x_2 \in (-4, -1)$. At $x_1 = 1$ for $x_2 \geq -1$, trajectories leave the region coming from the point at infinity.
- For $\beta > 2$, the dynamics are governed by a stable node with a behavior of trajectories similar to that of the previous case.

The analysis of the middle (interpolating) region dynamics is somehow more complex as the system is quadratic. First of all, apart from the equilibrium point at the origin (now, an actual equilibrium), if $\beta \neq 1$, there is another equilibrium point at

$$(\bar{x}_1, \bar{x}_2) = \left(\frac{1+\beta}{1-\beta}, -\frac{4\beta(1+\beta)}{(1-\beta)^2} \right), \tag{6.33}$$

which is a virtual equilibrium for $\beta > 0$, and an actual one for $\beta \leq 0$.

Note that for the complete system the value $\beta = 0$ clearly represents a bifurcation value, since the dynamics change at this value. When $\beta > 0$, the system has one equilibrium point. At $\beta = 0$, there appears a half-straight line of equilibrium points ($x_1 \geq 1$, with $x_2 = 0$) from which the system inherits one new equilibrium point at $(\bar{x}_1, \bar{x}_2)$ for $\beta < 0$. This bifurcation can be thought of as a degenerate saddle-node bifurcation.

The linearization of the equilibrium point at the origin is

$$J(0,0) = \begin{bmatrix} \frac{1-\beta}{2} & -\frac{1}{2} \\ \frac{1+\beta}{2} & 0 \end{bmatrix}, \tag{6.34}$$

so that

$$\text{trace} J(0,0) = \frac{1-\beta}{2}, \quad \det J(0,0) = \frac{1+\beta}{2}. \tag{6.35}$$

For $\beta < -1$ the origin is unstable (a saddle point). For $\beta = -1$ it is an unstable nonhyperbolic equilibrium, which is a necessary condition for a bifurcation to occur. The origin is also unstable (node or focus) for $-1 < \beta < 1$, becoming stable for $\beta > 1$. Further, when $\beta = 1$, this equilibrium is nonhyperbolic, since its linearization has a pair of pure imaginary eigenvalues, which might be associated to a Hopf bifurcation. To detect the character of this Hopf bifurcation, the Poincaré map can be applied for different values of β around $\beta = 1$ as shown in Fig. 6.11. Thus, it is concluded that for $\beta = 1$ there is a global nonlinear center (GC) that, for $|\beta - 1| \neq 0$ and $|\beta - 1|$ small, does not give rise to any periodic orbit. Notice that this can also be concluded from a mathematical analysis of the global system equations [171] as will be shown in Section 6.7. That analysis also yields that for $\beta = 1$ there is a global nonlinear center which, for $|\beta - 1| \neq 0$ and small, does not give rise to any periodic orbit.

To study the character of the second equilibrium point, it suffices to compute from (6.32) the corresponding linearization, namely,

$$\begin{aligned} J(\bar{x}_1, \bar{x}_2) &= \begin{bmatrix} \frac{(1-\beta)}{2} - (1+\beta)\bar{x}_1 & -\frac{1}{2} \\ \frac{(1+\beta)}{2} - (1-\beta)\bar{x}_1 & 0 \end{bmatrix} \\ &= \begin{bmatrix} -\frac{1+6\beta+\beta^2}{2(1-\beta)} & -\frac{1}{2} \\ -\frac{1+\beta}{2} & 0 \end{bmatrix}, \end{aligned} \tag{6.36}$$

and it should be remarked that

$$\text{trace}J(\bar{x}_1, \bar{x}_2) = -\frac{1+6\beta+\beta^2}{2(1-\beta)}, \quad \det J(\bar{x}_1, \bar{x}_2) = -\frac{1+\beta}{4}. \tag{6.37}$$

Thus, the point $(\bar{x}_1, \bar{x}_2)$ is a saddle point for $\beta > -1$. When $\beta = -1$, the system undergoes a bifurcation, since this equilibrium and the equilibrium at the origin coalesce.

It is clear that the sign of the trace in (6.37) is positive both for $\beta > 1$ and for $\beta \in (\beta_1, \beta_2)$, where β_1, β_2 are the roots of the quadratic equation $1 + 6\beta + \beta^2 = 0$, i.e.,

$$\beta_1 = -3 - 2\sqrt{2}, \quad \beta_2 = -3 + 2\sqrt{2}. \tag{6.38}$$

Therefore, at $\beta = -1$ a transcritical bifurcation takes place (two equilibrium points collide interchanging their stability properties).

Further, another bifurcation arises when $\beta = \beta_1$, since the trace changes its sign with a positive determinant. In fact, the system undergoes a subcritical Hopf bifurcation, denoted by H^{sub}, as the point $(\bar{x}_1, \bar{x}_2)$ (an unstable focus for $\beta > \beta_1$) becomes a stable focus for $\beta < \beta_1$, with one unstable limit cycle around it. According to bifurcation theory, the amplitude of this limit cycle evolves with the bifurcation parameter, and it can be approximated by $O(|\beta - \beta_1|^{\frac{1}{2}})$ for small $|\beta - \beta_1|$. Therefore, the unstable limit cycle will remain in the interpolating region for $\beta < \beta_1$ if $|\beta - \beta_1|$ is sufficiently small, representing the attraction basin of the point $(\bar{x}_1, \bar{x}_2)$, as shown in Fig. 6.10a.

When $|\beta - \beta_1|$ becomes larger (with $\beta < \beta_1$), the above unstable limit cycle begins to enter the left operating region, and approaches the origin from its right-hand side. Then, for a certain value of β, a global bifurcation appears when the limit cycle becomes a loop connecting one branch of the unstable manifold at the origin with one branch of its stable manifold (saddle connection or homoclinic bifurcation). After this critical value, the relative positions of these two manifold branches change, which results in the disappearance of the limit cycle. Now, the attraction basin of the point $(\bar{x}_1, \bar{x}_2)$ is no longer bounded and, depending on the situation of the initial conditions with respect to the stable manifold of the origin, trajectories go to infinity or to the stable point $(\bar{x}_1, \bar{x}_2)$.

It is obvious that for $\beta > 1$ the system exhibits a standard behavior, with no sensitivity to initial conditions and only one equilibrium, which is globally stable. When $\beta < \beta_1$, the system also possesses a stable equilibrium, but it can be said that it is not robust, since its attraction basin is limited. For the intermediate values of β, i.e., in the range $[\beta_1, 1]$, the system is unstable, and so it can be considered useless.

Thus, very different system behavior may be found depending on the actual value of β. It must be emphasized that the identification of the adequate value of β turns out to be a critical issue.

The whole analysis for the system (6.31) can be summarized in the bifurcation diagram of Fig. 6.9. Also, the corresponding phase portraits for several

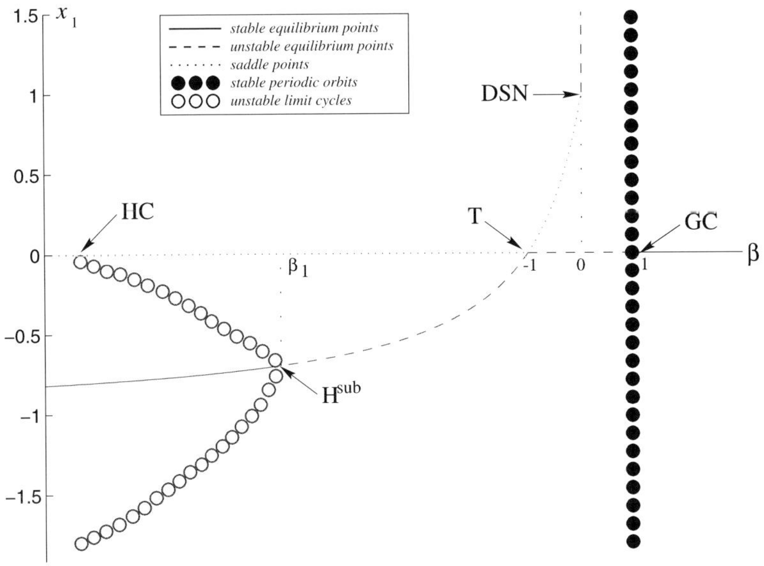

Fig. 6.9. Bifurcation diagram of Example 1 (HC = Homoclinic Connection; H^{sub} = Subcritical Hopf; T = Transcritical; DSN = Degenerated Saddle Node; GC = Global Center)

values of β are sketched in Fig. 6.10 and Fig. 6.11. Figure 6.11 shows the corresponding Poincaré maps for different values of β with the section S set at $x_1 = 0$. Notice that the slope of the curve in Fig. 6.11a is greater than 1 and the system is unstable. In Fig. 6.11b, the slope is always 1 which corresponds to a global center. On the other hand, in Fig. 6.11c the slope is always lower than 1 showing global stability.

6.7 Appendix: Bifurcation Analysis for $\beta = 1$

The standard techniques of Hopf bifurcation analysis do not provide any interesting information, apart from the fact that in this case the bifurcation is degenerate. Here, an alternative approach is proposed. For $\beta = 1$, system (6.32) is

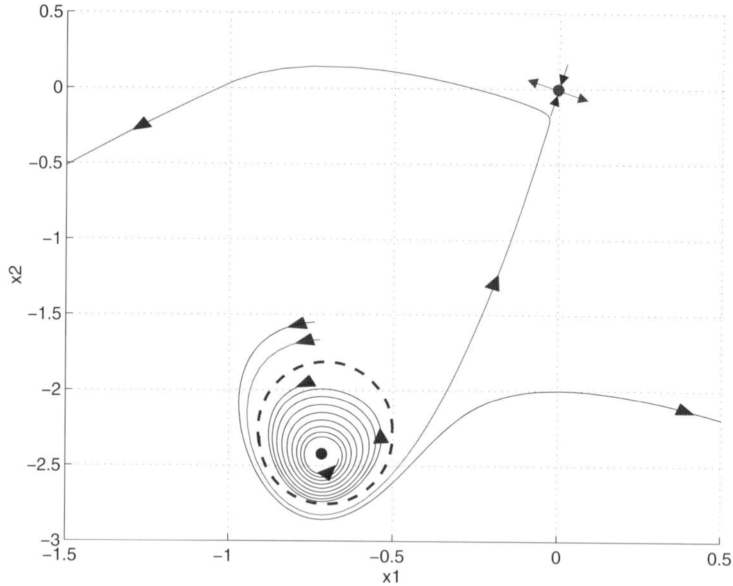

a) $\beta = -6$

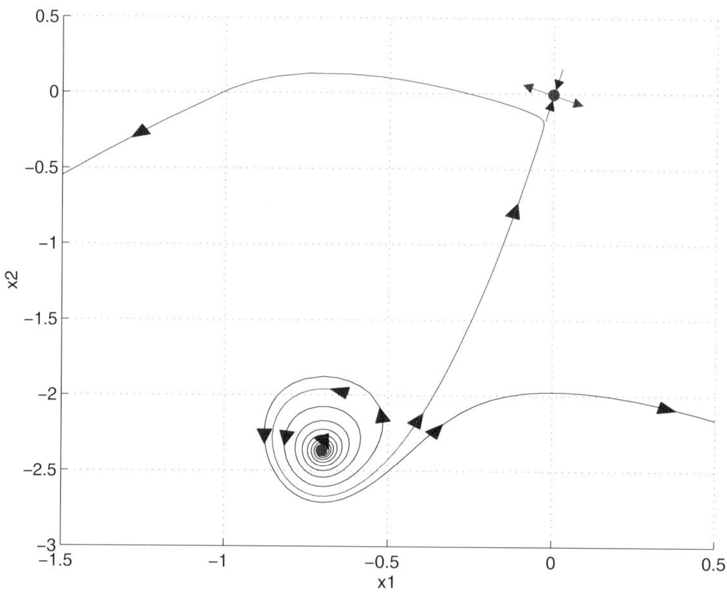

b) $\beta = -5.6$

Fig. 6.10. Phase portraits of Example 1 for different values of β: (**a**) for $\beta = -6$, there is an unstable limit cycle stating the attraction basin of a stable equilibrium (dashed line); (**b**) for $\beta = -5.6$, there are no stable equilibrium points (at $\beta = \beta_1$ the system undergoes a subcritical Hopf bifurcation within the interpolating region)

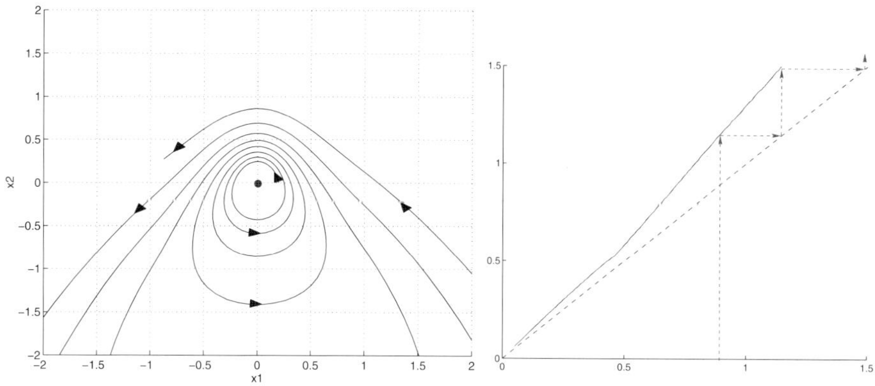

a) $\beta = 0.9$

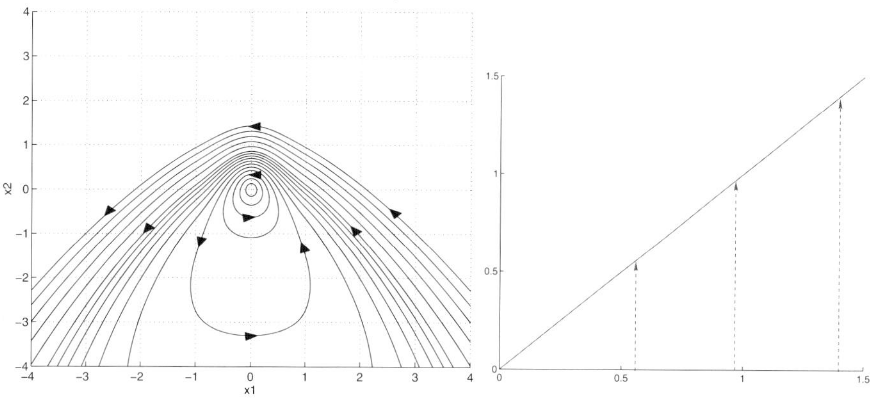

b) $\beta = 1$

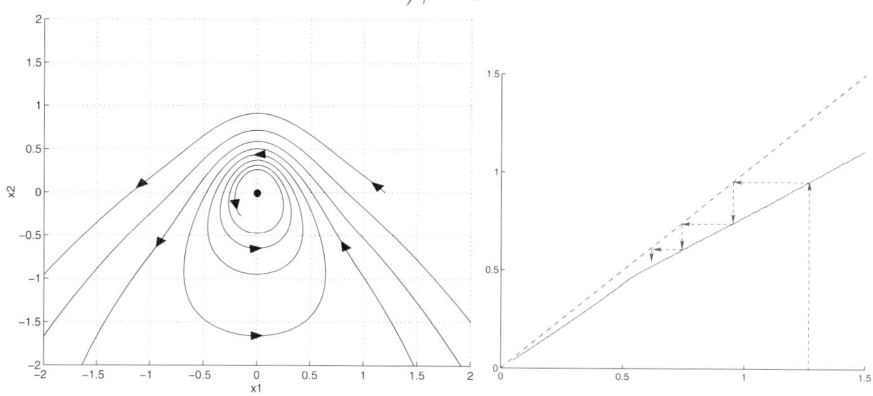

c) $\beta = 1.1$

Fig. 6.11. Phase portraits of Example 1 for different values of β and their corresponding Poincaré maps: (**a**) for $\beta = 0.9$, the system has an unstable equilibrium at the origin; (**b**) for $\beta = 1$, the system has a global center; (**c**) for $\beta = 1.1$, the origin is the global attractor of the system

$$\begin{cases} \dot{x}_1 = x_1 - \frac{1}{2}x_2 \\ \dot{x}_2 = x_1 \end{cases}, \quad \text{for } x_1 < -1,$$

$$\begin{cases} \dot{x}_1 = -x_1^2 - \frac{1}{2}x_2 \\ \dot{x}_2 = x_1 \end{cases}, \quad \text{for } |x_1| \leq 1, \tag{6.39}$$

$$\begin{cases} \dot{x}_1 = -x_1 - \frac{1}{2}x_2 \\ \dot{x}_2 = x_1 \end{cases}, \quad \text{for } x_1 > 1.$$

Lemma 6.5. *System (6.39) has a global nonlinear center.*

Proof. First, consider the quadratic system that coincides with system (6.39) in the middle region. Such quadratic system is not difficult to integrate by rewriting it as follows

$$(x_1^2 + \frac{x_2}{2})\frac{dx_2}{dx_1} + x_1 = 0, \tag{6.40}$$

and observing that e^{2x_2} is an integrating factor. Thus, trajectories are implicitly given by

$$e^{2x_2}(x_1^2 + \frac{x_2}{2} - \frac{1}{4}) = K \tag{6.41}$$

for each value of the constant K.

From (6.41) it is easily deduced that all trajectories of the quadratic system are symmetric with respect to the x_2-axis, since they are defined by

$$x_1 = \pm\sqrt{\frac{1}{4} - \frac{x_2}{2} + Ke^{-2x_2}}, \tag{6.42}$$

provided that there exists a range of values of x_2 where the expression in the radical is positive. In fact, that range always appears for $K \geq -\frac{1}{4}$. For instance, when $K = 0$, the radical is positive for $x_2 \leq \frac{1}{2}$, and the trajectory is given by the parabola $x_2 = \frac{1}{2} - 2x_1^2$.

For $K = -\frac{1}{4}$, the corresponding trajectory degenerates in a point (the origin), while for $K \in (-\frac{1}{4}, 0)$ trajectories form closed curves, see Fig. 6.12.

Obviously, from above trajectories only the portion in the interval $-1 \leq x_1 \leq 1$ represents actual trajectories for system (6.39). However, the complete system (6.39) is invariant under the following transformations

$$\begin{aligned} (x_1, x_2, t) &\to (-x_1, x_2, -t), \text{ for } |x_1| \leq 1, \\ (x_1, x_2, t) &\to (-x_1, -x_2, t), \text{ for } |x_1| \geq 1, \end{aligned}$$

what indicates that, apart from the aforementioned symmetry of trajectories with respect to the x_2−axis in the middle region, trajectories are symmetric with respect to the origin in the outer regions. In these regions the dynamics are of focus type (one unstable and one stable), and so all the trajectories are closed curves. The conclusion follows.

Lemma 6.6. *For $|\beta - 1| \neq 0$ and small, the system (6.32) has no periodic orbits.*

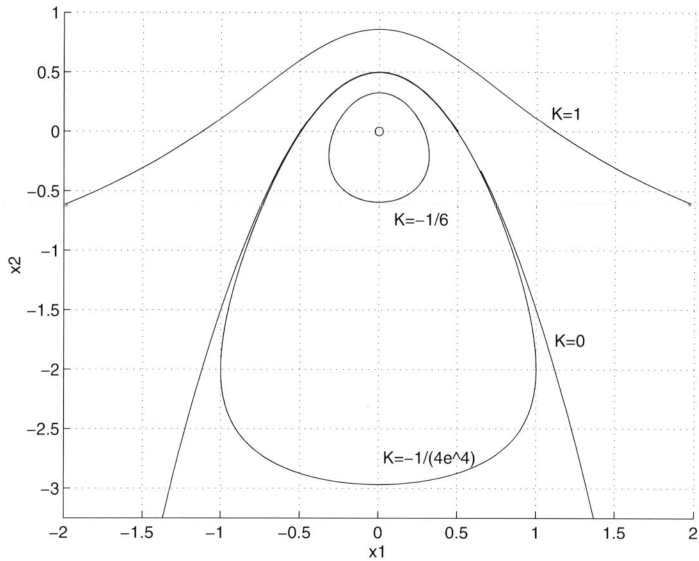

Fig. 6.12. Trajectories of the quadratic system (6.40) for different values of K; when $K \in (-\frac{1}{4}, 0)$, trajectories are closed curves, which completely belong to the middle region only for $K \in (-\frac{1}{4}, -\frac{1}{4e^4})$

Proof. For sake of brevity, technical details will be omitted. After an elemental scaling of time, system (6.32) can be written in Liénard form, i.e.,

$$\begin{aligned} \dot{x}_1 &= F(x_1) - x_2 \\ \dot{x}_2 &= g(x_2) \end{aligned} \tag{6.43}$$

For $|\beta - 1|$ small, it has only one equilibrium at the origin. Then, a necessary condition for the existence of periodic orbits, which is based upon Filippov transformations, is given in Theorem 5 of Cherkas [178] (see also the related remark therein). Basically, the condition needed is that there exist $u < 0 < v$ such that the system equation

$$\begin{aligned} F(u) &= F(v) \\ G(u) &= G(v) \end{aligned} \tag{6.44}$$

is fulfilled, where $G(u) = \int_0^u g(x)dx$. Detailed computations show that the above system cannot be compatible, and the conclusion follows.

7

Fuzzy Control of Chaotic Systems – I (Mamdani Model)

In this chapter, to stabilize chaotic dynamics a design method for fuzzy controllers based on the Mamdani model is introduced, which is also called a model-free approach. As illustrative examples, the chaotic Lorenz system and Chua's circuit will be controlled using this approach.

7.1 Introduction

As already known, many chaos control methods, such as OGY [30] or time-delay feedback control [97, 98], have so far been well derived and tested. So, why fuzzy control of chaos?

Fuzzy control has its intrinsic merits as described in Chapters 1 and 2, namely, there are no requirements to the exact mathematical model of a process or a system, when it is vaguely defined, nonlinear and complex, or when its dynamics are unknown and the sensors provide noisy and incomplete data; and domain experts' empirical knowlegde is employed in designing fuzzy logic controllers (FLC). In particular, fuzzy control can be useful together with one of these classic chaos control methods, as an extra layer of control, in order to improve the robustness to noise, and the ability to control long periodic orbits [33, 129].

The design principle of fuzzy logic controllers has been described in Chapter 2. Unfortunately, there is no standard or systematic procedure available for the design of an FLC, instead we can describe the steps involved and give hints to achieve a satisfactory design. This methodology follows the general guidelines described in [179]. It can be shown that an FLC design is not unique, is based on heuristic knowledge of the process to be controlled, and a trial-and-error process is employed until an adequate response is obtained in an iterative tuning procedure. So far, many methods have been proposed to automate the tuning procedure, which are based on genetic algorithms, neural networks or other optimization techniques.

Zhong Li: *Fuzzy Chaotic Systems*, StudFuzz **199**, 121–141 (2006)
www.springerlink.com

It is known that a fuzzy logic system involves fuzzification, inference, composition and defuzzification. A fuzzy controller can, thus, be designed to perform fuzzy logic operations on fuzzy sets representing linguistic variables in a qualitative set of control rules.

In this chapter, a design methodology for fuzzy logic controllers (although not systematic) will be introduced in detail, which includes to build up a fuzzy rule-base, design fuzzifier and defuzzifier, and perform compositions. Finally, two examples will be given to concretely show the process of the fuzzy controller design for chaotic systems.

7.2 Design of Fuzzy Logic Controllers

The architecture shown in Fig. 2.4 corresponds to a single-input-single-output (SISO) fuzzy controller. The input variable, usually an analogue signal, must be sampled and converted to a discrete signal for further processing. An FLC can be regarded as a special case of digital controllers with nonlinear behaviors.

In designing an FLC, the following design steps should be taken into account:

1. Determination of the input and output variables and the corresponding universes of discourse;
2. Adoption of a proper fuzzification strategy, including:
    a) Partitioning of input and output spaces;
    b) Selection of membership functions of primary fuzzy sets;
    c) Discretization and normalization of the universes of discourse;
3. Rule-base construction:
    a) Selection of input and control variables;
    b) Choice of the control rules;
4. Selection of decision-making logic:
    a) Implication definition;
    b) Definition of composition operator;
    c) Inference mechanism;
    d) Defuzzification strategy.

The steps are described in detail in the following.

7.2.1 Selection of Variables and the Universe of Discourse

The first step in designing a fuzzy control system is to specify the input and output variables and the corresponding universes of discourse. There are two different scenarios. If there exists *a priori* expertise, the FLC develops from human knowledge by mimicking the human expert; otherwise, it is necessary to select a control action (e.g., the output variable of the controller) and to monitor relevant variables, usually error signal and its first derivative.

7.2.2 Fuzzification

Fuzzification is related to vagueness and imprecision, and transforms a crisp value into a subjective one. Actually, it is a map from crisp measured data to grade-of-membership values for given fuzzy sets of the input universe of discourse. Therefore, it is necessary to define the fuzzy sets and their supports and the partitioning of the universes. Typical fuzzy sets are *Positive Small* (PS), *Negative Medium* (NM) etc. Note that increasing the number of fuzzy sets results in more powerful and flexible control.

The number of fuzzy rules increases exponentially as the partition of the universe of discourse increases, since it grows as the product of the number of fuzzy sets for each input. However, a large number of fuzzy rules results in time-consuming computation. Typical numbers of linguistic labels range from two to nine for each of input and output variable, but usually a trial-and-error procedure can generate an appropriate number.

In addition, it is noticed that there is no big difference between the performances of the control actions resulting from the use of different shapes of membership functions, such as triangular, trapezoidal, Gaussian, sigmoid etc. Most of the commercial software tools prefer the trapezoidal shape (and triangular as a special case) for its simple implementation.

In practice, a fuzzification unit in a fuzzy system, called a fuzzifier, converts a crisp value into a fuzzy singleton in a given universe of discourse. A singleton is not a real fuzzy variable in the sense that it represents an input x_0 as a fuzzy set A with a membership degree $\mu_A(x)$ equal to zero except at x_0, where it equals one. The partitioning process used for the inputs is also valid for the outputs, and the outputs sent to perform a control action (valve, current etc.) must be real numbers.

7.2.3 Discretization and Normalization of a Universe of Discourse

In a fuzzy system, uncertain information represented by fuzzy sets should be discretized for nowadays popular digital computer processing. Here, discretization of a universe is referred to quantization. If a universe of discourse is continuous, it should be discretized into segments. To ease processing, a continuous universe should also be normalized.

Quantization discretizes a universe into a certain number of segments (quantization levels), each one labelled by a generic element, to form a discrete universe. Assigning to each generic element of the discrete universe a grade-of-membership value defines a fuzzy set. In general, the number of quantization levels should be large enough to provide an adequate approximation, and yet be small to save memory storage. Table 7.1 shows an example of discretization, where a universe of discourse is discretized into 13 levels with 7 terms (fuzzy sets).

Normalization implies that the physical values of both inputs and outputs are mapped into a predetermined normalized domain. This process involves

Table 7.1. An example for discretization and partitioning

Level	Range	NB	NM	NS	ZE	PS	PM	PB
-6	$x_0 \leq -3.2$	1.0	0.3	0.0	0.0	0.0	0.0	0.0
-5	$-3.2 < x_0 \leq -1.6$	0.7	0.7	0.0	0.0	0.0	0.0	0.0
-4	$-1.6 < x_0 \leq -0.8$	0.3	1.0	0.3	0.0	0.0	0.0	0.0
-3	$-0.8 < x_0 \leq -0.4$	0.0	0.7	0.7	0.0	0.0	0.0	0.0
-2	$-0.4 < x_0 \leq -0.2$	0.0	0.3	1.0	0.3	0.0	0.0	0.0
-1	$-0.2 < x_0 \leq -0.1$	0.0	0.0	0.7	0.7	0.0	0.0	0.0
0	$-0.1 < x_0 \leq 0.1$	0.0	0.0	0.3	1.0	0.3	0.0	0.0
1	$0.1 < x_0 \leq 0.2$	0.0	0.0	0.0	0.7	0.7	0.0	0.0
2	$0.2 < x_0 \leq 0.4$	0.0	0.0	0.0	0.3	1.0	0.3	0.0
3	$0.4 < x_0 \leq 0.8$	0.0	0.0	0.0	0.0	0.7	0.7	0.0
4	$0.8 < x_0 \leq 1.6$	0.0	0.0	0.0	0.0	0.3	1.0	0.3
5	$1.6 < x_0 \leq 3.2$	0.0	0.0	0.0	0.0	0.0	0.7	0.7
6	$3.2 \leq x_0$	0.0	0.0	0.0	0.0	0.0	0.3	1.0

scaling or multiplication of the physical input by a normalization factor. The advantage is that fuzzification, rule firing and defuzzification can be designed independently of the physical domains of input and output variables. This facilitates the implementation of FLCs on microprocessors. In most cases, the fuzzy sets are symmetrical with equal support length.

7.2.4 Construction of the Rule-base

A fuzzy system is defined by a set of linguistic statements that embeds the knowledge of experts. The linguistic statements usually have the form of IF-THEN rules. Four schemes have been proposed to build up sets of fuzzy control rules [179]. It is remarked that the four schemes are not mutually exclusive, but beneficial to one another.

a) Based on the knowledge and experience of a control engineer on the process to be controlled: starting from a minimum set of rules as initial prototype, a complete rule-base is gradually established, mainly based on trial and error until the desired response is achieved.
b) Based on operators' control actions: in case that a system is too complex or has too many variables, it is difficult to derive its mathematical model. Yet a skilled operator can control such a system successfully without any quantitative models in mind by employing a set of fuzzy IF-THEN rules to control the process.
c) Based on a fuzzy model of a process: the linguistic description of the dynamics of a process to be controlled can be regarded as a fuzzy model of the process, based on which a set of fuzzy rules can be derived. Although this approach is more complicated, it results in better performance and reliability, and provides a more tractable structure for dealing theoretically with FLCs.

d) Based on learning: Mamdani proposed a self-organizing controller (SOC), consisting of two sets of rules [181, 182]. The first set is the standard FLC in charge of the control actions. The second one acts as a supervisor exhibiting a human-like behavior to tune the control rules. In this hierarchy of two sets of rules, the supervisory rules, known as "metarules", tune the rule-base according to the system response. This technique leads to adaptive FLCs, such as the neuro-fuzzy controllers ANFIS [183].

For instance, Fig. 7.1 shows the time evolution waveform of an input variable in a control process, where the input variables of the FLC are the error E and its derivative ΔE, and the output is the change of the control action Δu. The time evolution waveform can be discretized into segments, in each of which both E and ΔE keep their signs. Therefore, it is necessary to define the critical points (and, accordingly, the time instants) where either E or ΔE becomes zero. Denote the critical points by $a, b, c, d, \ldots$ and the zones between them by $i, ii, iii, iv, \ldots$ as shown in Fig. 7.1. The objective of the control action is to reduce the overshoot and the rise time.

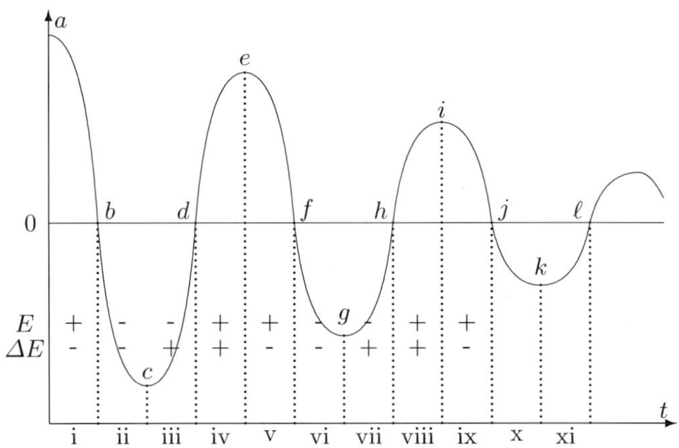

Fig. 7.1. Time evolution waveform of the input variable

To build up a primary fuzzy rule-base, assign E and ΔE with three linguistic variables, {negative (N), zero (Z) and positive (P)}. In Zone i, the rule can be formulated as Rule R_i to shorten the rise time, and in Zone ii, Rule R_{ii} reduces the overshoot of the system response.

$$R_i : \text{IF } E \text{ is } P \text{ and } \Delta E \text{ is } N \text{ THEN } u \text{ is } P;$$
$$R_{ii} : \text{IF } E \text{ is } N \text{ and } \Delta E \text{ is } N \text{ THEN } u \text{ is } N.$$

In this way a coarse fuzzy rule-base can be built up, which may, however, lead to a limit circle around the stable point. Increasing the number of linguistic terms of E and ΔE can resolve this undesired situation, where, for instance, seven linguistic variables, {negative and big (NB), negative and medium (NM), negative and small (NS), zero (Z), positive and small (PS), positive and medium (PM), positive and big (PB)}, can be employed to yield 49 rules.

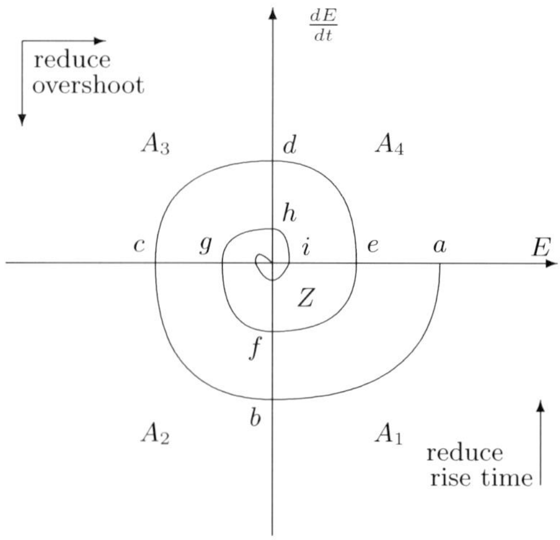

Fig. 7.2. Rule-base derivation by phase plane partitioning

Figure 7.1 can be redrawn in a phase plane as Fig. 7.2. Then, the fuzzy rules can be generated according to the following general rules:

1. If $E = \Delta E$, keep the output of the FLC, i.e., $\Delta u = 0$;
2. If E approaches zero at a right speed, do nothing;
3. The sign and magnitude of Δu depend on E and ΔE according to the following rules:
   a) At the critical points, when the waveform (in Fig. 7.1) or trajectory (in Fig. 7.2) crosses the zero axis ($E = 0$), (b, d, f, ...), $\text{sgn}(\Delta u) = \text{sgn}(\Delta E)$;
   b) At peaks and valleys ($\Delta E = 0$) (c, e, g, ...), $\text{sgn}(\Delta u) = \text{sgn}(\Delta E)$;
   c) In A_1, when E is big, one needs to shorten the rise time; and in A_2, when E is small, one needs to prevent from the overshoot, then:

$$\Delta u > 0, \text{ when the trajectory is far from zero in Fig. 7.2,}$$
$$\Delta u \leq 0, \text{ when the trajectory approaches zero.}$$

   d) In A_2, to prevent from overshoot, one has $\Delta u < 0$;
   e) In A_3 (reverse to that in A_1)

Table 7.2. Initial rule-base

Rule	E	ΔE	Δu	Reference Value
1	PB	ZE	PB	a
2	PM	Z	PM	e
3	PS	Z	PS	i
4	Z	NB	NB	b
5	Z	NM	NM	f
6	Z	NS	NS	j
7	NB	Z	NB	c
8	NM	Z	NM	g
9	NS	Z	NS	k
10	Z	PB	PB	d
11	Z	PM	PM	h
12	Z	PS	PS	l
13	Z	Z	Z	Desired Value
14	PB	NS	PM	i (rise time)
15	PS	NB	NM	i (overshoot)
16	NB	PS	NM	iii
17	NS	PB	PM	iii
18	PS	NS	Z	ix
19	NS	PS	ZE	xi

$$\Delta u \geq 0 \text{ when the trajectory approaches zero,}$$
$$\Delta u < 0 \text{ when } |E| \text{ is big.}$$

f) In A_4, to reduce the overshoot in the valley, one has $\Delta u > 0$.
A primary set of fuzzy rules is concluded in Table 7.2, which is reformed in Table 7.3a), based on which a complete set of fuzzy rules is generated in Table 7.3b).
As the magnitude of the control action is concerned, consider that when $\Delta E \approx 0$, then $u = u_0$, where u_0 takes the following values:

E	NB	NM	NS	Z	PS	PM	PB
u_0	NB	NM	NS	Z	PS	PM	PB

When $\Delta E \neq 0$, then $u = (u_0 + \Delta E) + C$, where C is a compensation factor, which is usually zero.

The sum is defined as a *linguistic sum*, i.e.,

$$PM + PS = PB, \quad PS + PB = PB, \quad PS + NM = NS, \ldots.$$

When E is big (NM, NL, PM, or PL), ΔE has little influence, then $C \neq 0$ for speeding up the response.

Table 7.3. Fuzzy rule-base: a) primary; b) complete

	NB	NM	NS	Z	PS	PM	PB
PB			PM	PB			
PM				PM			
PS	NM		Z	PS			
Z	NB	NM	NS	Z	PS	PM	PB
NS				NS	Z		
NM				NM			
NB				NB	NM		

a)

	NB	NM	NS	Z	PS	PM	PB
PB	Z	PS	PM	PB	PB	PB	PB
PM	NS	Z	PS	PM	PB	PB	PB
PS	NM	NS	Z	PS	PM	PB	PB
ZE	NB	NM	NS	Z	PS	PM	PB
NS	NB	NB	NM	NS	Z	PS	PM
NM	NB	NB	NB	NM	NS	Z	PS
NB	NB	NB	NB	NB	NM	NS	Z

b)

When $|E|$ is small (Z, PS, or NS), ΔE has a big influence, then C is set to be small to prevent overshoots.

The relation between the phase plane and the fuzzy rule-base is depicted in Table 7.4.

Table 7.4. Relation of phase plane and inference matrix

$\Delta E/E$	NB NM NS	Z	PS PM PB
PB PM PS	A_3	d	A_4
Z	c	Z	a,e
NS NM NB	A_2	b	A_1

7.2.5 Defuzzification

The result of rule inference is a fuzzy set. However, a physical actuator cannot be driven by a fuzzy signal. For instance, if the actuator is a valve, the controller's output, acting over a servomotor to control the valve, does not support a command such as *apply a small positive voltage*. It is, therefore, necessary to convert the fuzzy signal into a crisp value, which is executed by the so-called defuzzifier. The most popular defuzzification methods are Maximum (MAX), Mean of Maximum (MOM), and Center of Area (COA), as mentioned in Chapter 2 and shown in Fig. 7.3.

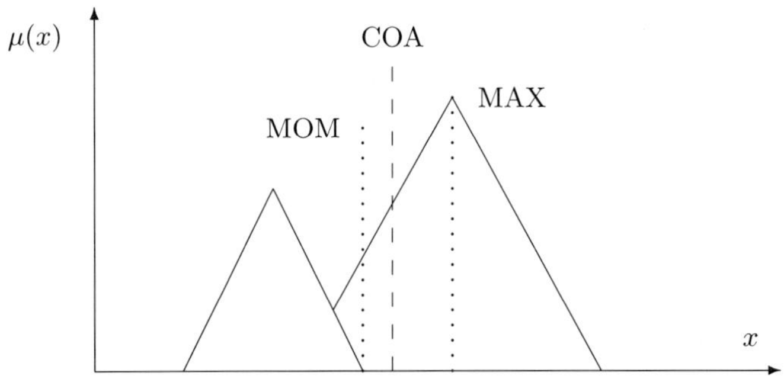

Fig. 7.3. Comparison of defuzzification strategies

7.2.6 Fuzzy Inference Mechanisms

Mamdani proposed an inference mechanism [181] as shown in Fig. 7.4, where Mamdani's minimum operator is employed.

Another popular fuzzy inference mechanism due to Sugeno [180] adopts singletons in the consequents of the rules instead of the fuzzy set, as shown in Fig. 7.5. This greatly eases the defuzzification process and the implementation in computers. The Sugeno systems are classified by order. For instance, a typical rule of order zero is given as:

$$\text{IF } x \text{ is } A \text{ and } y \text{ is } B \text{ THEN } z = k,$$

where A and B are fuzzy sets and k is a constant. Likewise, a Sugeno rule of order one is written as:

$$\text{IF } x \text{ is } A \text{ and } y \text{ is } B \text{ THEN } z = px + qy + r,$$

where A and B are fuzzy sets, and p, q and r are constants [183].

7.3 Fuzzy Control of Chua's Chaotic Circuit

It is well known that Chua's circuit [184, 185] can exhibit chaotic behavior, and its dynamics have been well studied [186]. Occasional proportional feedback

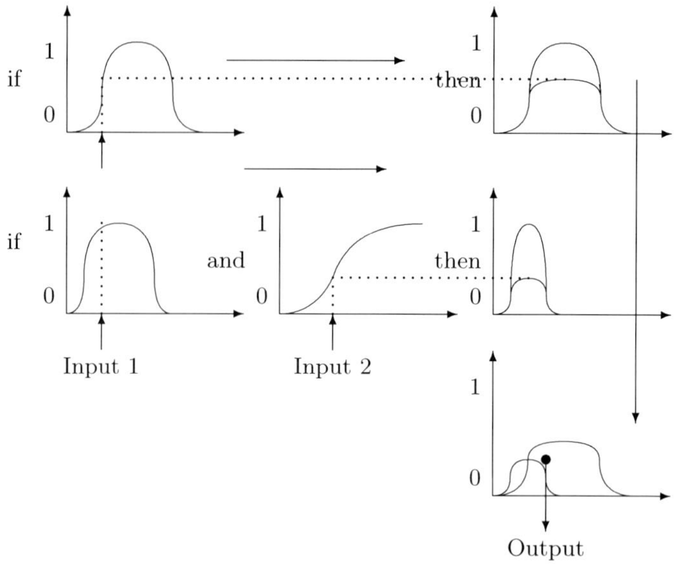

Fig. 7.4. Inference mechanisms for two inputs and single output

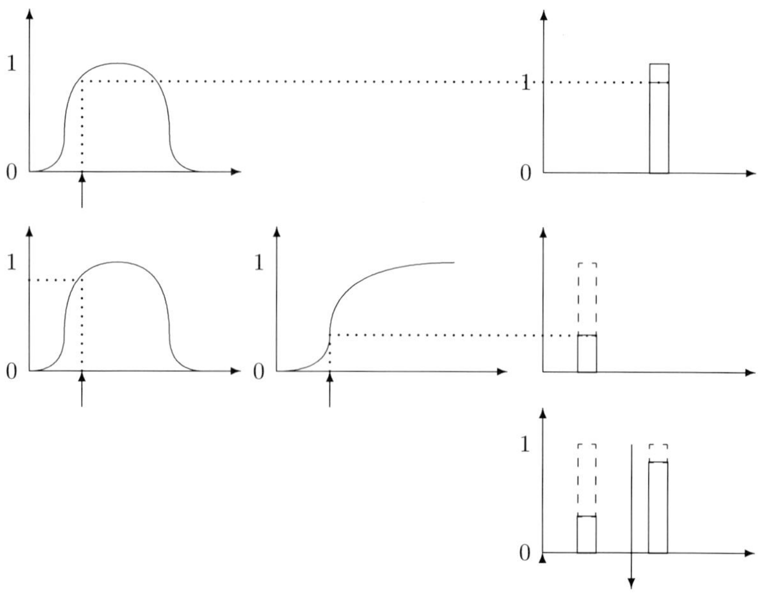

Fig. 7.5. Sugeno's inference method

control has been proposed to control the circuit [187]. This method uses an electronic circuit to sample the peaks of the voltage across the negative resistance, and when it falls into the neighborhood of a set-point value a, the slope of the negative resistance is modified by an amount proportional to the difference between the set-point and the peak value. The nonlinear nature of the system and the heuristic approach used to find the best set of parameters to drive the system to a given periodic orbit suggest that a fuzzy controller is much better suited than the occasional proportional feedback method.

A fuzzy controller is proposed to control the nonlinearity of the nonlinear element (a three-segment nonlinear resistance) within Chua's circuit. The block diagram of the fuzzy controller is like the one shown in Fig. 2.4. It consists of four components, viz., fuzzy rule-base, fuzzifier, inference engine and defuzzifier. The input and output membership functions are shown in Fig. 7.6, which provide a basis for the fuzzification, defuzzification and inference mechanisms. The rule-base is made up of a set of linguistic rules mapping inputs to control actions. The fuzzifier converts the input signals e and Δe into fuzzified signals with grade-of-membership values assigned to linguistic sets. The inference mechanisms operate on all fuzzy rules to generate fuzzy outputs. Finally, the defuzzifier converts the fuzzy outputs to crisp control signals, which control the slope of the negative resistance Δa in Chua's circuit as shown in Fig. 7.7. There, scaling and quantification operations are first applied to the inputs. Table 7.5 shows the quantified levels and the linguistic labels used for inputs and outputs. The knowledge rules (Table 7.6) are represented as control statements such as:

IF e is NB and Δe is NS
THEN Δa is NB.

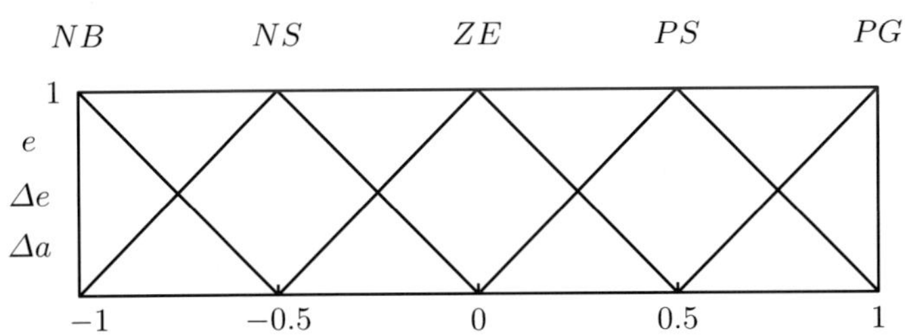

Fig. 7.6. Membership functions of the input and output variables, e, Δe and Δa

Chua's circuit can be formalized as:

$$\begin{cases} \dot{x} = \alpha(y - x - f(x)) \\ \dot{y} = x - y + z \\ \dot{z} = -\beta y \end{cases}, \tag{7.1}$$

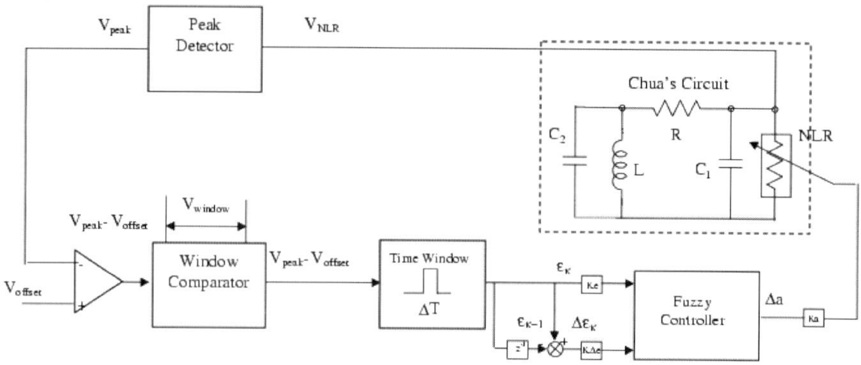

Fig. 7.7. The whole controller and control system in the form of a block diagram, including fuzzy controller, peak detector, window comparator, and Chua's circuit system being controlled

Table 7.5. Quantification levels and membership functions

Linguistic Labels	Membership Functions								
Positive Big (PB)	0	0	0	0	0	0	0	0.5	1
Positive Small (PS)	0	0	0	0	0	0.25	0.5	0.75	1
Approximately Zero (AZ)	0	0	0	0.5	1	0.5	0	0	0
Negative Small (NS)	0	0.5	1	0.5	0	0	0	0	0
Negative Big (NB)	1	0.5	0	0	0	0	0	0	0
Error (e)	-1	-0.75	-0.5	-0.25	0	0.25	0.5	0.75	1
Change in error (Δe)	-1	-0.75	-0.5	-0.25	0	0.25	0.5	0.75	1
Control (Δa)	-1	-0.75	-0.5	-0.25	0	0.25	0.5	0.75	1
Quantification level	-4	-3	-2	-1	0	1	2	3	4

Table 7.6. Rule-table for the linguistic variables defined in Table 7.5

$\Delta e \backslash e$	NB	NS	AZ	PS	PB
NB	NB	NS	NS	AZ	AZ
NS	NS	AZ	AZ	PS	
AZ	NS	AZ	PS	PS	
PS	AZ	AZ	PS	PB	
PB	AZ	PS	PS	PB	

in which

$$f(x) = bx + \frac{1}{2}(a - b)(|x^2 - 1|),$$

where $f(x)$ represents the nonlinear element of the circuit. Changes in the negative resistance are made by changing a by an amount

$$\Delta a = \text{Fuzzy Controller Output} \times \text{Gain} \times a.$$

Figure 7.7 shows the whole control system, including Chua's circuit, a fuzzy controller, a peak detector and a window comparator. Figure 7.8 shows a simulation result of using the fuzzy controller to stabilize an unstable period-1 orbit, where a single correction pulse per cycle of oscillation is applied. Changing the control parameters, one can stabilize orbits of different periods, as shown in Fig. 7.9, where more complex higher-period orbits are stabilized by the fuzzy controller. Furthermore, one can tune the fuzzy controller over

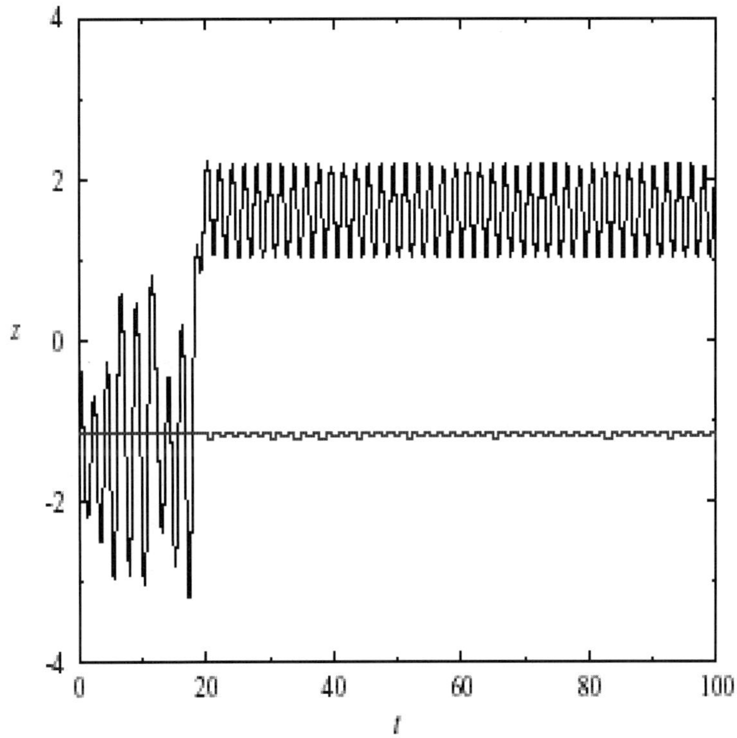

Fig. 7.8. The fuzzy controller stabilizes a previously unstable period-1 orbit; control is switched on at time 20; the lower trace shows the correction pulses applied by the controller

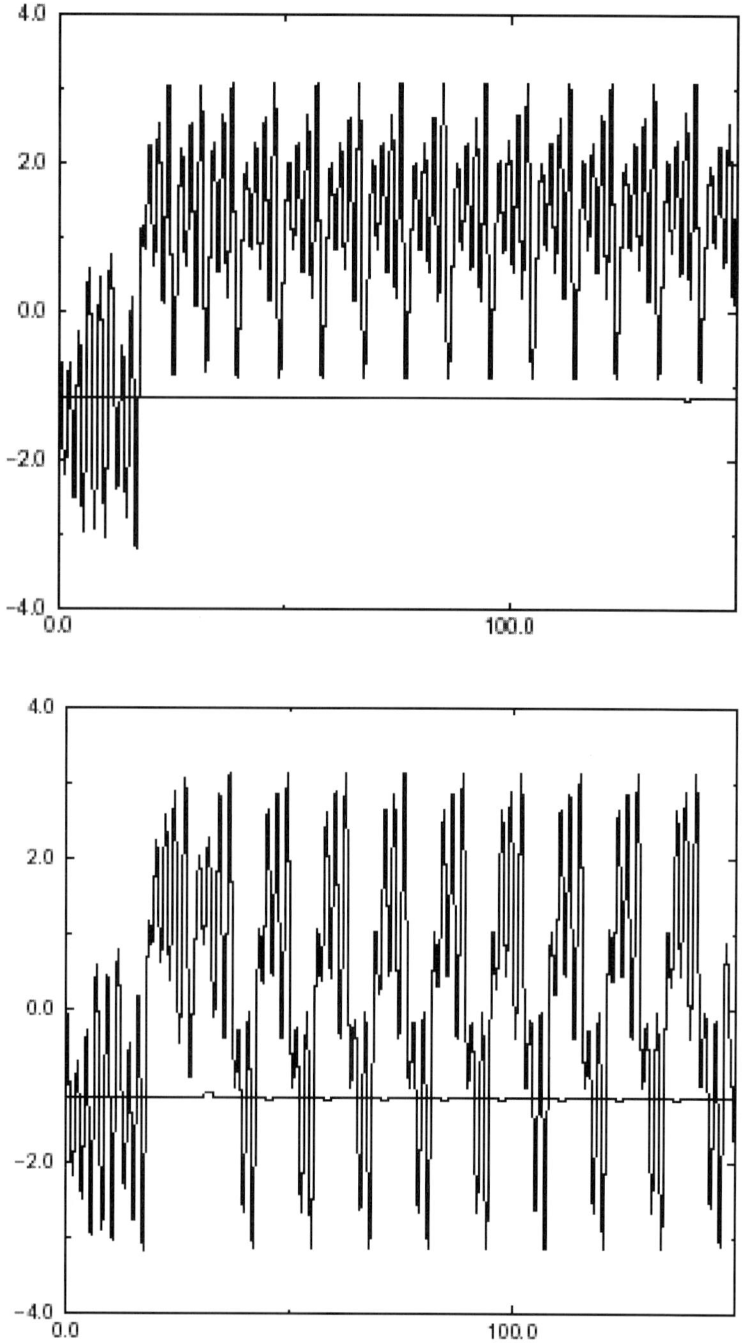

Fig. 7.9. The fuzzy controller stabilizes a previously unstable period-1 orbit; control is switched on at time 20; the lower trace shows the correction pulses applied by the controller

the circuit to achieve various types of responses required in a given situation by modifying some or all of the rules in the system's rule-base.

It is remarked that when the system equations of Chua's circuit are available, fuzzy control is essentially not necessary, but the simple example shows the possibility of using fuzzy control.

7.4 Fuzzy Control of Chaotic Lorenz System

The Lorenz system is described as:

$$\begin{cases} \dot{x} = \sigma(y - x) \\ \dot{y} = \rho x - y - xz \ , \\ \dot{z} = -\beta z + xy \end{cases} \tag{7.2}$$

where σ, ρ and β are parameters. Here, ρ is chosen as the control parameter to be adjusted by the fuzzy controller.

The fuzzy logic controller consists of two inputs and one output with 9 rules as shown in Table 7.7. The membership functions of the input and output variables are shown in Fig. 7.10. The block diagram of the fuzzy controller is shown in Fig. 7.11, in which x, z are the linguistic variables; X and Z are the scaled variables via K_x and K_z, respectively. The FLC is to adjust ρ by $\Delta\rho$:

$$\Delta\rho = \rho \times \text{Fuzzy Controller Output} \times \text{Gain}.$$

Table 7.7. Rule-table for the linguistic variables x and z

$x\backslash z$	N	Z	P
N	NB	NM	ZE
Z	NM	ZE	PM
P	ZE	PM	PB

With the parameter setting $(\sigma, \rho, \beta) = (10, 20, 1)$, the phase portrait and time evolution of the uncontrolled Lorenz system are shown in Fig. 7.12 (a) and (b), respectively.

The control signal u is the fuzzy controller's output, and U is the one scaled by K_u. For $K_x = 0.5$, $K_z = 0.5$ and $K_u = 0.75$, the simulation results are shown in Fig. 7.13. It is shown that the fuzzy controller drives the chaotic behavior to a stable periodic orbit. While for $K_x = 0.15$, $K_z = 0.15$ and $K_u = 0.32$, the Lorenz system is stabilized to the equilibrium point (constant solution) as shown in Fig. 7.14.

Remark 7.1. A fuzzy logic controller can be applied when there is no mathematical model available for the process, which provides robustness of the

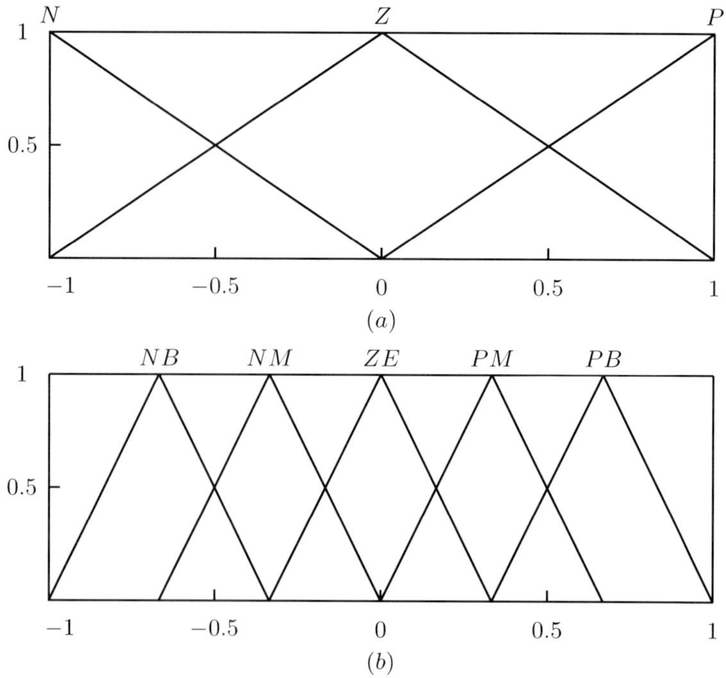

Fig. 7.10. Membership functions of the (a) inputs x and z and (b) output u

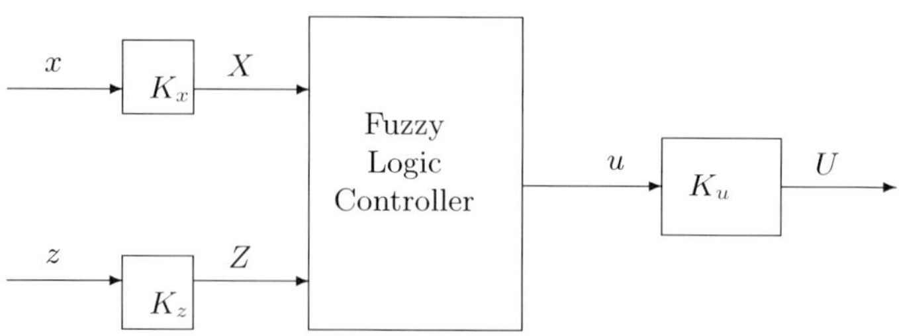

Fig. 7.11. Block diagram of the fuzzy logic controller

proposed fuzzy logic controller design. This is proved by a systematic study of the control error versus the parameters of the nonlinear dynamical system as shown in Figs. 7.15–7.17 for the Lorenz attractor and the same result for Chua's circuit. Figure 7.15 shows the control error when σ changes to 10 (Curve a), to 15 (Curve b), to 20 (Curve c) and to 5 (Curve d). On the other hand, Fig. 7.16 shows the control error when β is changed, Curve a for $\beta = 1$, Curve b for $\beta = 0.5$ and Curve c for $\beta = 1.5$. In addition, Fig. 7.17 shows the control error when both σ and β are changed, Curve a corresponds to the

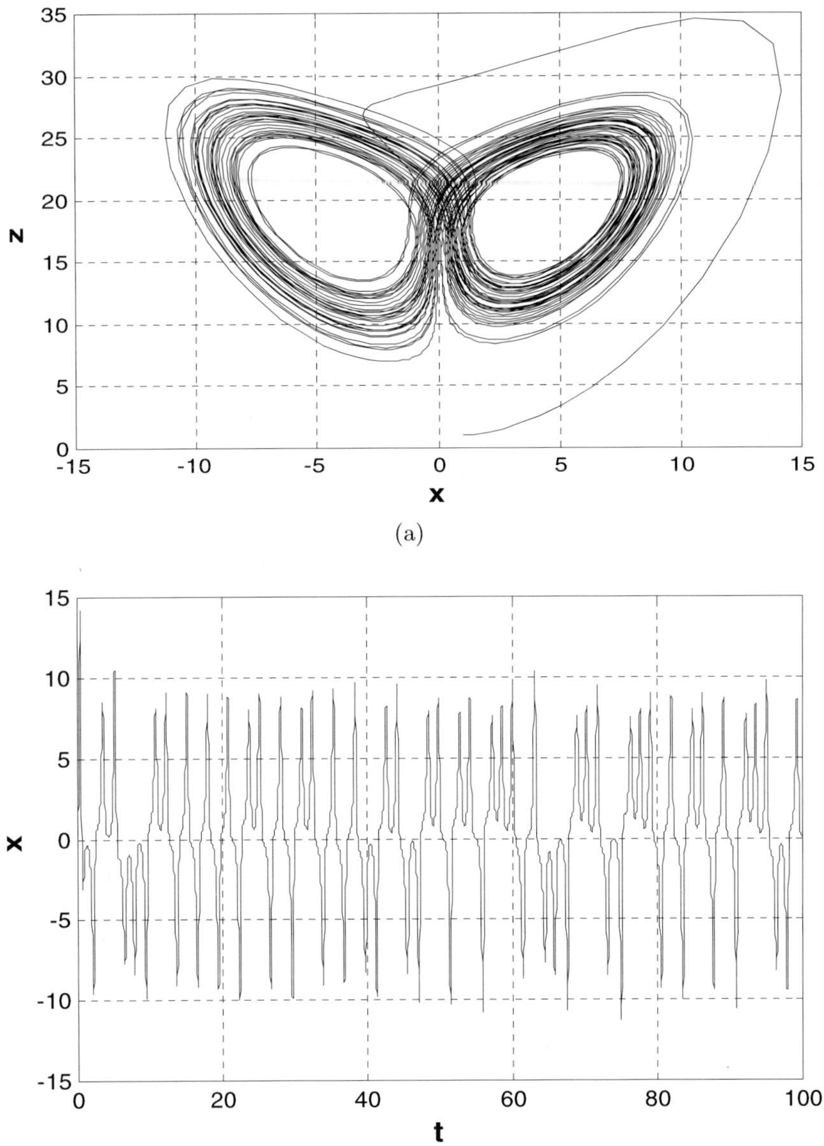

Fig. 7.12. Uncontrolled Lorenz system

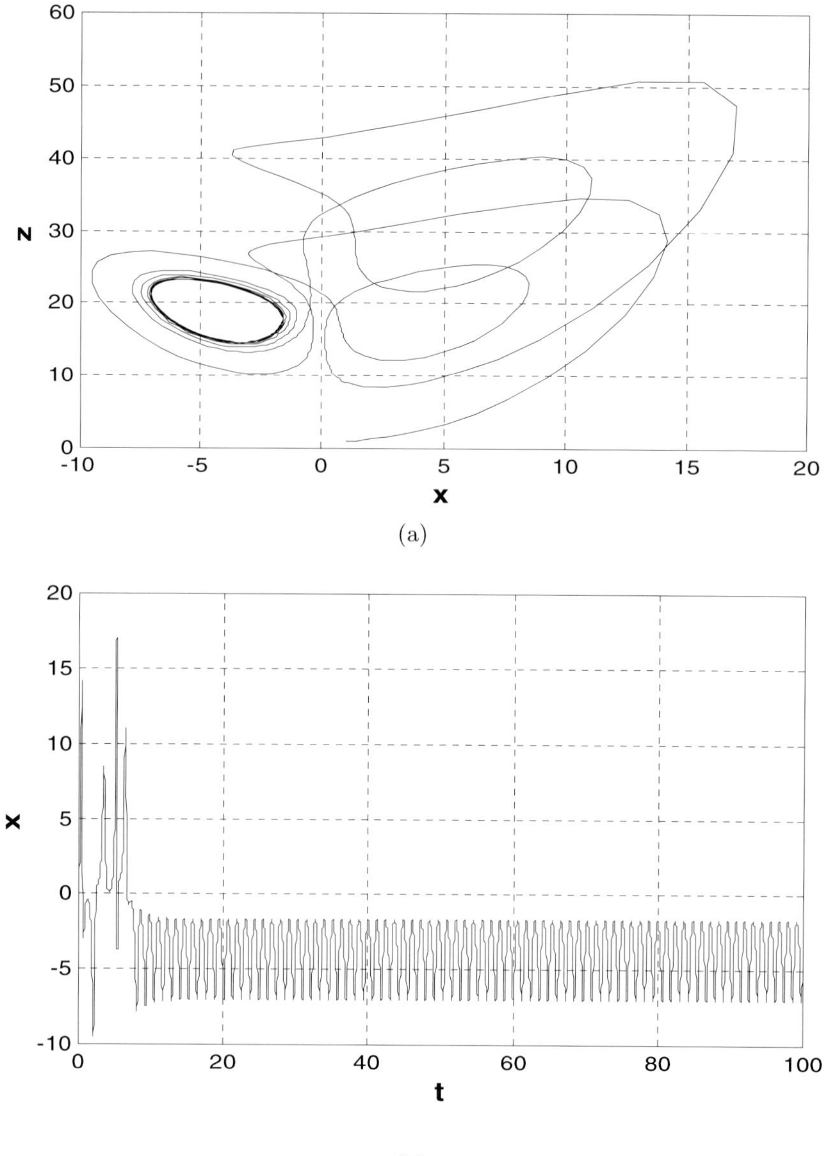

Fig. 7.13. Stabilizing the Lorenz system to a periodic solution

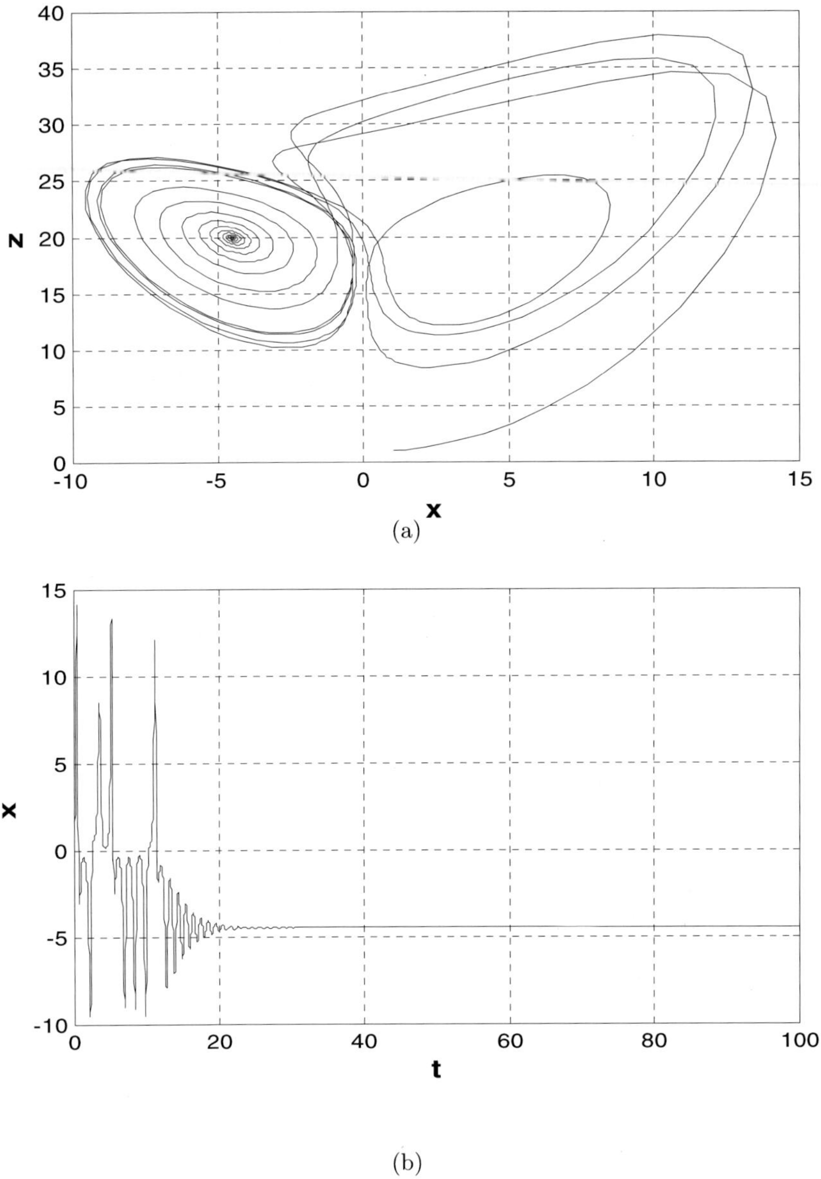

Fig. 7.14. Stabilizing the Lorenz system to an equilibrium point

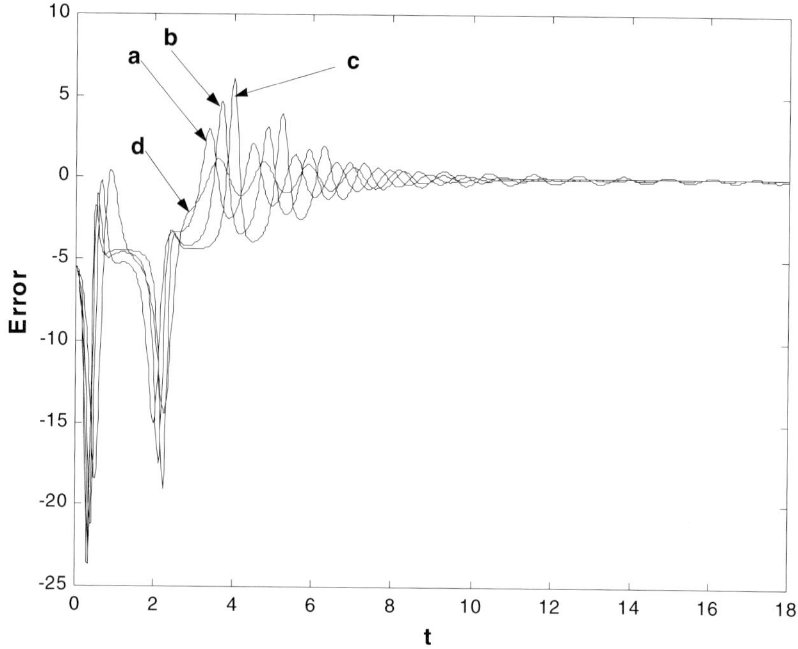

Fig. 7.15. Control error as σ is changed

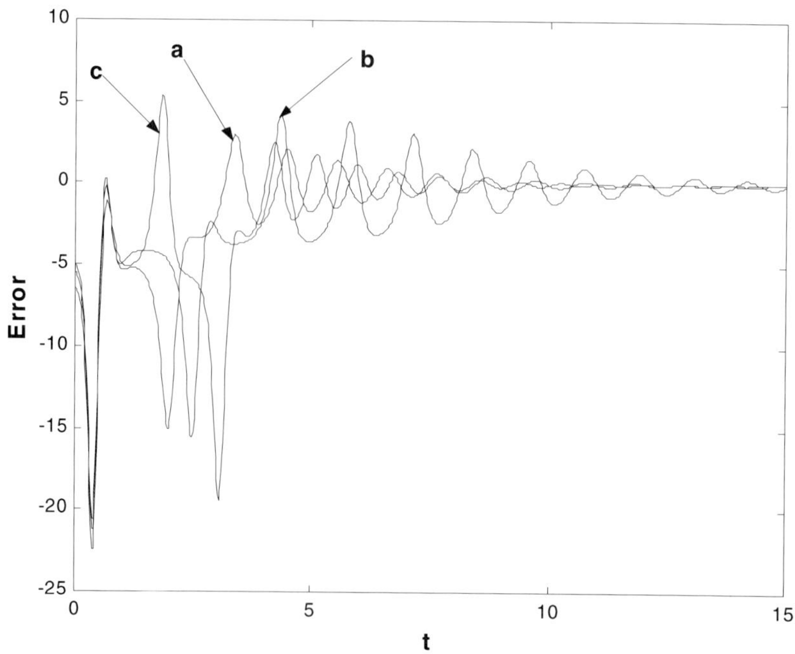

Fig. 7.16. Control error as β is changed

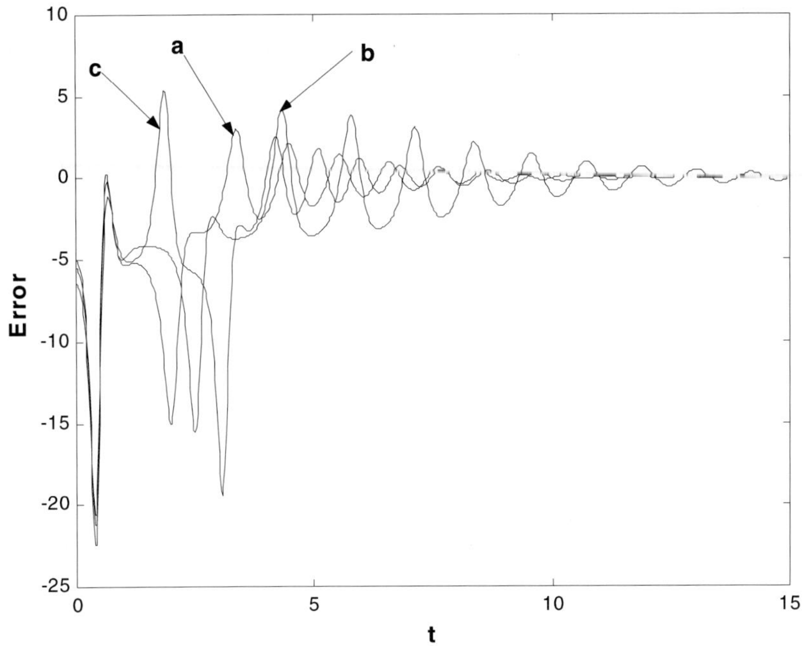

Fig. 7.17. Control error as σ and β are changed

control error when $\sigma = 15$ and $\beta = 1.5$, Curve b to $\sigma = 12$ and $\beta = 1.3$, and Curve c to $\sigma = 8$ and $\beta = 0.5$. Therefore, it follows that the proposed fuzzy logic controller is robust under parameter variations.

8

Adaptive Fuzzy Control of Chaotic Systems (Mamdani Model)

In this chapter, methodologies of adaptive fuzzy control for chaotic systems will be introduced, and some illustrative examples will be presented.

8.1 Introduction

Adaptive control is one of the main approaches in control engineering that deal with uncertain systems. System uncertainties come from unknown or changing system parameters, nonparametric uncertainties, including high-frequency unmodeled dynamics (e.g., actuator dynamics and structural vibrations), low-frequency unmodeled dynamics (e.g., friction and stiffness), measurement noise, and computational round-off errors as well as sampling delays. Uncertainties often cause performance degradation and instability, so that an intended control task such as target-tracking becomes impossible.

An adaptive system is capable of adapting to a changing environment as well as varying internal parameters. Such a system is usually made adaptive by a feedback controller, called an *adaptive controller*. When the system to be controlled is unknown or uncertain, methods leading to accomplishing control tasks, via system identification plus controller design based on the identified model in either an on-line or off-line manner, are also referred to as *adaptive control methods*. Generally speaking, an adaptive controller is one that has adjustable parameters, such as control gains, and the capability of self-adjusting these parameters in response to changes within the dynamics and environment of the controlled system. Many kinds of controllers can be considered adaptive. Typically, adaptive control methods include the popular model-referenced adaptive control, self-tuning regulation, gain scheduling and dual control.

To develop a fuzzy-logic-based control system, it is often beneficial to integrate adaptive capability so that the systems developed cannot only control complex dynamics but also adapt to environmental changes. Compared with

Zhong Li: *Fuzzy Chaotic Systems*, StudFuzz **199**, 143–151 (2006)
www.springerlink.com

conventional adaptive control, the prominent advantage of adaptive fuzzy control lies in the ability to incorporate linguistic fuzzy information from domain experts. This is of special significance for systems with a high degree of uncertainty, such as chemical processes or aircrafts as, although they are difficult to be controlled from a control-theoretical viewpoint, they are often successfully controlled by domain experts [59]. Adaptive fuzzy control provides a tool for making use of fuzzy information in a systematic and efficient manner.

Adaptive fuzzy controllers can be classified into directly and indirectly adaptive fuzzy controllers. In directly adaptive fuzzy control, the controller parameters are directly adjusted to reduce the difference between plant output and output of the reference model; whereas, in indirectly adaptive fuzzy control, the plant parameters are estimated and the controller is designed assuming that the estimated parameters represent the true values of the plant parameters. Further, adaptive fuzzy controllers can be classified into of first and second type according to the linearity or nonlinearity of their adjustable parameters. In this chapter, we only introduce first-type directly adaptive fuzzy control of chaotic systems of the following form [59, 188, 189]:

$$f(x) = \sum_{i=1}^{q} \omega_i \xi_i(x) = \varpi^T \xi(x), \tag{8.1}$$

where $\varpi = [\omega_1, \ldots, \omega_q]^T$, and fuzzy basis function vector, $\xi(x) = [\xi_1(x), \ldots, \xi_q]^T$, defined by

$$\xi_i(x) = \frac{\prod_{j=1}^{n} \mu_{\Gamma_j^i}(x_i)}{\sum_{k=1}^{q} \prod_{j=1}^{n} \mu_{\Gamma_i^k}(x_i)}, \quad i = 1, \ldots, q, \tag{8.2}$$

in which $\{\mu_{\Gamma_i^j}(x_i)\}$ are given membership functions, which can be Gaussian, triangular, or of any other type of membership functions.

8.2 Stable Directly Adaptive Fuzzy Control of Chaotic Systems

Parameters in a directly adaptive fuzzy controller are those of the membership functions of the linguistic variables given in the fuzzy rule-base, or the number or contents of the rule-base. In adaptive control, the parameters are automatically tuned during the control process by the adaptation law [25, 59, 190].

Consider a general, nth-order nonlinear system

$$\begin{cases} x^{(n)}(t) = f(x(t), \dot{x}(t), \ldots, x^{(n-1)}(t)) + bu(t), \\ y(t) = x(t), \qquad n = 1, 2, \cdots, \end{cases} \tag{8.3}$$

where f is an unknown function, $b > 0$ is an unknown constant, and $u \in \Re$ and $y \in \Re$ are the system input and output, respectively.

Assume that the state vector $x = [x, \dot{x}, \ldots, x^{(n-1)}]^T$ is measurable. The control objective is to determine a fuzzy feedback $u = u(x|\varpi)$ and an adaptation law for tuning ϖ such that the following constraints are satisfied:

i) The closed-loop system is globally bounded-input, bounded-output stable:

$$\sup_{0 \le t < \infty} \|x\| \le M_x < \infty,$$
$$\|\varpi\| \le M_\varpi < \infty,$$
$$\sup_{0 \le t < \infty} |u(x|\varpi)| \le M_u < \infty,$$

where M_x, M_ϖ, and M_u are acceptable bounds specified by the designer;

ii) $y(t)$ approaches a given reference signal $y_m(t)$, or say, the tracking error $e \equiv y_m(t) - y(t)$ should be as small as possible under the constraints in i).

For this purpose, consider a controller

$$u = u_c + u_s = u_c(x|\varpi) + u_s(x), \tag{8.4}$$

where $u_c(x|\varpi)$ is a fuzzy controller in the form of (8.1), and $u_s(x)$ is a supervisory controller to be determined. Substituting (8.4) into (8.3) yields

$$x^{(n)} = f(x) + b(u_c(x|\varpi) + u_s(x)). \tag{8.5}$$

After some straightforward manipulations, the error dynamics of the closed-loop control system can be obtained as

$$\dot{\mathbf{e}} = A_{\mathbf{e}}\mathbf{e} + \mathbf{b}_c(u^* - u_c(x|\varpi) - u_s(x)), \tag{8.6}$$

where

$$A_{\mathbf{e}} = \begin{bmatrix} 0 & 1 & 0 & \cdots & 0 & 0 \\ 0 & 0 & 1 & \cdots & 0 & 0 \\ \vdots & \vdots & \vdots & & \vdots & \vdots \\ 0 & 0 & 0 & \cdots & 0 & 1 \\ -a_n & -a_{n-1} & -a_{n-2} & \cdots & -a_2 & -a_1 \end{bmatrix}, \quad \mathbf{b}_c = \begin{bmatrix} 0 \\ \vdots \\ 0 \\ b \end{bmatrix},$$

and

$$u^* = \frac{1}{b}(-f(x) + y_m^{(n)} + \mathbf{a}^T\mathbf{e}), \tag{8.7}$$

in which $\mathbf{a} = [a_n, \ldots, a_1]^T$, and $\mathbf{e} = [e, \dot{e}, \ldots, e^{(n-1)}]^T$.

If all the eigenvalues of $A_{\mathbf{e}}$ have negative real parts, the Lyapunov equation

$$A_{\mathbf{e}}^T P + P A_{\mathbf{e}} = -Q < 0 \tag{8.8}$$

has a unique positive-definite and symmetric matrix solution, P, for any given positive-definite and symmetric matrix Q (usually, taking the identity matrix $Q = I$).

8.2.1 Design of the Supervisory Controller u_s

Define a Lyapunov function, $V_{\mathbf{e}} = \frac{1}{2}\mathbf{e}^T P\mathbf{e}$, where P is a positive-definite and symmetric matrix derived from (8.8).

Using (8.8) and the error equation (8.6), one has

$$\begin{aligned} \dot{V}_{\mathbf{e}} =& -\frac{1}{2}\mathbf{e}^T Q\mathbf{e} + \mathbf{e}^T P\mathbf{b}_c(u^* - u_c(x|\varpi) - u_s(x)) \\ \leq& -\frac{1}{2}\mathbf{e}^T Q\mathbf{e} + |\mathbf{e}^T P\mathbf{b}_c| \cdot (|u^*| + |u_c(x|\varpi)|) - \mathbf{e}^T P\mathbf{b}_c u_s(x). \end{aligned} \tag{8.9}$$

The supervisory controller u_s can be chosen as following, such that $\dot{V}_{\mathbf{e}} \leq 0$,

$$u_s(x) = I^* \operatorname{sgn}(\mathbf{e}^T P\mathbf{b}_c) \left[|u_c| + \beta^{-1}(f_b(x) + |y_m^{(n)}| + |\mathbf{a}^T\mathbf{e}|) \right], \tag{8.10}$$

where $0 < \beta \leq b$ is a constant, f_b is a predetermined function satisfying $|f(x)| \leq f_b(x)$, and

$$I^* = \begin{cases} 1 & \text{if } V_{\mathbf{e}} > \bar{V}, \\ 0 & \text{if } V_{\mathbf{e}} \leq \bar{V}, \end{cases}$$

in which $\bar{V}$ is a constant specified by the designer.

8.2.2 Design of the Controller u_c

The controller $u_c(x|\varpi)$ adopts the fuzzy logic system of the form (8.1). The aim is to develop an adaptation law to adjust the parameter vector ϖ. Let the optimal vector be

$$\varpi^* = \arg \min_{\|\varpi\| \leq M_\varpi} \left(\sup_{\|x\| \leq M_x} |u_c(x|\varpi) - u^*| \right) \tag{8.11}$$

and the mininum approximation error be

$$e^* = u_c(x|\varpi^*) - u^*. \tag{8.12}$$

Then, the error equation (8.6) becomes

$$\begin{aligned} \dot{\mathbf{e}} =& A_{\mathbf{e}}\mathbf{e} + \mathbf{b}_c\left(u_c(x|\varpi^*) - u_c(x|\varpi)\right) - \mathbf{b}_c u_s(x) - \mathbf{b}_c e^* \\ =& A_{\mathbf{e}}\mathbf{e} + \mathbf{b}_c\mathbf{h}^T\mathbf{g}(x) - \mathbf{b}_c u_s - \mathbf{b}_c e^*, \end{aligned} \tag{8.13}$$

where $\mathbf{h} := \varpi^* - \varpi$ and the fuzzy basis function is give by Eq. (8.2).

Consider a Lyapunov function candidate

$$V_{\mathbf{e},\mathbf{h}} = \frac{1}{2}\mathbf{e}^T P\mathbf{e} + \frac{b}{2\alpha}\mathbf{h}^T\mathbf{h}, \tag{8.14}$$

where $\alpha > 0$ is an adjustable parameter to be determined. From the Lyapunov equation (8.8) and the error equation (8.13), one has

$$\begin{aligned}\dot{V}_{\mathbf{e},\mathbf{h}} &= -\frac{1}{2}\mathbf{e}^T Q\mathbf{e} + \mathbf{e}^T P\mathbf{b}_c(\mathbf{h}^T\mathbf{g}(x) - u_s - e^*) + \frac{b}{\alpha}\mathbf{h}^T\mathbf{h} \\ &= -\frac{1}{2}\mathbf{e}^T Q\mathbf{e} + \frac{b}{\alpha}\mathbf{h}^T(\alpha\mathbf{e}^T\mathbf{p}_n\mathbf{g}(x) + \dot{\mathbf{h}}) - \mathbf{e}^T P\mathbf{b}_c(u_s + e^*),\end{aligned} \quad (8.15)$$

where $\mathbf{p}_n$ is the last column of P and $\mathbf{e}^T P\mathbf{b}_c = \mathbf{e}^T\mathbf{p}_n b$. If the basic adaptation law is chosen as

$$\dot{\varpi} = \alpha\mathbf{e}^T\mathbf{p}_n\mathbf{g}(x), \quad (8.16)$$

then, since e^* is ideally zero, (8.15) becomes

$$\dot{V}_{\mathbf{e},\mathbf{h}} \le -\frac{1}{2}\mathbf{e}^T Q\mathbf{e} - \mathbf{e}^T P\mathbf{b}_c e^* < 0.$$

In order to guarantee $\|\varpi\| \le M_{\varpi}$, the projection estimation algorithm to be introduced in the next section can be used to modify the basic adaptation law (8.16). A fuzzy logic rule-base, to be incorporated into the directly adaptive fuzzy controller, has the form

$$\begin{aligned}R_c^{(r)} :\ & \text{IF } x_1 \text{ is } A_{r1} \text{ and } \cdots \text{ and } x_n \text{ is } A_{rn} \\ & \text{THEN } u \text{ is } C^r.\end{aligned} \quad (8.17)$$

8.3 Design of Directly Adaptive Fuzzy Controllers

In this section, the detailed steps to design directly adaptive fuzzy controllers will be introduced, and then their properties will be investigated.

Step 1. Off-line preprocessing

1.1 Specify $\{a_k\}_{k=1}^n$ such that the roots of $s^n + a_1 s^{n-1} + \cdots + a_n = 0$ are located in the left half-plane.

1.2 Specify a positive-definite $n \times n$ matrix Q (usually, $Q = I_n$).

1.3 Specify the bounds $M_{\varpi} > 0$, M_x and M_u.

1.4 Solve the Lyapunov equation (8.8) to obtain the unique positive-definite and symmetric matrix P.

Step 2. Controller construction

2.1 Define m_i fuzzy sets $\Gamma_i^{k_i}$, whose membership functions $\mu_{\Gamma_i^{k_i}}$ uniformly cover U_i, the projection of U onto the ith coordinate, where $k_i = 1, \ldots, m_i$ and $i = 1, 2, \ldots, n$. The $U = U_1 \times \cdots \times U_n$ can be chosen as $U = \{x \in \Re^n : \|x\| \le M_x\}$.

2.2 Construct the fuzzy logic rule-base for the controller $u_c(x|\varpi)$, which consists of $m_1 \times \cdots \times m_n$ rules whose IF parts comprise all possible combinations of $\Gamma_i^{k_i}$ for $i = 1, 2, \ldots, n$. Specifically, the fuzzy rule-base of $u_c(x|\varpi)$ consists of rules

$$R^{(k_1,\cdots,k_n)} : \text{IF } x_1 \text{ is } \Gamma_1^{k_1} \text{ and } \ldots \text{ and } x_n \text{ is } \Gamma_n^{k_n}, \quad \text{THEN } u_c \text{ is } G^{(k_1,\ldots,k_n)}, \tag{8.18}$$

where $k_i = 1, 2, \ldots, m_i$ and $i = 1, 2, \ldots, n$. The fuzzy sets, $G^{(k_1,\ldots,k_n)} \subset \Re$, are chosen to be equal to the corresponding C^r if the IF part of (8.18) agrees with that of (8.17); otherwise, they are set arbitrarily with the only constraint that the centers of $G^{(k_1,\ldots,k_n)}$ are inside the constraint sets $\{\varpi : \|\varpi\| \le M_\varpi\}$.

2.3 Construct the basis functions

$$\xi^{(k_1,\ldots,k_n)}(x) = \frac{\prod_{i=1}^{n} \mu_{\Gamma_i^{k_i}(x_i)}}{\sum_{k_1=1}^{m_1} \cdots \sum_{k_n=1}^{m_n} \left(\prod_{i=1}^{n} \mu_{\Gamma_i^{k_i}(x_i)} \right)} \tag{8.19}$$

and collect them into a $\prod_{i=1}^{n} m_i$-dimensional vector, $\mathbf{g}(x)$, with $k_1 = 1, \ldots, m_1, \ldots, k_n = 1, \ldots, m_n$. Collect the points, at which $\mu_{G^{(k_1,\ldots,k_n)}}$ achieve their maximum values, into ϖ. Then, the basis controller is

$$u_c(x|\varpi) = \varpi^T \mathbf{g}(x). \tag{8.20}$$

Step 3. On-line adaptation

3.1 Apply $u = u_c + u_s$ to the plant (8.3).

3.2 Adjust the parameter vector ϖ by the adaptation law

$$\dot{\varpi} = \begin{cases} \alpha \mathbf{e}^T \mathbf{p}_n \mathbf{g}(x) & \text{if } \|\varpi\| < M_\varpi \\ & \text{or } (\|\varpi\| = M_\varpi \text{ and } \mathbf{e}^T \mathbf{p}_n \varpi^T \mathbf{g}(x) \ge 0) \\ P_o\{\alpha \mathbf{e}^T \mathbf{p}_n \mathbf{g}(x)\} & \text{if } \|\varpi\| = M_\varpi \text{ and } \mathbf{e}^T \mathbf{p}_n \varpi^T \mathbf{g}(x) < 0, \end{cases} \tag{8.21}$$

where the projection operator $P_o\{\cdot\}$ is defined as

$$P_o\{\alpha \mathbf{e}^T \mathbf{p}_n \mathbf{g}(x)\} = \alpha \mathbf{e}^T \mathbf{p}_n \left(1 - \frac{\varpi \varpi^T}{\|\varpi\|^2} \mathbf{g}(x) \right).$$

The overall schematic diagram of a directly adaptive fuzzy control system is shown in Fig. 8.1.

As an example, this adaptive fuzzy control method will be applied to control a typical chaotic system, the Duffing oscillator.

8.4 Adaptive Fuzzy Control of the Duffing Oscillator

Consider the controlled Duffing oscillator

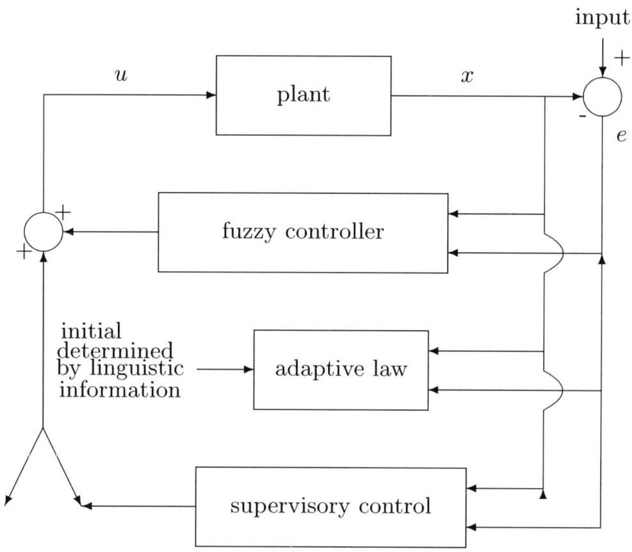

Fig. 8.1. Configuration of a directly adaptive fuzzy control system

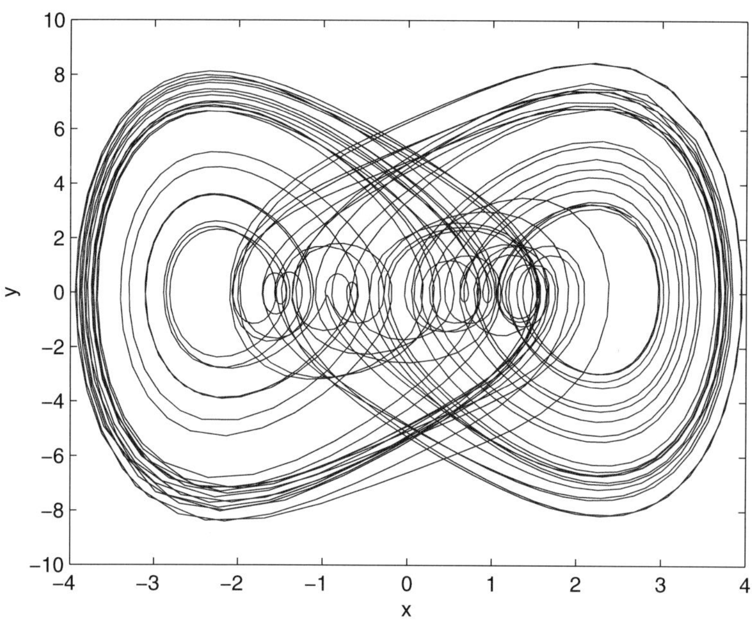

Fig. 8.2. Trajectory of the uncontrolled Duffing system

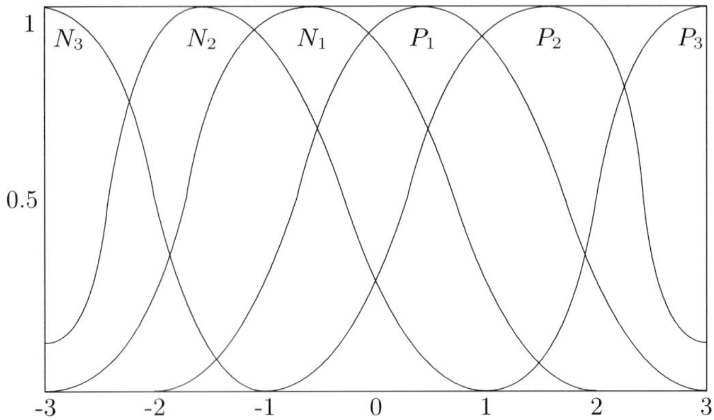

Fig. 8.3. Fuzzy membership functions used in the fuzzy adaptive controller

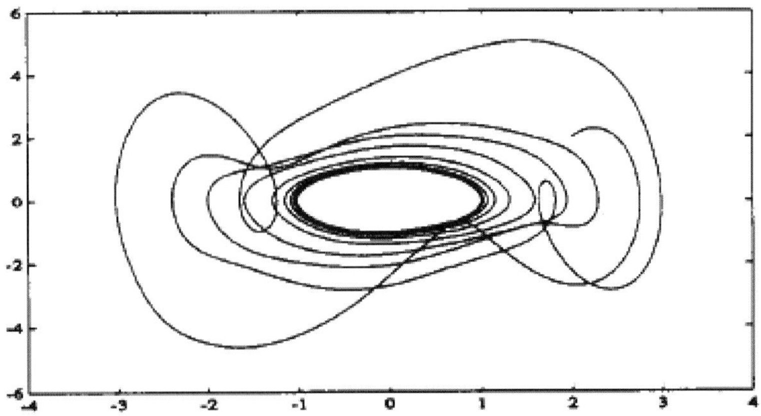

(a)

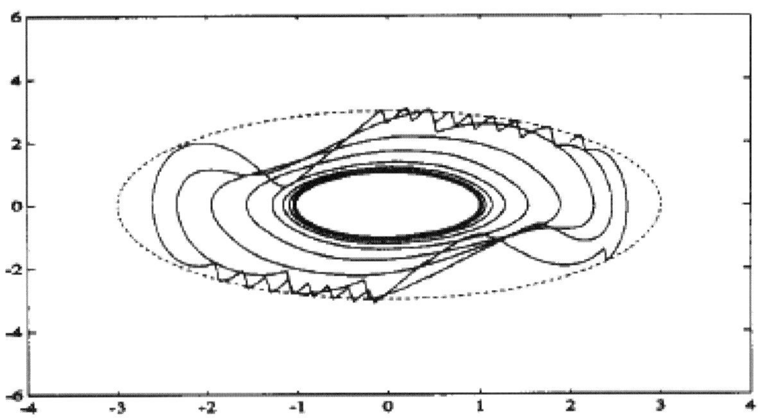

(b)

Fig. 8.4. Trajectories of the controlled Duffing system: (a) $M_x = 10$; (b) $M_x = 3$

$$\begin{cases} \dot{x} = y \\ \dot{y} = -x^3 - 0.1y + 12\cos(t) + u(t), \end{cases} \tag{8.22}$$

which is chaotic when $u(t) = 0$. The trajectory of the uncontrolled system is shown in Fig. 8.2. The aim is to use a directly adaptive fuzzy controller to control the system state x to track a given reference trajectory $y_m(t) = \sin t$.

In the phase plane, the reference trajectory is the unit circle $y_m^2(t) + \dot{y}_m^2(t) = 1$. In terms of the above design procedure the controller parameters are chosen as

$$k_1 = 2, k_2 = 1, \alpha = 2, Q = \begin{bmatrix} 10 & 0 \\ 0 & 10 \end{bmatrix},$$
$$M_{\varpi} = 30, M_x = 3, \beta = 1, f_b(x, y) = 12 + |x|^3.$$

Six fuzzy sets, shown in Fig. 8.3, are used, and assume that there are no fuzzy control rules.

Figure 8.4 (a) and (b) show that the directly adaptive fuzzy controller drives the chaotic Duffing system's trajectory to track the reference one, where in (a) $M_x = 10$ and u_s is not activated, because the trajectory never hits the boundary $x^2 + y^2 = 10$; in (b) $M_x = 3$ and u_s is activated to force the trajectory to be within the constraint $x^2 + y^2 \leq 3$. Here, the initial condition is set as $(x(0), y(0)) = (2, 2)$, and the time period is from $t_0 = 0$ to $t_f = 60$.

9

Fuzzy Control of Chaotic Systems – II (TS Model)

In this chapter, we introduce a TS-model-based fuzzy control method to stabilize chaotic dynamics with parametric uncertainties, also called model-based approach, by using the Linear Matrix Inequalities (LMI) techniques. In the end, this approach will be applied to control the chaotic Lorenz system and Chua's chaotic circuit.

9.1 Introduction

Most real plants in industry have severe nonlinearities and uncertainties, which often cause performance degradation and instability, and post additional difficulties in stability analysis and controllers design. Fuzzy control methodology provides an effective solution to the control of plants that are complex, uncertain, or ill-defined by incorporating qualitative knowledge from domain experts for their controllers' design.

LMI techniques have been serving as powerful design tools in control engineering, system identification and structural design [191]. The LMI techniques have three prominent merits:

1. A variety of design specifications and constraints can be expressed as LMIs;
2. Once formulated in terms of LMIs, a problem can be solved exactly by efficient convex optimization algorithms;
3. Although most problems with multiple constraints or objectives lack analytical solutions in terms of matrix equations, they often remain tractable in the LMI framework. This makes LMI-based design a valuable alternative to classical "analytical" methods.

In this chapter, the parametric uncertainties in nonlinear systems will be considered. Some sufficient conditions in the LMI format and a systematic controller design procedure for general nonlinear systems with parametric uncertainties, for both continuous-time and discrete-time TS fuzzy systems, will

Zhong Li: *Fuzzy Chaotic Systems*, StudFuzz **199**, 153–187 (2006)
www.springerlink.com

be proposed. In particular, the robust stabilization problem for a class of nonlinear systems with time-varying but norm-bounded parametric uncertainties will be solved to some extent.

Finally, the chaotic Lorenz system and a flexible-joint robot arm system are used to illustrate the effectiveness of the proposed fuzzy control techniques.

9.2 Preliminaries

First, some preliminaries about LMIs need to be introduced.

Lemma 9.1. [192] *Given constant symmetric matrices N, O and L of appropriate dimensions, the following two inequalities are equivalent:*

$$(a) \qquad O > 0, \quad N + L^T O L < 0\,,$$

$$(b) \qquad \begin{bmatrix} N & L^T \\ L & -O^{-1} \end{bmatrix} < 0 \quad or \quad \begin{bmatrix} -O^{-1} & L \\ L^T & N \end{bmatrix} < 0\,.$$

Lemma 9.1 is called *Schur complements*, which is one of the most basic tool in converting nonlinear matrix inequalities to LMIs.

Lemma 9.2. [192] *Given constant matrices D and E, and a symmetric constant matrix S of appropriate dimensions, the following inequality holds:*

$$S + DFE + E^T F^T D^T < 0\,,$$

where F satisfies $F^T F \leq I$, if and only if for some $\epsilon > 0$,

$$S + \begin{bmatrix} \epsilon^{-1} E^T & \epsilon D \end{bmatrix} \begin{bmatrix} \epsilon^{-1} E \\ \epsilon D^T \end{bmatrix} < 0\,.$$

Lemma 9.2 eliminates the theoretical difficulties in dealing with uncertain systems.

The interior-point algorithm can quite efficiently find solutions to LMI problems, and many fuzzy-model-based control problems can be recast as LMI problems.

Definition 9.3. [191] *A linear matrix inequality (LMI) is a matrix inequality of the form:*

$$J(x) = J_0 + \sum_{i=1}^{m} x_i J_i > 0\,, \tag{9.1}$$

where $x = [x_1\,, x_2\,, \cdots, x_m]^T$ is the variable and symmetric matrices J_i are given. The symbol "> 0" means positive-definite.

9.3 Parallel-Distributed Compensation

The so-called parallel-distributed compensation (PDC) technique is employed to determine the structure of a fuzzy controller for a given TS fuzzy model. Each control rule in the PDC is constructed from the corresponding rule of the TS fuzzy model. The designed fuzzy controller shares the same fuzzy sets with the fuzzy model in the premise parts. The PDC provides the following fuzzy control rule structure from the fuzzy model (6.2) or (6.5):

$$\begin{aligned}
&\text{Control Rule } i: \\
&\text{IF } x_1(t) \text{ is } \Gamma_1^i \text{ and } \dots \text{ and } x_n(t) \text{ is } \Gamma_n^i \\
&\text{THEN } u_i(t) = F_i x(t), \\
&i = 1, 2, \dots, q.
\end{aligned} \tag{9.2}$$

where $\{F_i\}_{i=1}^q$ are constant gain matrices to be designed.

The fuzzy control rules have linear state-feedback laws in the consequent parts. The overall fuzzy controller is represented by

$$u(t) = \frac{1}{\sum_{i=1}^{q} \omega_i(t)} \sum_{i=1}^{q} \omega_i(t) F_i x(t) \equiv \sum_{i=1}^{q} h_i(t) F_i x(t). \tag{9.3}$$

To be practical, the control-gain matrices $\{F_i\}_{i=1}^q$ are required to be uniformly bounded:

$$\sup_{1 \le i \le q} \|F_i\| \le M < \infty, \tag{9.4}$$

for some positive constant M.

Substituting (9.3) into (6.3) or its discrete counterpart yields

$$sx(t) = \sum_{i=1}^{q} \sum_{j=1}^{q} h_i(t) h_j(t) \{A_i + B_i F_j\} x(t). \tag{9.5}$$

where $sx(t)$ denotes $\dot{x}(t)$ for continuous-time TS fuzzy systems (CFS), or $x(t+1)$ for discrete-time TS fuzzy systems (DFS).

System (9.5) can also be written as

$$
\begin{aligned}
sx(t) &= \sum_{i=1}^{q} h_i(t)h_i(t)\{A_i + B_iF_i\}x(t) \\
&\quad + 2\sum_{i<j}^{q} h_i(t)h_j(t)\frac{\{A_i + B_iF_j\} + \{A_j + B_jF_i\}}{2}x(t) \\
&= \sum_{i=1}^{q} h_i(t)h_i(t)G_{ii}x(t) \\
&\quad + 2\sum_{i<j}^{q} h_i(t)h_j(t)\left\{\frac{G_{ij} + G_{ji}}{2}\right\}x(t),
\end{aligned} \tag{9.6}
$$

where $G_{ij} = A_i + B_iF_j$ for all $i, j = 1, 2, \ldots, q$.

9.4 Lyapunov Stability of TS Fuzzy Systems

Stability conditions for the fuzzy control system (9.6) can be derived in terms of the Lyapunov direct method [193, 194, 195, 196, 197, 198, 153].

Theorem 9.4. [Continuous-time TS fuzzy systems (CFS)]
The equilibrium of the continuous TS fuzzy system (9.6) *is asymptotically stable in the large if there exists a common positive-definite matrix P such that*

$$
G_{ii}^T P + PG_{ii} < 0, \tag{9.7}
$$

for all i and

$$
\left(\frac{G_{ij} + G_{ji}}{2}\right)^T P + P\left(\frac{G_{ij} + G_{ji}}{2}\right) \leq 0,
$$

for $i < j$, except the pairs (i, j) such that $\mu_i(x(t))\mu_j(x(t)) = 0$, $\forall\, t$.

Theorem 9.5. [Discrete-time TS fuzzy systems (DFS)]
The equilibrium of the discrete TS fuzzy systems (9.6) *is asymptotically stable in the large if there exists a common positive-definite matrix P such that*

$$
G_{ii}^T PG_{ii} - P < 0, \tag{9.8}
$$

for all i and

$$
\left(\frac{G_{ij} + G_{ji}}{2}\right)^T P\left(\frac{G_{ij} + G_{ji}}{2}\right) - P \leq 0,
$$

for $i < j$, except the pairs (i, j) such that $\mu_i(x(t))\mu_j(x(t)) = 0$, $\forall\, t$.

The task of fuzzy controller design is to determine F_j ($j = 1, 2, \ldots, q$) satisfying the conditions in Theorems 9.4 or 9.5 for a common positive-definite matrix P.

9.5 Stability Analysis and Controller Design Based on LMIs

The LMI (9.1) can represent various convex constraints on x such as Lyapunov stability conditions and convex quadratic matrix inequalities. In this chapter, the stability conditions for TS fuzzy model are formulated in the form of LMIs, since this LMI-based controller design is a very efficient, systematic and powerful technique.

9.5.1 Continuous-time Case

Consider a continuous-time TS fuzzy model, with parametric uncertainties, described by the following state-space equation:

$$\dot{x}(t) = \sum_{i-1}^{q} \mu_i(x(t))((A_i + \Delta A_i)x(t) + (B_i + \Delta B_i)u(t)) . \tag{9.9}$$

where ΔA_i and ΔB_i are time-varying matrices with appropriate dimensions, which represent parametric uncertainties in the plant model.

The closed-loop system of (9.9) and (9.3) is

$$\begin{aligned}\dot{x}(t) &= \sum_{i=1}^{q}\sum_{j=1}^{q} \mu_i(x(t))\mu_j(x(t))(A_i + \Delta A_i + (B_i + \Delta B_i)F_j)x(t)\\ &= \sum_{i=1}^{q} \mu_i^2(x(t))(A_i + \Delta A_i + (B_i + \Delta B_i)F_i)x(t)\\ &\quad + 2\sum_{i<j}^{q} \mu_i(x(t))\mu_j(x(t))\\ &\quad \times \left(\frac{A_i + \Delta A_i + (B_i + \Delta B_i)F_j + A_j + \Delta A_j + (B_j + \Delta B_j)F_i}{2}\right) x(t) .\end{aligned} \tag{9.10}$$

Since there are time-varying uncertainty matrices, it is not easy to design the controller gain matrices. In order to find these gain matrices, F_i, the uncertainty matrices should be removed under some reasonable assumptions. Therefore, we assume, as usual, that the uncertainty matrices ΔA_i and ΔB_i are admissibly norm-bounded and structured.

Assumption 1 *The parameter uncertainties considered here are norm-bounded, in the form*

$$\left[\Delta A_i \ \Delta B_i\right] = D_i K_i(t) \left[E_{1i} \ E_{2i}\right] ,$$

where D_i, E_{1i} and E_{2i} are known real constant matrices of appropriate dimensions, and $K_i(t)$ is an unknown matrix function with Lebesgue-measurable elements and satisfying $K_i(t)^T K_i(t) \leq I$, in which I is the identity matrix of appropriate dimension.

Thus, the global asymptotic stability of the continuous-time TS fuzzy model, with parametric uncertainties, is summarized in the following theorem:

Theorem 9.6. *If there exist a symmetric and positive-definite matrix P, some matrices F_i, and some scalars $\epsilon_{ij}, (i,j = 1,\ldots,q)$, such that the following LMIs are satisfied, then the continuous-time TS fuzzy system (9.9) is asymptotically stabilizable via the PDC controller* (9.3)*:*

$$(a) \qquad \begin{bmatrix} \Psi_{ii} & * & * \\ E_{1i}Q + E_{2i}M_i & -\epsilon_{ii}I & * \\ D_i^T & 0 & -\epsilon_{ii}^{-1}I \end{bmatrix} < 0 \,, \qquad (1 \le i \le q)\,, \tag{9.11}$$

$$(b) \qquad \begin{bmatrix} \Upsilon_{ij} & * & * & * & * \\ E_{1i}Q + E_{2i}M_j & -\epsilon_{ij}I & * & * & * \\ E_{1j}Q + E_{2j}M_i & 0 & -\epsilon_{ij}I & * & * \\ D_i^T & 0 & 0 & -\epsilon_{ij}^{-1}I & * \\ D_j^T & 0 & 0 & 0 & -\epsilon_{ij}^{-1}I \end{bmatrix} < 0 \,, \qquad (1 \le i < j \le q)\,, \tag{9.12}$$

where

$$\Psi_{ii} = QA_i^T + A_iQ + M_i^TB_i^T + B_iM_i \,,$$
$$\Upsilon_{ij} = QA_i^T + A_iQ + QA_j^T + A_jQ + M_j^TB_i^T + B_iM_j + M_i^TB_j^T + B_jM_i \,,$$

and $Q = P^{-1}$ and $M_i = F_iP^{-1}$, where $*$ denotes the transposed elements in the symmetric positions.

Proof. Consider the Lyapunov function candidate

$$V(t) = x(t)^TPx(t) \,, \tag{9.13}$$

where P is a time-invariant, symmetric and positive-definite matrix. Clearly, $V(t)$ is positive-definite and radially unbounded. The time derivative of $V(t)$ is

$$\dot{V}(t) = \dot{x}(t)^TPx(t) + x(t)^TP\dot{x}(t) \,. \tag{9.14}$$

By substituting (9.10) into (9.14), one has

$$
\begin{aligned}
\dot{V}(t) = &\sum_{i=1}^{q} \mu_i(x(t)) \sum_{j=1}^{q} \mu_j(x(t)) \Big(\big(A_i + \Delta A_i + (B_i + \Delta B_i)F_j)x(t)\big)^T Px(t)\Big) \\
&+ x(t)^T P \left(A_i + \Delta A_i + (B_i + \Delta B_i)F_j\right) x(t)\Big) \\
= &\sum_{i=1}^{q} (\mu_i^2(x(t))x(t)^T \Big((A_i + \Delta A_i + (B_i + \Delta B_i)F_i)^T P \\
&+ P(A_i + \Delta A_i + (B_i + \Delta B_i)F_i)\big)\, x(t) \\
&+ 2\sum_{i<j}^{q} \mu_i(x(t))\mu_j(x(t)) \\
&\times x(t)^T \Bigg(\left(\frac{A_i + \Delta A_i + (B_i + \Delta B_i)F_j + A_j + \Delta A_j + (B_j + \Delta B_j)F_i}{2} \right)^T P \\
&+ P \left(\frac{A_i + \Delta A_i + (B_i + \Delta B_i)F_j + A_j + \Delta A_j + (B_j + \Delta B_j)F_i}{2} \right) \Bigg)\, x(t)\,.
\end{aligned}
\tag{9.15}
$$

If the time derivative of (9.13) is uniformly negative-definite for all $x(t)$ and for all $t \geq 0$ except at $x(t) = 0$, then the controlled fuzzy system (9.10) is asymptotically stable about its zero equilibrium. Therefore, if it is possible to assume each sum of the second equation in (9.15) to be negative-definite, respectively, then the controlled continuous-time TS fuzzy system is asymptotically stable.

First, assume that the first sum of the last equation in (9.15) is negative-definite, i.e.,

$$
(A_i P + \Delta A_i + (B_i + \Delta B_i)F_i)^T P + P(A_i + \Delta A_i + (B_i + \Delta B_i)F_i) < 0\,, \tag{9.16}
$$

$$
1 \leq i \leq q\,.
$$

Then, applying Assumption 1 to (9.16) yields

$$
\Phi_{ii} + PD_i K_i(t)(E_{1i} + E_{2i}F_i) + (E_{1i} + E_{2i}F_i)^T K_i(t)^T D_i^T P < 0\,, \tag{9.17}
$$

where

$$
\Phi_{ii} = A_i^T P + PA_i + F_i^T B_i^T P + PB_i F_i\,.
$$

According to Lemma 9.2, the above matrix inequality (9.17) holds for all $K_i(t)$ satisfying $K_i(t)^T K_i(t) \leq I$ if and only if there exists a constant $\epsilon_{ii}^{1/2} > 0$ such that

$$
\begin{aligned}
&\Phi_{ii} + \begin{bmatrix} \epsilon_{ii}^{-1/2}(E_{1i} + E_{2i}F_i)^T & \epsilon_{ii}^{1/2} PD_i \end{bmatrix} \begin{bmatrix} \epsilon_{ii}^{-1/2}(E_{1i} + E_{2i}F_i) \\ \epsilon_{ii}^{1/2}(PD_i)^T \end{bmatrix} \\
&= \Phi_{ii} + \begin{bmatrix} (E_{1i} + E_{2i}F_i)^T & PD_i \end{bmatrix} \begin{bmatrix} \epsilon_{ii}^{-1} I & 0 \\ 0 & \epsilon_{ii} I \end{bmatrix} \begin{bmatrix} E_{1i} + E_{2i}F_i \\ (PD_i)^T \end{bmatrix} < 0\,.
\end{aligned}
\tag{9.18}
$$

Applying Lemma 9.1 to (9.18) results in

$$\begin{bmatrix} \Phi_{ii} & * & * \\ E_{1i} + E_{2i}F_i & -\epsilon_{ii}I & * \\ D_i^T P & 0 & -\epsilon_{ii}^{-1}I \end{bmatrix} < 0 \,. \tag{9.19}$$

The matrix inequality (9.19) is not an LMI but a quadratic matrix inequality (QMI). In order to use the convex optimization technique, the QMI must be converted to an LMI via some variable changes or transformations. For this purpose, define the following transformation matrix as

$$\begin{bmatrix} P^{-1} & 0 & 0 \\ 0 & I & 0 \\ 0 & 0 & I \end{bmatrix},$$

and take a congruence transformation. This yields

$$\begin{aligned} &\begin{bmatrix} P^{-1} & 0 & 0 \\ 0 & I & 0 \\ 0 & 0 & I \end{bmatrix} \begin{bmatrix} \Phi_{ii} & * & * \\ E_{1i} + E_{2i}F_i & -\epsilon_{ii}I & * \\ D_i^T P & 0 & -\epsilon_{ii}^{-1}I \end{bmatrix} \begin{bmatrix} P^{-1} & 0 & 0 \\ 0 & I & 0 \\ 0 & 0 & I \end{bmatrix}^T \\ &= \begin{bmatrix} P^{-1}\Phi_{ii}P^{-1} & * & * \\ E_{1i}P^{-1} + E_{2i}F_iP^{-1} & -\epsilon_{ii}I & * \\ D_i^T & 0 & -\epsilon_{ii}^{-1}I \end{bmatrix} < 0 \,. \end{aligned} \tag{9.20}$$

Denoting $Q = P^{-1}$ and $M_i = F_iP^{-1}$ yields the first LMI, (9.11) in Theorem 9.6.

The second LMI (9.12) can be established through a similar procedure. Assume that

$$\begin{aligned} &\left(\frac{A_i + \Delta A_i + (B_i + \Delta B_i)F_j + A_j + \Delta A_j + (B_j + \Delta B_j)F_i}{2}\right)^T P \\ &\quad + P\left(\frac{A_i + \Delta A_i + (B_i + \Delta B_i)F_j + A_j + \Delta A_j + (B_j + \Delta B_j)F_i}{2}\right) < 0 \,. \end{aligned} \tag{9.21}$$

$$1 \le i < j \le q \,.$$

Then, using Assumption 1, (9.21) can be represented as

$$\begin{aligned} &\Theta_{ij} + \begin{bmatrix} PD_i & PD_j \end{bmatrix} \begin{bmatrix} K_i(t) & 0 \\ 0 & K_j(t) \end{bmatrix} \begin{bmatrix} E_{1i} + E_{2i}F_j \\ E_{1j} + E_{2j}F_i \end{bmatrix} \\ &\quad + \begin{bmatrix} E_{1i} + E_{2i}F_j \\ E_{1j} + E_{2j}F_i \end{bmatrix}^T \begin{bmatrix} K_i(t) & 0 \\ 0 & K_j(t) \end{bmatrix}^T \begin{bmatrix} PD_i & PD_j \end{bmatrix}^T < 0 \,, \end{aligned} \tag{9.22}$$

where

$$\Theta_{ij} = A_i^T P + PA_i + A_j^T P + PA_j + F_j^T B_i^T P + PB_iF_j + F_i^T B_j^T P + PB_jF_i \,.$$

Using Lemma 9.2 repeatedly, the above matrix inequality (9.22) holds for all $K_i(t)$ satisfying

$$\begin{bmatrix} K_i(t) & 0 \\ 0 & K_j(t) \end{bmatrix}^T \begin{bmatrix} K_i(t) & 0 \\ 0 & K_j(t) \end{bmatrix} \leq I\,,$$

if and only if there exists a constant $\epsilon_{ij}^{1/2} > 0$ such that

$$\begin{aligned} &\Theta_{ij} + \left[\epsilon_{ij}^{-1/2}(E_{1i} + E_{2i}F_j)^T \ \epsilon_{ij}^{-1/2}(E_{1j} + E_{2j}F_i)^T \ \epsilon_{ij}^{1/2}PD_i \ \epsilon_{ij}^{1/2}PD_j\right] \\ &\quad \times \begin{bmatrix} \epsilon_{ij}^{-1/2}(E_{1i} + E_{2i}F_j) \\ \epsilon_{ij}^{-1/2}(E_{1j} + E_{2j}F_i) \\ \epsilon_{ij}^{1/2}(PD_i)^T \\ \epsilon_{ij}^{1/2}(PD_j)^T \end{bmatrix} \\ &= \Theta_{ij} + \left[(E_{1i} + E_{2i}F_j)^T \ (E_{1j} + E_{2j}F_i)^T \ PD_i \ PD_j\right] \\ &\quad \times \begin{bmatrix} \epsilon_{ij}^{-1}I & 0 & 0 & 0 \\ 0 & \epsilon_{ij}^{-1}I & 0 & 0 \\ 0 & 0 & \epsilon_{ij}I & 0 \\ 0 & 0 & 0 & \epsilon_{ij}I \end{bmatrix} \begin{bmatrix} E_{1i} + E_{2i}F_j \\ E_{1j} + E_{2j}F_i \\ (PD_i)^T \\ (PD_j)^T \end{bmatrix} < 0\,. \end{aligned} \tag{9.23}$$

Applying Lemma 9.1 to (9.23) yields

$$\begin{bmatrix} \Theta_{ij} & * & * & * & * \\ E_{1i} + E_{2i}F_j & -\epsilon_{ij}I & * & * & * \\ E_{1j} + E_{2j}F_i & 0 & -\epsilon_{ij}I & * & * \\ D_i^T P & 0 & 0 & -\epsilon_{ij}^{-1}I & * \\ D_j^T P & 0 & 0 & 0 & -\epsilon_{ij}^{-1}I \end{bmatrix} < 0\,. \tag{9.24}$$

Define the transformation matrix as

$$\begin{bmatrix} P^{-1} & 0 & 0 & 0 & 0 \\ 0 & I & 0 & 0 & 0 \\ 0 & 0 & I & 0 & 0 \\ 0 & 0 & 0 & I & 0 \\ 0 & 0 & 0 & 0 & I \end{bmatrix}.$$

Then, it can easily be verified that

$$\begin{bmatrix} P^{-1} & 0\,0\,0\,0 \\ 0 & I\,0\,0\,0 \\ 0 & 0\,I\,0\,0 \\ 0 & 0\,0\,I\,0 \\ 0 & 0\,0\,0\,I \end{bmatrix} \begin{bmatrix} \Theta_{ij} & * & * & * & * \\ E_{1i}+E_{2i}F_j & -\epsilon_{ij}I & * & * & * \\ E_{1j}+E_{2j}F_i & 0 & -\epsilon_{ij}I & * & * \\ D_i^T P & 0 & 0 & -\epsilon_{ij}^{-1}I & * \\ D_j^T P & 0 & 0 & 0 & -\epsilon_{ij}^{-1}I \end{bmatrix} \begin{bmatrix} P^{-1} & 0\,0\,0\,0 \\ 0 & I\,0\,0\,0 \\ 0 & 0\,I\,0\,0 \\ 0 & 0\,0\,I\,0 \\ 0 & 0\,0\,0\,I \end{bmatrix}^T$$

$$= \begin{bmatrix} P^{-1}\Theta_{ij}P^{-1} & * & * & * & * \\ E_{1i}P^{-1}+E_{2i}F_jP^{-1} & -\epsilon_{ij}I & * & * & * \\ E_{1j}P^{-1}+E_{2j}F_iP^{-1} & 0 & -\epsilon_{ij}I & * & * \\ D_i^T & 0 & 0 & -\epsilon_{ij}^{-1}I & * \\ D_j^T & 0 & 0 & 0 & -\epsilon_{ij}^{-1}I \end{bmatrix} < 0 \,. \tag{9.25}$$

Letting $Q = P^{-1}$ and $M_i = F_iP^{-1}$ yields the second LMI, (9.12). This completes the proof of the theorem.

Q. E. D.

Remark 9.7. The second inequalities in Theorem 9.6, for the pair (i, j) such that $\mu_i(x(t))\mu_j(x(t)) = 0$, for all $t \geq 0$, do not have to be solved in determining the system stability.

9.5.2 Discrete-time Case

This section deals with the controller design problem for the discrete-time TS fuzzy model with parametric uncertainties. The state-space representation of the fuzzy system and its corresponding PDC controller can be described as follows:

$$x(t+1) = \sum_{i=1}^{q} \mu_i(x(t))(G_i + \Delta G_i)x(t) + (H_i + \Delta H_i)u(t)), \tag{9.26}$$

$$u(t) = \sum_{i=1}^{q} \mu_i(x(t))F_ix(t) \,. \tag{9.27}$$

The closed-loop system of (9.26) and (9.27) is given by

$$\begin{aligned} x(t+1) &= \sum_{i=1}^{q}\sum_{j=1}^{q} \mu_i(x(t))\mu_j(x(t))(G_i + \Delta G_i + (H_i + \Delta H_i)F_j)x(t) \\ &= \sum_{i=1}^{q} \mu_i^2(x(t))(G_i + \Delta G_i + (H_i + \Delta H_i)F_i)x(t) \\ &\quad + 2\sum_{i<j}^{q} \mu_i(x(t))\mu_j(x(t)) \\ &\qquad \times \left(\frac{G_i + \Delta G_i + (H_i + \Delta H_i)F_j + G_j + \Delta G_j + (H_j + \Delta H_j)F_i}{2} \right) . \end{aligned} \tag{9.28}$$

The following theorem provides a sufficient condition for robust stabilization of the controlled discrete-time TS fuzzy system (9.28), in the presence of parametric uncertainties.

Theorem 9.8. *If there exist a symmetric and positive-definite matrix P, some matrices F_i, and some scalars ϵ_{ij}, $(i,j = 1,\ldots,q)$, such that the following LMIs are satisfied, then the discrete-time TS fuzzy system (9.26) is asymptotically stabilizable by the PDC controller (9.27):*

$$(a) \quad \begin{bmatrix} -Q & * & * & * \\ G_iQ + H_iM_i & -Q & * & * \\ E_{1i}Q + E_{2i}M_i & 0 & -\epsilon_{ii}I & * \\ 0 & D_i^T & 0 & -\epsilon_{ii}^{-1}I \end{bmatrix} < 0\,, \quad (1 \le i \le q)\,, \tag{9.29}$$

$$(b) \quad \begin{bmatrix} -4Q & * & * & * & * & * \\ \begin{pmatrix} G_iQ + H_iM_j \\ +G_jQ + H_jK_i \end{pmatrix} & -Q & * & * & * & * \\ E_{1i}Q + E_{2i}M_j & 0 & -\epsilon_{ij}I & * & * & * \\ E_{1j}Q + E_{2j}M_i & 0 & 0 & -\epsilon_{ij}I & * & * \\ 0 & D_i^T & 0 & 0 & -\epsilon_{ij}^{-1}I & * \\ 0 & D_j^T & 0 & 0 & 0 & -\epsilon_{ij}^{-1}I \end{bmatrix} < 0\,, \quad (1 \le i < j \le q)\,, \tag{9.30}$$

where $Q = P^{-1}$, $M_i = F_iP^{-1}$, and $*$ denotes the transposed elements in the symmetric positions.

Proof. The proof is analogous to that of Theorem 9.6. Consider the Lyapunov function candidate

$$V(t) = x(t)^T Px(t)\,, \tag{9.31}$$

which is positive-definite. The rate with which $V(t)$ increases is

$$\begin{aligned}
\Delta V(t) &= V(t+1) - V(t) \\
&= x(t+1)^T Px(t+1) - x(t)^T Px(t) \\
&= \sum_{i=1}^{q}\sum_{j=1}^{q}\sum_{k=1}^{q}\sum_{l=1}^{q} \mu_i(x(t))\mu_j(x(t))\mu_k(x(t))\mu_l(x(t))x(t)^T \\
&\quad \times \Big((G_i + \Delta G_i + (H_i + \Delta H_i)F_j)^T P(G_k + \Delta G_k + (H_k + \Delta H_k)F_l) - P\Big)\, x(t) \\
&= \frac{1}{4}\sum_{i=1}^{q}\sum_{j=1}^{q}\sum_{k=1}^{q}\sum_{l=1}^{q} \mu_i(x(t))\mu_j(x(t))\mu_k(x(t))\mu_l(x(t))x(t)^T \\
&\quad \times \Big((G_i + \Delta G_i + (H_i + \Delta H_i)F_j + G_j + \Delta G_j + (H_j + \Delta H_j)F_i)^T P \\
&\quad \times (G_k + \Delta G_k + (H_k + \Delta H_k)F_l + G_l + \Delta G_l + (H_l + \Delta H_l)F_k) - 4P)\, x(t)
\end{aligned}$$

$$
\begin{aligned}
&\leq \frac{1}{4} \sum_{i=1}^{q} \sum_{j=1}^{q} \mu_i(x(t))\mu_j(x(t))x(t)^T \\
&\quad \times \Big((G_i + \Delta G_i + (H_i + \Delta H_i)F_j + G_j + \Delta G_j + (H_j + \Delta H_j)F_i)^T P \\
&\quad \times (G_i + \Delta G_i + (H_i + \Delta H_i)F_j + G_j + \Delta G_j + (H_j + \Delta H_j)F_i) - 4P \Big) x(t) \\
&= \sum_{i=1}^{q} \mu_i^2(x(t))x(t)^T ((G_i + \Delta G_i + (H_i + \Delta H_i)F_i)^T P \\
&\quad \times (G_i + \Delta G_i + (H_i + \Delta H_i)F_i) - P)x(t) \\
&+ 2 \sum_{i<j}^{q} \mu_i(x(t))\mu_j(x(t))x(t)^T \\
&\quad \times \left(\left(\frac{G_i + \Delta G_i + (H_i + \Delta H_i)F_j + G_j + \Delta G_j + (H_j + \Delta H_j)F_i}{2} \right)^T P \right. \\
&\quad \times \left. \left(\frac{G_i + \Delta G_i + (H_i + \Delta H_i)F_j + G_j + \Delta G_j + (H_j + \Delta H_j)F_i}{2} \right) - P \right) x(t) \,.
\end{aligned}
\tag{9.32}
$$

Therefore, if the two sums above are both negative-definite uniformly for all $x(t)$ and for all $t \geq 0$, then $\Delta V(t)$ is negative-definite so that the controlled system is asymptotically stable.

Assume that the first sum of the last equality in (9.32) is negative-definite:

$$
\begin{aligned}
(G_i + \Delta G_i + (H_i + \Delta H_i)F_i)^T P (G_i + \Delta G_i + \\
(H_i + \Delta H_i)F_i) - P < 0 \,, \qquad 1 \leq i \leq q \,.
\end{aligned}
\tag{9.33}
$$

By applying Lemma 9.1 and Assumption 1, (9.33) is equivalent to

$$
\begin{aligned}
&\begin{bmatrix} -P & * \\ G_i + \Delta G_i + (H_i + \Delta H_i)F_i & -P^{-1} \end{bmatrix} = \Omega_{ii} + \begin{bmatrix} 0 & * \\ \Delta G_i + \Delta H_i F_i & 0 \end{bmatrix} \\
&= \Omega_{ii} + \begin{bmatrix} 0 \\ D_i \end{bmatrix} K_i(t) \begin{bmatrix} E_{1i} + E_{2i}F_i & 0 \end{bmatrix} + \begin{bmatrix} E_{1i} + E_{2i}F_i & 0 \end{bmatrix}^T K_i(t)^T \begin{bmatrix} 0 \\ D_i \end{bmatrix}^T < 0 \,,
\end{aligned}
\tag{9.34}
$$

where

$$
\Omega_{ii} = \begin{bmatrix} -P & * \\ G_i + H_i F_i & -P^{-1} \end{bmatrix} .
$$

Using Lemma 9.2, (9.34) holds for all $K_i(t)$ satisfying $K_i(t)^T K_i(t) \leq I$ if and only if there exists a constant $\epsilon_{ii}^{1/2} > 0$ such that

$$
\begin{aligned}
&\Omega_{ii} + \begin{bmatrix} \epsilon_{ii}^{-1/2}(E_{1i} + E_{2i}F_i)^T & 0 \\ 0 & \epsilon_{ii}^{1/2} D_i \end{bmatrix} \begin{bmatrix} \epsilon_{ii}^{-1/2}(E_{1i} + E_{2i}F_i) & 0 \\ 0 & \epsilon_{ii}^{1/2} D_i^T \end{bmatrix} \\
&= \Omega_{ii} + \begin{bmatrix} (E_{1i} + E_{2i}F_i)^T & 0 \\ 0 & D_i \end{bmatrix} \begin{bmatrix} \epsilon_{ii}^{-1} I & 0 \\ 0 & \epsilon_{ii} I \end{bmatrix} \begin{bmatrix} E_{1i} + E_{2i}F_i & 0 \\ 0 & D_i^T \end{bmatrix} < 0 \,.
\end{aligned}
\tag{9.35}
$$

Applying Lemma 9.1 to (9.35) gives

$$\begin{bmatrix} -P & * & * & * \\ G_i + H_i F_i & -P^{-1} & * & * \\ E_{1i} + E_{2i} F_i & 0 & -\epsilon_{ii} I & * \\ 0 & D_i^T & 0 & -\epsilon_{ii}^{-1} I \end{bmatrix} < 0\,. \tag{9.36}$$

Define the following transformation matrix:

$$\begin{bmatrix} P^{-1} & 0 & 0 & 0 \\ 0 & I & 0 & 0 \\ 0 & 0 & I & 0 \\ 0 & 0 & 0 & I \end{bmatrix}$$

and take the congruence transformation. This results in

$$\begin{aligned} &\begin{bmatrix} P^{-1} & 0 & 0 & 0 \\ 0 & I & 0 & 0 \\ 0 & 0 & I & 0 \\ 0 & 0 & 0 & I \end{bmatrix} \begin{bmatrix} -P & * & * & * \\ G_i + H_i F_i & -P^{-1} & * & * \\ E_{1i} + E_{2i} F_i & 0 & -\epsilon_{ii} I & * \\ 0 & D_i^T & 0 & -\epsilon_{ii}^{-1} I \end{bmatrix} \begin{bmatrix} P^{-1} & 0 & 0 & 0 \\ 0 & I & 0 & 0 \\ 0 & 0 & I & 0 \\ 0 & 0 & 0 & I \end{bmatrix}^T \\ &= \begin{bmatrix} -P^{-1} & * & * & * \\ G_i P^{-1} + H_i F_i P^{-1} & -P^{-1} & * & * \\ E_{1i} P^{-1} + E_{2i} F_i P^{-1} & 0 & -\epsilon_{ii} I & * \\ 0 & D_i^T & 0 & -\epsilon_{ii}^{-1} I \end{bmatrix} < 0\,. \end{aligned} \tag{9.37}$$

Denoting $Q = P^{-1}$ and $M_i = F_i P^{-1}$ yields (9.29).

The second LMI can be established through a similar procedure. Using Assumption 1, applying Lemmata 9.1 and 9.2, and then taking the congruence transformation, we obtain (9.30). If the following inequality is assumed:

$$\begin{aligned} &\left(\left(\frac{G_i + \Delta G_i + (H_i + \Delta H_i) F_j + G_j + \Delta G_j + (H_j + \Delta H_j) F_i}{2} \right)^T P \right. \\ &\times \left. \left(\frac{G_i + \Delta G_i + (H_i + \Delta H_i) F_j + G_j + \Delta G_j + (H_j + \Delta H_j) F_i}{2} \right) - P \right) < 0\,, \end{aligned} \tag{9.38}$$

$$1 \le i < j \le q\,,$$

then (9.38) is equivalent to

$$\begin{aligned} &\begin{bmatrix} -4P & * \\ G_i + \Delta G_i + (H_i + \Delta H_i) F_j + G_j + \Delta G_j + (H_j + \Delta H_j) F_i & -P^{-1} \end{bmatrix} \\ &= \Xi_{ij} + \begin{bmatrix} 0 & * \\ \Delta G_i + \Delta H_i F_j + \Delta G_j + \Delta H_j F_i & 0 \end{bmatrix} < 0\,, \end{aligned}$$

or

$$\Xi_{ij} + \begin{bmatrix} 0 & 0 \\ D_i & D_j \end{bmatrix} \begin{bmatrix} K_i(t) & 0 \\ 0 & K_j(t) \end{bmatrix} \begin{bmatrix} E_{1i} + E_{2i}F_j & 0 \\ E_{1j} + E_{2j}F_i & 0 \end{bmatrix} + \begin{bmatrix} E_{1i} + E_{2i}F_j & 0 \\ E_{ij} + E_{2j}F_i & 0 \end{bmatrix}^T \begin{bmatrix} K_i(t) & 0 \\ 0 & K_j(t) \end{bmatrix}^T \begin{bmatrix} 0 & 0 \\ D_i & D_j \end{bmatrix}^T < 0 \,. \tag{9.39}$$

where

$$\Xi_{ij} = \begin{bmatrix} -4P & * \\ G_i + H_iF_j + G_j + H_jF_i & -P^{-1} \end{bmatrix} .$$

The above inequality (9.39) holds for all $F_i(t)$ satisfying

$$\begin{bmatrix} K_i(t) & 0 \\ 0 & K_j(t) \end{bmatrix}^T \begin{bmatrix} K_i(t) & 0 \\ 0 & K_j(t) \end{bmatrix} \leq I \,,$$

if and only if there exists a constant $\epsilon_{ii}^{1/2} > 0$ such that

$$\begin{aligned} &\Xi_{ij} + \begin{bmatrix} \epsilon_{ij}^{-1/2}(E_{1i} + E_{2i}F_j)^T & \epsilon_{ij}^{-1/2}(E_{1j} + E_{2j}F_i)^T & 0 & 0 \\ 0 & 0 & \epsilon_{ij}^{1/2}D_i & \epsilon_{ij}^{1/2}D_j \end{bmatrix} \\ &\times \begin{bmatrix} \epsilon_{ij}^{-1/2}(E_{1i} + E_{2i}F_j) & 0 \\ \epsilon_{ij}^{-1/2}(E_{1j} + E_{2j}F_i) & 0 \\ 0 & \epsilon_{ij}^{1/2}D_i^T \\ 0 & \epsilon_{ij}^{1/2}D_j^T \end{bmatrix} \\ &= \Xi_{ij} + \begin{bmatrix} (E_{1i} + E_{2i}F_j)^T & (E_{1j} + E_{2j}F_i)^T & 0 & 0 \\ 0 & 0 & D_i & D_j \end{bmatrix} \begin{bmatrix} \epsilon_{ij}^{-1}I & 0 & 0 & 0 \\ 0 & \epsilon_{ij}^{-1}I & 0 & 0 \\ 0 & 0 & \epsilon_{ij}I & 0 \\ 0 & 0 & 0 & \epsilon_{ij}I \end{bmatrix} \\ &\times \begin{bmatrix} E_{1i} + E_{2i}F_j & 0 \\ E_{1j} + E_{2j}F_i & 0 \\ 0 & D_i^T \\ 0 & D_j^T \end{bmatrix} < 0 \,, \end{aligned} \tag{9.40}$$

Applying Lemma 9.1 to (9.40) yields

$$\begin{bmatrix} -4P & * & * & * & * & * \\ \begin{pmatrix} G_i + H_iF_j \\ +G_j + H_jF_i \end{pmatrix} & -P^{-1} & * & * & * & * \\ E_{i1} + E_{2i}F_j & 0 & -\epsilon_{ij}I & * & * & * \\ E_{1j} + E_{2j}F_i & 0 & 0 & -\epsilon_{ij}I & * & * \\ 0 & D_i^T & 0 & 0 & -\epsilon_{ij}^{-1}I & * \\ 0 & D_j^T & 0 & 0 & 0 & -\epsilon_{ij}^{-1}I \end{bmatrix} < 0 \,. \tag{9.41}$$

Introduce the following congruence transformation matrix,

$$\begin{bmatrix} P^{-1} & 0 & 0 & 0 & 0 & 0 \\ 0 & I & 0 & 0 & 0 & 0 \\ 0 & 0 & I & 0 & 0 & 0 \\ 0 & 0 & 0 & I & 0 & 0 \\ 0 & 0 & 0 & 0 & I & 0 \\ 0 & 0 & 0 & 0 & 0 & I \end{bmatrix},$$

then, one has

$$\begin{bmatrix} P^{-1} & 0 & 0 & 0 & 0 & 0 \\ 0 & I & 0 & 0 & 0 & 0 \\ 0 & 0 & I & 0 & 0 & 0 \\ 0 & 0 & 0 & I & 0 & 0 \\ 0 & 0 & 0 & 0 & I & 0 \\ 0 & 0 & 0 & 0 & 0 & I \end{bmatrix} \begin{bmatrix} -4P & * & * & * & * & * \\ \begin{pmatrix} G_i + H_iF_j \\ +G_j + H_jF_i \end{pmatrix} & -P^{-1} & * & * & * & * \\ E_{i1} + E_{2i}F_j & 0 & -\epsilon_{ij}I & * & * & * \\ E_{1j} + E_{2j}F_i & 0 & 0 & -\epsilon_{ij}I & * & * \\ 0 & D_i^T & 0 & 0 & -\epsilon_{ij}^{-1}I & * \\ 0 & D_j^T & 0 & 0 & 0 & -\epsilon_{ij}^{-1}I \end{bmatrix}$$

$$\times \begin{bmatrix} P^{-1} & 0 & 0 & 0 & 0 & 0 \\ 0 & I & 0 & 0 & 0 & 0 \\ 0 & 0 & I & 0 & 0 & 0 \\ 0 & 0 & 0 & I & 0 & 0 \\ 0 & 0 & 0 & 0 & I & 0 \\ 0 & 0 & 0 & 0 & 0 & I \end{bmatrix}^T$$

$$= \begin{bmatrix} -4P^{-1} & * & * & * & * & * \\ \begin{pmatrix} G_iP^{-1} + H_iF_jP^{-1} \\ +G_jP^{-1} + H_jF_iP^{-1} \end{pmatrix} & -P^{-1} & * & * & * & * \\ E_{i1}P^{-1} + E_{2i}F_jP^{-1} & 0 & -\epsilon_{ij}I & * & * & * \\ E_{1j}P^{-1} + E_{2j}F_iP^{-1} & 0 & 0 & -\epsilon_{ij}I & * & * \\ 0 & D_i^T & 0 & 0 & -\epsilon_{ij}^{-1}I & * \\ 0 & D_j^T & 0 & 0 & 0 & -\epsilon_{ij}^{-1}I \end{bmatrix} < 0. \tag{9.42}$$

To this end, a change of variables, $P^{-1} = Q$ and $F_iP^{-1} = M_i$, yields (9.30), which completes the proof of the theorem.

Q. E. D.

Remark 9.9. The second inequalities in Theorem 9.8 for the pair (i, j), such that $\mu_i(x(t))\mu_j(x(t)) = 0$, for all t, do not have to be solved in determining the system stability.

Remark 9.10. The matrices D_i and E_{ij} in Theorems 9.6 and 9.8 can be chosen arbitrarily to express parametric uncertainties. Note, however, that the choices of D_i and E_{ij} generally influence the controller performance.

9.6 Simulations

In this section, to show the effectiveness of the proposed system modeling and controller design techniques, we simulate the control of the chaotic Lorenz system and the flexible-joint robot arm both with parametric uncertainties in continuous-time and discrete-time cases, respectively. The control objective is to drive their trajectories to the origin.

9.6.1 Continuous-time Case

This subsection presents examples of controller design for continuous-time TS fuzzy models.

Chaotic Lorenz System

First, the continuous-time chaotic Lorenz system is simulated. The input matrices B_1 and B_2 are arbitrarily chosen as

$$B_1 = B_2 = \begin{bmatrix} 1 \\ 0 \\ 0 \end{bmatrix},$$

which preserve the system's controllability. The TS fuzzy-model-based controller has the following structure:

Controller Rules:

$$\text{Rule 1: IF } x(t) \text{ is about } M_1 \text{ THEN } u(t) = F_1 \begin{bmatrix} x(t) \\ y(t) \\ z(t) \end{bmatrix},$$

$$\text{Rule 2: IF } x(t) \text{ is about } M_2 \text{ THEN } u(t) = F_2 \begin{bmatrix} x(t) \\ y(t) \\ z(t) \end{bmatrix}.$$

Choose $\left[M_1\,, M_2\right]$ to be $\left[-20\,, 30\right]$, because $x(t)$ is seemingly bounded within $\left[-20\,, 30\right]$, as shown in Fig. 6.1. The membership functions for the plant rules and the controller rules are shown in Fig. 9.1.

If we assume that the uncertainty parameters for σ, r, and b are all bounded within 30% of their nominal values, then the uncertainty matrices can be represented as follows:

$$\Delta A_1 = \Delta A_2 = 0.3 \times \begin{bmatrix} -\sigma\delta_1(t) & \sigma\delta_1(t) & 0 \\ r\delta_2(t) & 0 & 0 \\ 0 & 0 & -b\delta_3(t) \end{bmatrix}, \quad \Delta B_1 = \Delta B_2 = \begin{bmatrix} 0 \\ 0 \\ 0 \end{bmatrix},$$

just for simplicity. There, $|\delta_i(t)|^2 \leq 1, (i = 1, 2, 3)$. Based on Assumption 1, the uncertainty matrices $\Delta A_1, \Delta A_2, \Delta B_1$ and ΔB_i can be decomposed as follows:

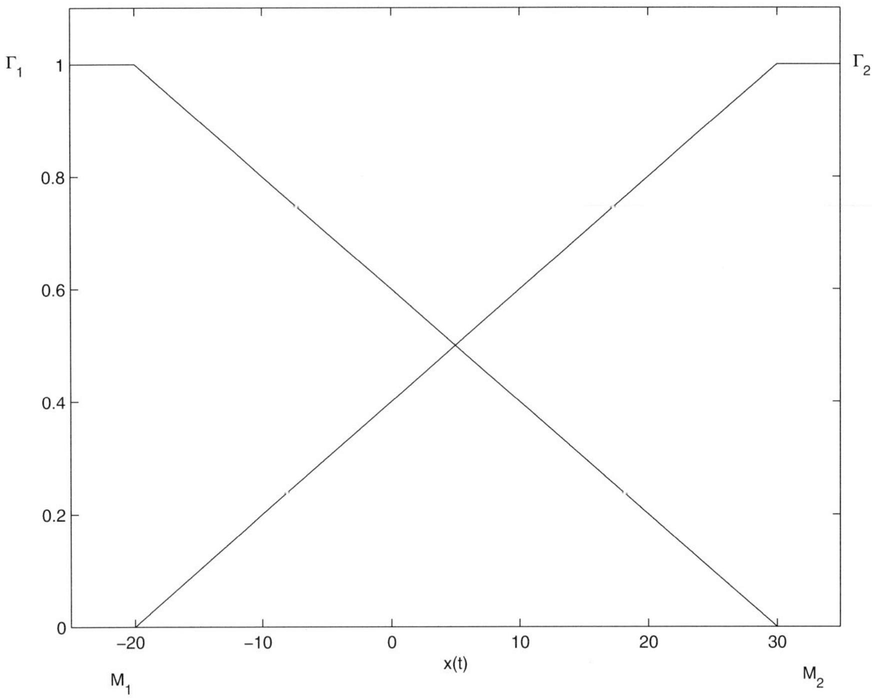

Fig. 9.1. Membership functions for the TS fuzzy model of the Lorenz system

$$D_1 = D_2 = \begin{bmatrix} -0.3 & 0 & 0 \\ 0 & 0.3 & 0 \\ 0 & 0 & 0.3 \end{bmatrix}, \quad E_{11} = E_{12} = \begin{bmatrix} \sigma & -\sigma & 0 \\ r & 0 & 0 \\ 0 & 0 & b \end{bmatrix},$$

$$E_{21} = E_{22} = \begin{bmatrix} 0 \; 0 \; 0 \end{bmatrix}^T, \quad K_1(t) = K_2(t) = \begin{bmatrix} \delta_1(t) & 0 & 0 \\ 0 & \delta_2(t) & 0 \\ 0 & 0 & \delta_3(t) \end{bmatrix}.$$

By applying Theorem 9.6 and solving the corresponding LMIs, we obtain the following controller gain matrices:

$$F_1 = \begin{bmatrix} -295.7653 & -137.2603 & -8.0866 \end{bmatrix},$$
$$F_2 = \begin{bmatrix} -443.0647 & -204.8089 & 12.6930 \end{bmatrix}.$$

Checking the global stability of the TS fuzzy-model-based control system results in the common positive-definite matrix P as

$$P = \begin{bmatrix} 0.0087 & -0.0129 & 0.0001 \\ -0.0129 & 0.0303 & -0.0002 \\ 0.0001 & -0.0002 & 0.0311 \end{bmatrix}.$$

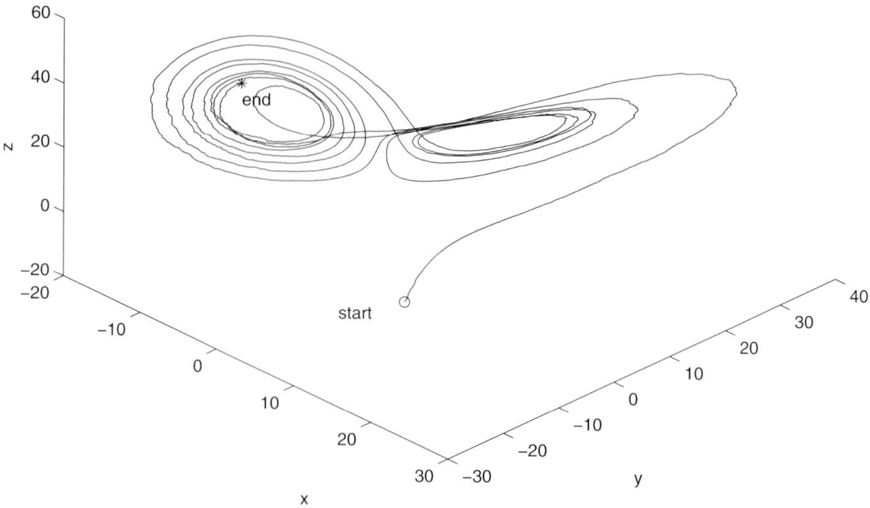

Fig. 9.2. Phase trajectory of the chaotic Lorenz system with parametric uncertainties (all system parameters are randomly varied within 30% of their nominal values)

The initial values of states are $\left[x(0)\ y(0)\ z(0)\right]^T = \left[10\ -10\ -10\right]^T$. The system parameters σ, r, and b are randomly varied within 30% of their nominal values. The simulation time is 10 sec.

Figure 9.2 shows the phase trajectory of the Lorenz system, with parametric uncertainties but without control inputs. Although the system parameters are varied during the simulation process, the system trajectory has only a slight distortion and remains bounded in the state space. The control result is shown in Fig. 9.3. The control input is activated at $t = 3.89$ sec for comparison. Before the control input was activated, the phase trajectory of the Lorenz system was chaotic. However, after the control input is activated, the phase trajectory is quickly directed to the origin. Figure 9.4 shows the simulation result for the Lorenz system without parametric uncertainties but with the same feedback controller. The control input is also activated at $t = 3.89$ sec. Similarly to the case of the parametrically uncertain system, after control input was activated, the system state is rapidly guided to the origin. Two simulation results show that the TS fuzzy-model-based controller, designed by using Theorem 9.6, is robust against norm-bounded parametric uncertainties.

Flexible-joint Robot Arm

Designing a controller for the flexible-joint robot arm is exemplified in this subsection. The structure of the TS fuzzy-model-based controller is as follows:

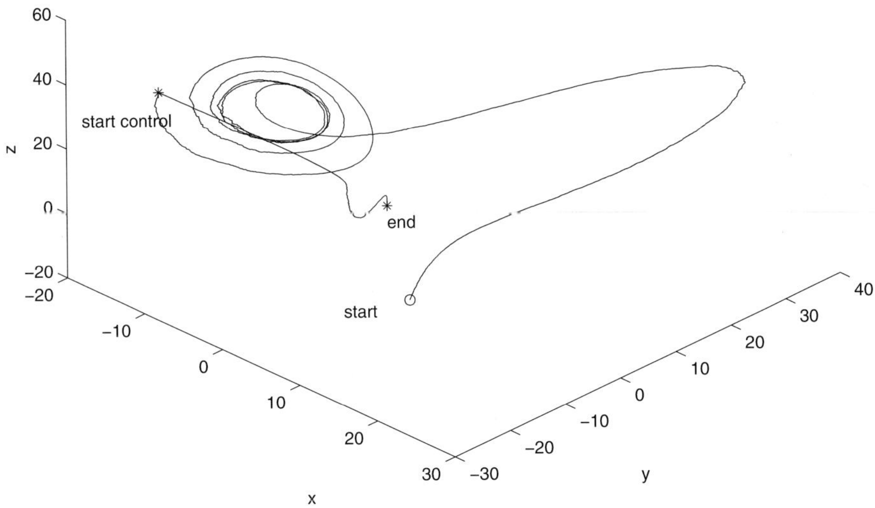

Fig. 9.3. Controlled phase trajectory of the Lorenz system with parametric uncertainties (all system parameters are randomly varied within 30% of their nominal values)

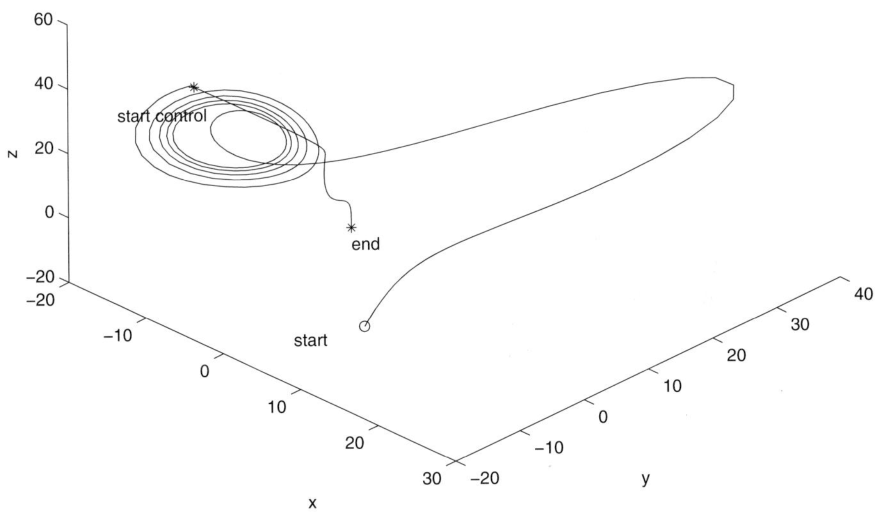

Fig. 9.4. Phase trajectory of the controlled Lorenz system without parametric uncertainties

Controller Rules:

Rule 1: IF $x(t)$ is about M_1 THEN $u(t) = F_1 x(t)$,

Rule 2: IF $x(t)$ is about M_2 THEN $u(t) = F_2 x(t)$.

In this simulation, $[M_1\,,\, M_2]$ is chosen as $[-2.85\,,\, 2.85]$, and the design parameter α is selected as 0.1 such that Γ_i is positive-semidefinite for all t. The membership functions for the plant rules and the controller rules are depicted in Fig. 9.5.

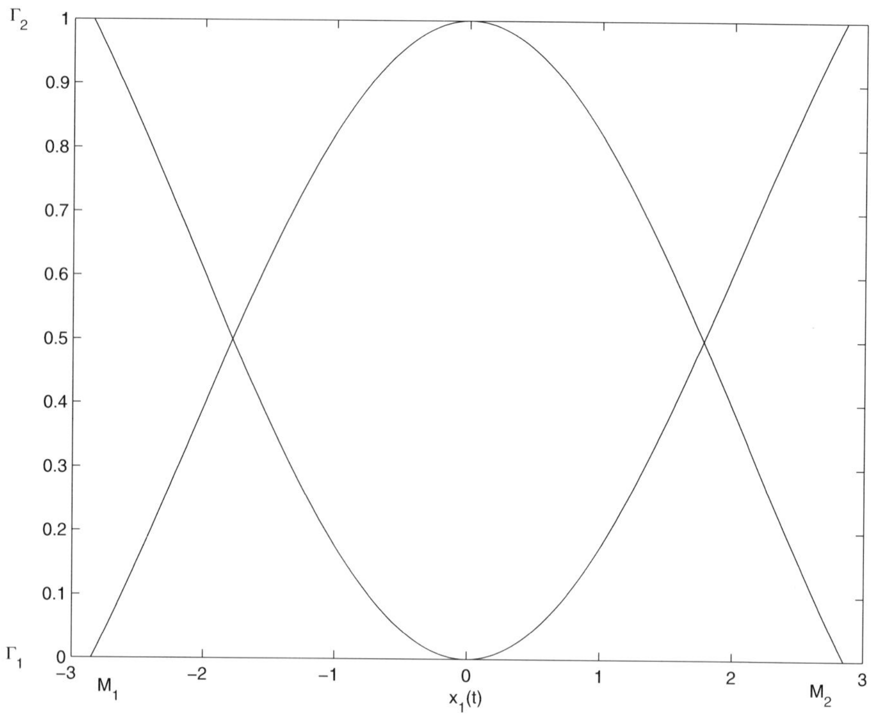

Fig. 9.5. Membership functions for the TS fuzzy model of the flexible-joint robot arm

The system parameters $\frac{k}{I}$, $\frac{k}{J}$, $\frac{1}{J}$ and $\frac{Mgl}{I}$ are all assumed to be bounded within 30% of their nominal values. The associated uncertainty matrices can be represented as follows:

$$\Delta A_1 = 0.3 \times \begin{bmatrix} 0 & 0 & 0 & 0 \\ -\frac{Mgl}{I}\delta_1(t) - \frac{k}{I}\delta_2(t) & 0 & \frac{k}{I}\delta_2(t) & 0 \\ 0 & 0 & 0 & 0 \\ \frac{k}{J}\delta_3(t) & 0 & -\frac{k}{J}\delta_3(t) & 0 \end{bmatrix},$$

$$\Delta A_2 = 0.3 \times \begin{bmatrix} 0 & 0 & 0 & 0 \\ -\frac{\alpha Mgl}{I}\delta_1(t) - \frac{k}{I}\delta_2(t) & 0 & \frac{k}{I}\delta_2(t) & 0 \\ 0 & 0 & 0 & 0 \\ \frac{k}{J}\delta_3(t) & 0 & -\frac{k}{J}\delta_3(t) & 0 \end{bmatrix},$$

$$\Delta B_1 = \Delta B_2 = 0.3 \times \begin{bmatrix} 0 \\ 0 \\ 0 \\ \frac{k}{J}\delta_4(t) \end{bmatrix}.$$

where $\delta_i(t)\,(i = 1, 2, 3, 4)$ is an unknown function satisfying $|\delta_i(t)|^2 \leq 1$. Applying Assumption 1, the uncertainty matrices $\Delta A_1, A_2, B_1$ and B_2 can be decomposed as

$$D_1 = \begin{bmatrix} 0 & 0 & 0 & 0 \\ \frac{70k}{I} & \frac{70Mgl}{I} & 0 & 0 \\ 0 & 0 & 0 & 0 \\ 0 & 0 & \frac{70K}{J} & \frac{70}{J} \end{bmatrix}, \quad D_2 = \begin{bmatrix} 0 & 0 & 0 & 0 \\ \frac{70k}{I} & \frac{70\alpha Mgl}{I} & 0 & 0 \\ 0 & 0 & 0 & 0 \\ 0 & 0 & \frac{70k}{J} & \frac{70}{J} \end{bmatrix},$$

$$E_{11} = E_{12} = \begin{bmatrix} -\frac{1}{70} & 0 & \frac{1}{70} & 0 \\ -\frac{1}{70} & 0 & 0 & 0 \\ \frac{1}{70} & 0 & -\frac{1}{70} & 0 \\ 0 & 0 & 0 & 0 \end{bmatrix}, \quad E_{21} = E_{22} = \begin{bmatrix} 0 \\ 0 \\ 0 \\ \frac{1}{70} \end{bmatrix},$$

$$K_1(t) = K_2(t) = \begin{bmatrix} \delta_1(t) & 0 & 0 & 0 \\ 0 & \delta_2(t) & 0 & 0 \\ 0 & 0 & \delta_3(t) & 0 \\ 0 & 0 & 0 & \delta_4(t) \end{bmatrix}.$$

By applying Theorem 9.6 and solving the associated LMIs, the following controller gain matrices are obtained:

$$F_1 = \begin{bmatrix} 11.7696 & -1.9279 & -44.0304 & -0.3549 \end{bmatrix},$$
$$F_2 = \begin{bmatrix} 11.5981 & -1.8931 & -43.2852 & -0.3485 \end{bmatrix}.$$

The global stability of the controlled TS fuzzy model (9.28) is guaranteed by finding the common positive-definite matrix P as

$$P = 1.0\mathrm{e} - 3 \times \begin{bmatrix} 0.0343 & -0.0012 & -0.0419 & -0.0003 \\ -0.0012 & 0.0004 & 0.0072 & 0.0000 \\ -0.0419 & 0.0072 & 0.1622 & 0.0011 \\ -0.0003 & 0.0000 & 0.0011 & 0.0000 \end{bmatrix}.$$

During the simulation process, the uncertain system parameters are randomly varied within the bound of 30% of their nominal values. The initial values of system states are $x(0) = \left[\frac{\pi}{2} \; -\frac{\pi}{6} \; -\frac{\pi}{2} \; \frac{\pi}{6}\right]^T$.

The simulation results are depicted in Figs. 9.6 and 9.7. The control input is activated at $t = 1$ sec for the purpose of more clearly visualizing the effectiveness of the proposed control method. Before control input is applied, the trajectories of the system state do not go to zero. Here, it should be noted that the state $x_1(t)$ is indeed bounded within $\left[-2.85\,,\, 2.85\right]$. This fact means that μ_1 and μ_2 are positive-semidefinite for all t. Therefore, the design parameter $\alpha = 0.1$ is a reasonable choice. As soon as the control input is applied, the system trajectories are rapidly guided to zero. Figure 9.7 shows the simulation result of the case of the flexible-joint robot arm without uncertainties but with the same feedback controller. The control input is applied at $t = 1$ sec. After the control input is applied to the system, the system trajectory is immediately directed to zero as expected. From these simulation results, it is obvious that the designed controller is robust against norm-bounded parametric uncertainties.

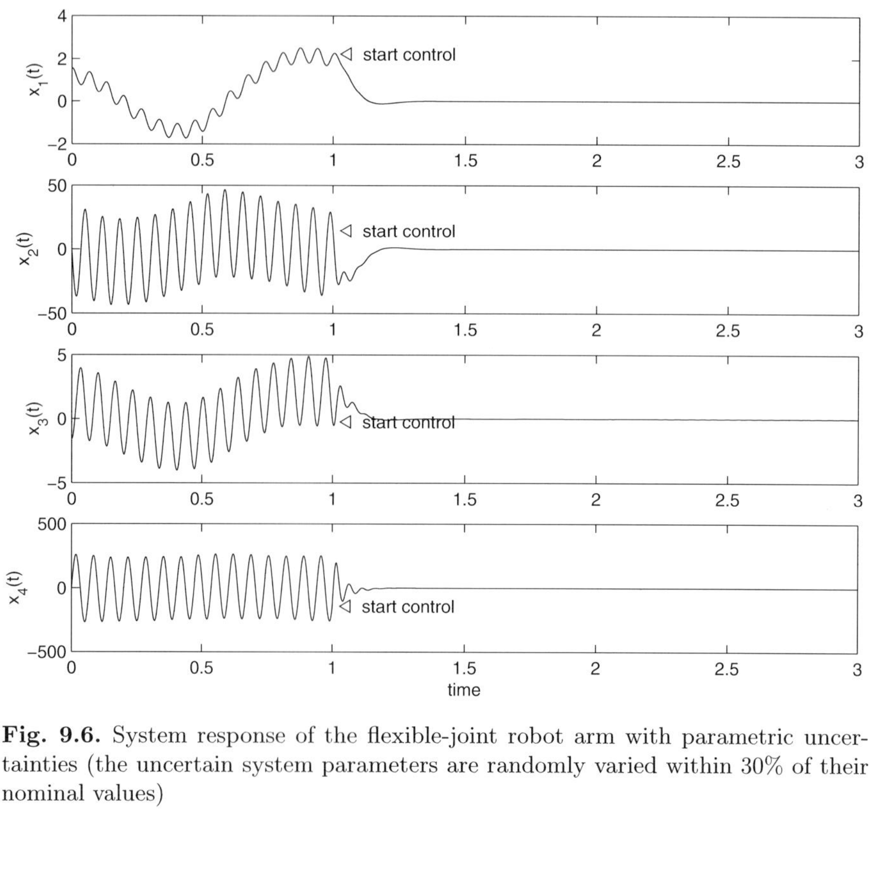

Fig. 9.6. System response of the flexible-joint robot arm with parametric uncertainties (the uncertain system parameters are randomly varied within 30% of their nominal values)

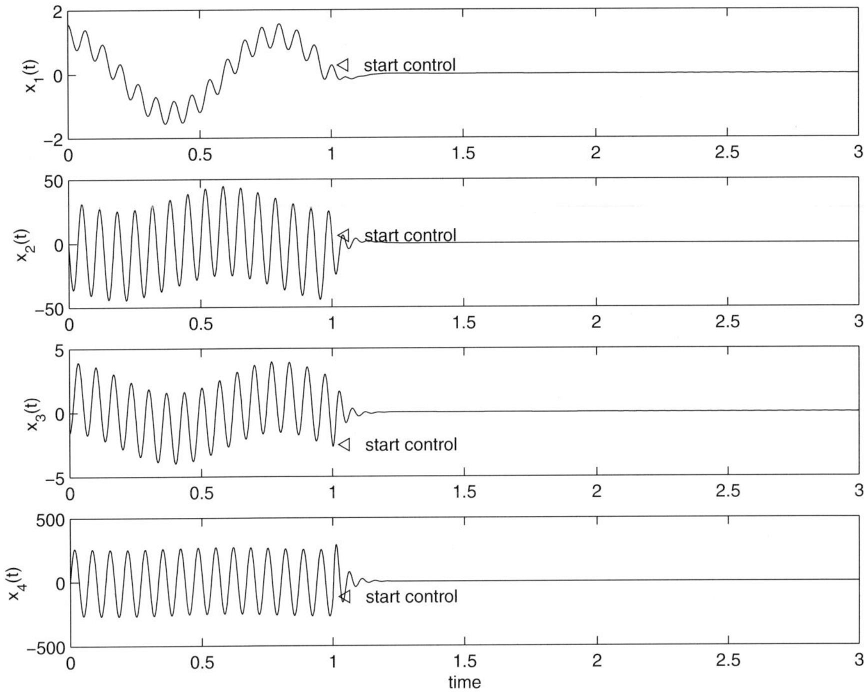

Fig. 9.7. System response of the flexible-joint robot arm without parametric uncertainties

9.6.2 Discrete-time Case

In this section, discrete-time PDC controllers are to be designed, based on the scheme described in Section 9.5.2.

Discretized TS Fuzzy Model of the Chaotic Lorenz System

Without loss of controllability, the input matrices are chosen as

$$H_1 = H_2 = \begin{bmatrix} 1 \\ 0 \\ 0 \end{bmatrix} .$$

In this case, the TS fuzzy-model-based controller structure, for the discretized chaotic Lorenz system, is as follows:

Controller Rules:

$$\text{Rule 1: IF } x(t) \text{ is about } M_1 \text{ THEN } u(t) = F_1 \begin{bmatrix} x(t) \\ y(t) \\ z(t) \end{bmatrix},$$

$$\text{Rule 2: IF } x(t) \text{ is about } M_2 \text{ THEN } u(t) = F_2 \begin{bmatrix} x(t) \\ y(t) \\ z(t) \end{bmatrix}.$$

The system parameters σ, r, b are the same as those in the continuous-time case. Also, the parameter uncertainties are assumed to be bounded within 30% of their nominal values. The uncertainty matrices are represented as follows:

$$\Delta G_1 = \Delta G_2 = 0.3 \times \begin{bmatrix} \sigma T_s \delta_1(t) & \sigma T_s \delta_1(t) & 0 \\ r T_s \delta_2(t) & 0 & 0 \\ 0 & 0 & -b T_s \delta_3(t) \end{bmatrix}, \quad \Delta H_1 = \Delta H_2 = \begin{bmatrix} 0 \\ 0 \\ 0 \end{bmatrix},$$

where $\delta_i(t)$ $(i = 1, 2, 3)$ is an unknown function satisfying $|\delta_i(t)|^2 \le 1$. Based on Assumption 1, the matrices D_1, D_2, E_{11}, E_{12}, E_{21}, and E_{22} are defined as

$$D_1 = D_2 = \begin{bmatrix} 0.006 & 0 & 0 \\ 0 & 0.006 & 0 \\ 0 & 0 & 0.006 \end{bmatrix}, \quad E_{11} = E_{12} = \begin{bmatrix} -\sigma & \sigma & 0 \\ b & 0 & 0 \\ 0 & 0 & -r \end{bmatrix},$$

$$E_{21} = E_{22} = \begin{bmatrix} 0 & 0 & 0 \end{bmatrix}^T, \quad K_1(t) = K_2(t) = \begin{bmatrix} \delta_1(t) & 0 & 0 \\ 0 & \delta_2(t) & 0 \\ 0 & 0 & \delta_3(t) \end{bmatrix}.$$

From Theorem 9.8, the controller gain matrices are obtained as

$$F_1 = \begin{bmatrix} -1.0024 & -0.4328 & -0.0158 \end{bmatrix},$$
$$F_2 = \begin{bmatrix} -1.0023 & -0.4338 & 0.0045 \end{bmatrix}.$$

The common positive-definite matrix P is found to be

$$P = \begin{bmatrix} 0.0079 & -0.0002 & -0.0000008 \\ -0.0002 & 0.0006 & 0.0000008 \\ -0.0000008 & 0.0000008 & 0.0006 \end{bmatrix},$$

which guarantees the global stability of the controlled discrete-time TS fuzzy system (9.28).

Figures 9.8, 9.9 and 9.10 show the the results of computer simulations. The simulation time is 10 sec, and the trajectory of each simulation starts with $\begin{bmatrix} x(0) & y(0) & z(0) \end{bmatrix}^T = \begin{bmatrix} 10 & -10 & -10 \end{bmatrix}^T$. During the simulation time, all system parameters are randomly varied within the bound of 30% of their nominal values.

The phase trajectory, in the presence of parametric uncertainties, is shown in Fig. 9.8. Although the system parameters are varied throughout the simulation process, the system trajectory is always bounded inside the state

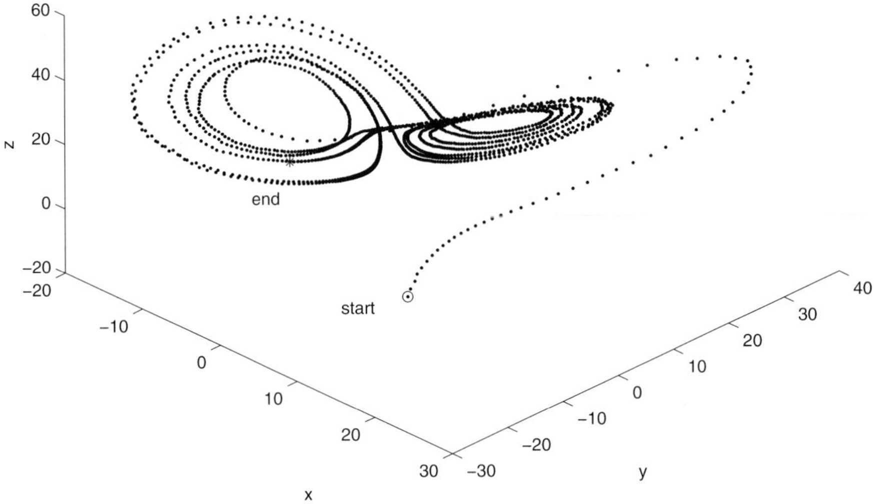

Fig. 9.8. Phase trajectory of the discretized chaotic Lorenz system with parametric uncertainties (all system parameters are varied within 30% of their nominal values)

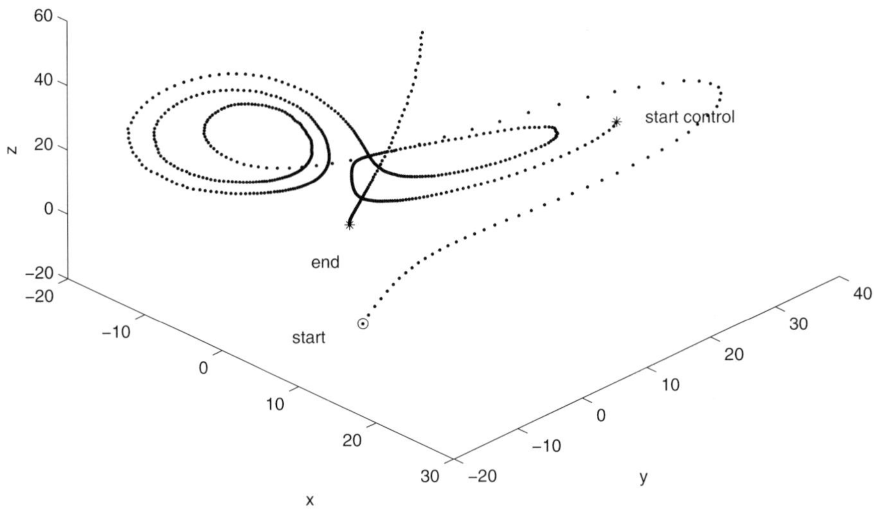

Fig. 9.9. Phase trajectory of the controlled discretized system of the chaotic Lorenz system in the presence of parametric uncertainties (all system parameters are varied within 30% of their nominal values)

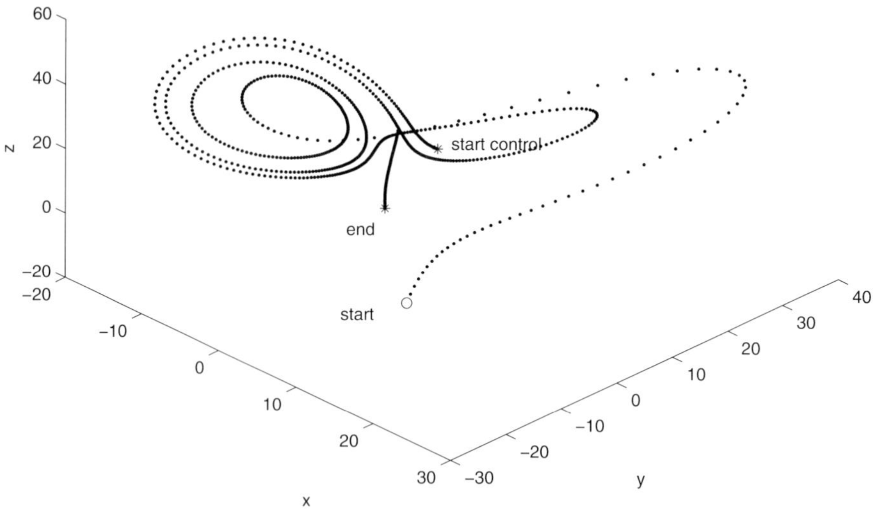

Fig. 9.10. Phase trajectory of the controlled discretized system of the chaotic Lorenz system without parameter uncertainties

space, and remains to be chaos-like. The controlled trajectory is shown in Fig. 9.9. Similar to the continuous-time case, the control input is activated at $t = 3.89$ sec for the purpose of comparison. Before $t = 3.89$ sec, the trajectory does not go to the origin, but exhibits a chaos-like, irregular behavior. After $t = 3.89$ sec, the trajectory is controlled very quickly to the origin. The simulation result, without parametric uncertainties, is depicted in Fig. 9.10. As soon as the control input is activated, at $t = 3.89$ sec, the system phase trajectory is well guided to the origin by the controller designed according to Theorem 9.8. From the two simulation results, the designed controller is quite robust against admissible and norm-bounded uncertainties.

Discretized TS Fuzzy Model of the Flexible-joint Robot Arm

Finally, the design procedure of the discrete-time TS fuzzy-model-based controller is presented for the discretized TS fuzzy model of the flexible-joint robot arm, which is derived in Section 6.5. In this case, the TS fuzzy-model-based controller rules are as follows:

Controller Rules:

Rule 1: IF $x(t)$ is about M_1 THEN $u(t) = F_1 x(t)$,

Rule 2: IF $x(t)$ is about M_2 THEN $u(t) = F_2 x(t)$.

All system parameters are the same as those in the continuous-time case including the property of the uncertain system parameters. The uncertainty matrices are expressed as follows:

$$\Delta G_1 = 0.3 \times \begin{bmatrix} 0 & 0 & 0 & 0 \\ -\frac{MglT_s}{I}\delta_1(t) - \frac{kT_s}{I}\delta_2(t) & 0 & \frac{kT_s}{I}\delta_2(t) & 0 \\ 0 & 0 & 0 & 0 \\ \frac{kT_s}{J}\delta_3(t) & 0 & -\frac{kT_s}{J}\delta_3(t) & 0 \end{bmatrix},$$

$$\Delta G_2 = 0.3 \times \begin{bmatrix} 0 & 0 & 0 & 0 \\ -\frac{\alpha MglT_s}{I}\delta_1(t) - \frac{kT_s}{I}\delta_2(t) & 0 & \frac{kT_s}{I}\delta_2(t) & 0 \\ 0 & 0 & 0 & 0 \\ \frac{kT_s}{J}\delta_3(t) & 0 & -\frac{kT_s}{J}\delta_3(t) & 0 \end{bmatrix},$$

$$\Delta H_1 = \Delta H_2 = 0.3 \times \begin{bmatrix} 0 \\ 0 \\ 0 \\ \frac{kT_s}{J}\delta_4(t) \end{bmatrix},$$

where $\delta_i(t)$ $(i = 1, 2, 3, 4)$ is an unknown function satisfying $|\delta_i(t)|^2 \leq 1$. Based on Assumption 1, the associated matrices are defined as

$$D_1 = \begin{bmatrix} 0 & 0 & 0 & 0 \\ \frac{0.0006k}{I} & \frac{0.0006Mgl}{I} & 0 & 0 \\ 0 & 0 & 0 & 0 \\ 0 & 0 & \frac{0.0006K}{J} & \frac{0.0006}{J} \end{bmatrix}, \quad D_2 = \begin{bmatrix} 0 & 0 & 0 & 0 \\ \frac{0.0006k}{I} & \frac{0.0006\alpha Mgl}{I} & 0 & 0 \\ 0 & 0 & 0 & 0 \\ 0 & 0 & \frac{0.0006k}{J} & \frac{0.0006}{J} \end{bmatrix},$$

$$E_{11} = E_{12} \begin{bmatrix} -0.0002 & 0 & 0.0002 & 0 \\ -0.0002 & 0 & 0 & 0 \\ 0.0002 & 0 & -0.0002 & 0 \\ 0 & 0 & 0 & 0 \end{bmatrix}, \quad E_{21} = E_{22} \begin{bmatrix} 0 \\ 0 \\ 0 \\ 0.0002 \end{bmatrix},$$

$$K_1(t) = K_2(t) = \begin{bmatrix} \delta_1(t) & 0 & 0 & 0 \\ 0 & \delta_2(t) & 0 & 0 \\ 0 & 0 & \delta_3(t) & 0 \\ 0 & 0 & 0 & \delta_4(t) \end{bmatrix}.$$

From Theorem 9.8, the controller gain matrices are found by

$$F_1 = \begin{bmatrix} 9.4424 & -0.6647 & -19.5414 & -0.2365 \end{bmatrix},$$
$$F_2 = \begin{bmatrix} 7.2697 & -0.5076 & -14.9979 & -0.1851 \end{bmatrix}.$$

The common positive-definite matrix P is found to be

$$P = 1.0\mathrm{e} - 4 \times \begin{bmatrix} 0.1467 & -0.0029 & -0.1469 & -0.0012 \\ -0.0029 & 0.0005 & 0.0110 & 0.0001 \\ -0.1469 & 0.0110 & 0.3149 & 0.0024 \\ -0.0012 & 0.0001 & 0.0024 & 0.0000 \end{bmatrix},$$

which guarantees the global stability of the controlled discrete-time TS fuzzy system (9.28). The simulation results are depicted in Fig. 9.11 and 9.12. The system trajectory of each simulation starts from $x(0) = \begin{bmatrix} \frac{\pi}{2} & -\frac{\pi}{6} & -\frac{\pi}{2} & \frac{\pi}{6} \end{bmatrix}^T$. During the simulation time, uncertain time-varying system parameters within the

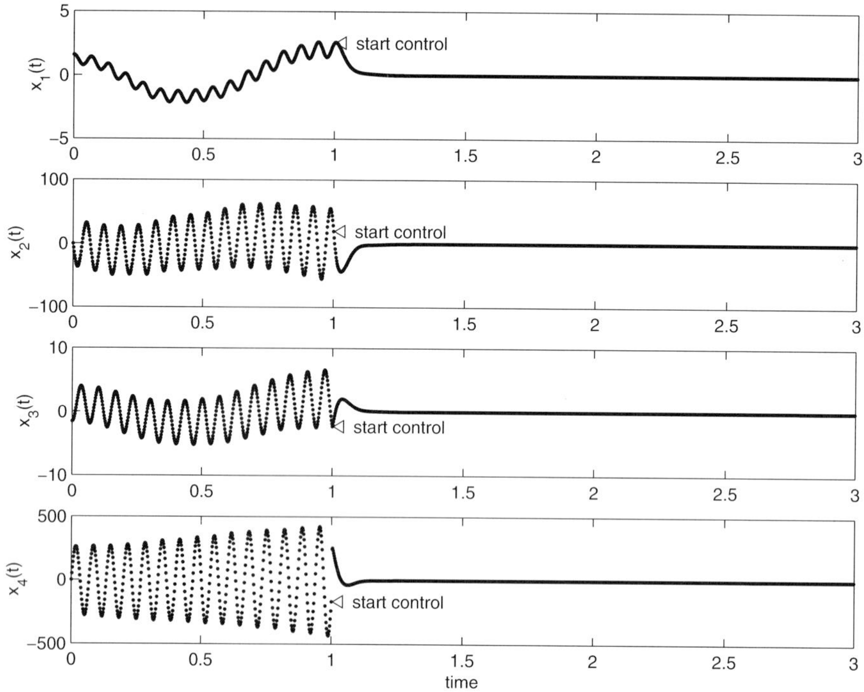

Fig. 9.11. System response of the discretized flexible-joint robot arm with parametric uncertainties (uncertain system parameters are varied within 30% of their nominal values)

bounds of 30% of their nominal values are introduced. Figure 9.11 shows the system response in the presence of parametric uncertainties. The control input is activated at $t = 1$ sec. Before the control input is activated, the system trajectories do not go to zero, but oscillate. However, it is observed that the oscillation of $x_1(t)$ occurs within the bound $[-2.85\,, 2.85]$. Once again, it should be noted that design parameter $\alpha = 0.1$ is a feasible choice. Therefore, $\mu_1(x(t))$ and $\mu_2(x(t))$ are positive-semidefinite for the entire simulation time. As soon as the control input is activated, even if the system contains time-varying uncertain parameters, the system state is instantly guided to zero. The simulation result, with no parametric uncertainties, is reported in Fig. 9.12. As expected, after the control input is applied, the state responses quickly converge to zero. From the two simulation results, it is obvious that the designed TS fuzzy-model-based state-feedback controller not only can stabilize the nonlinear systems, but exhibits strong robustness against admissible parametric uncertainties.

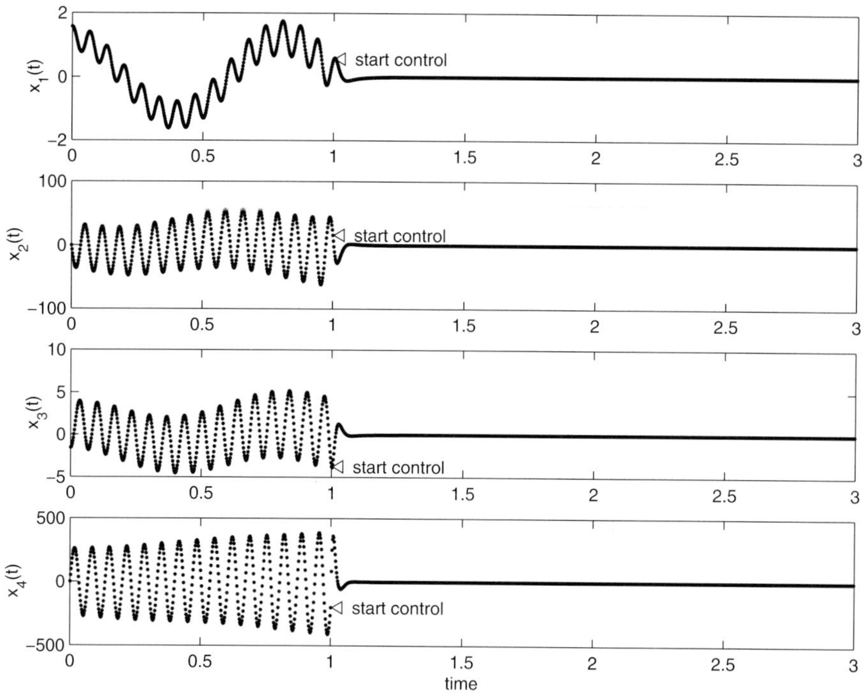

Fig. 9.12. System response of the discretized flexible-joint robot arm without parametric uncertainties

9.7 TS Fuzzy-model-based Adaptive Control

Corresponding to Chapter 8, fuzzy adaptive control based on TS fuzzy model is introduced in this section, taking only the CFS case into account.

Recall the TS fuzzy model described in Chapter 6 and in this chapter,

$$\dot{x}(t) = \frac{\sum_{i=1}^{q} \mu(x(t))\{A_i x(t) + B_i u(t)\}}{\sum_{i=1}^{q} \mu(x(t))}, \tag{9.43}$$

where $x(t) \in \Re^n$ is measurable, $A_i \in \Re^{n \times n}$, $B_i \in \Re^{n \times l}$ $(i = 1, \ldots, q)$ are unknown constant matrices and (A_i, B_i) are controllable. The control objective is to determine the control input $u(t) \in \Re^l$ such that all the signals in the closed-loop system are bounded and the trajectory follows that of a stable reference model (SRM), $x_m(t) \in \Re^n$, specified by the system [199, 200]

$$\dot{x}_m = \frac{\sum_{i=1}^{q} \sum_{j=1}^{q} \mu_i(x_m(t)\mu_j(x_m(t))\{(A_m)_{ij} x_m(t) + (B_m)_{ij} r(t)\}}{\sum_{i=1}^{q} \sum_{j=1}^{q} \mu_i(x_m(t))\mu_j(x_m(t))}, \tag{9.44}$$

where $(A_m)_{ij} \in \Re^{n \times n}$, $i, j = 1, \ldots, q$, satisfy the stability condition (9.7) given in Theorem 9.4, $(B_m)_{ij} \in \Re^{n \times l}$ and $r(t) \in \Re^l$ is a bounded reference input vector.

In terms of the PDC, the control law has the following form

$$\begin{aligned} R^i : &\text{IF } x_1(t) \text{ is } \Gamma_1^i \text{ and } \cdots \text{ and } x_n(t) \text{ is } \Gamma_n^i \\ &\text{THEN } u = -F_i^* x(t) + L_j^* r(t), \end{aligned} \tag{9.45}$$

where $x(t) = [x_1(t), x_2(t), \ldots, x_n(t)]^T$, $r(t) = [r_1(t), r_2(t), \ldots, r_l(t)]^T$ and $i = 1, \ldots, q$.

It is inferred as

$$u(t) = \frac{\sum_{i=1}^{q} \mu_i(x(t))(-F_i^* x(t) + L_i^* r(t))}{\sum_{i=1}^{q} \mu_i(x(t))}. \tag{9.46}$$

If A_i and B_i are known, substituting (9.46) into (9.43) yields

$$\dot{x}(t) = \frac{\sum_{i=1}^{q} \sum_{j=1}^{q} \mu_i(x(t))\mu_j(x(t))\{(A_i - B_i F_j^*)x(t) + B_i L_j^* r(t)\}}{\sum_{i=1}^{q} \sum_{j=1}^{q} \mu_i(x(t))\mu_j(x(t))}. \tag{9.47}$$

Thus, if $F_i^* \in \Re^{l \times n}$ and $L_i^* \in \Re^{l \times l}$ can be chosen to satisfy

$$A_i - B_i F_j^* = (A_m)_{ij}, \qquad B_i L_j^* = (B_m)_{ij}, \tag{9.48}$$

then the closed-loop system (9.47) is the same as the SRM (9.44) and $x(t) \to x_m(t)$ for any bounded reference input $r(t)$.

However, if the system parameters are unknown, the controller design becomes impossible. To solve this problem, the following adaptive fuzzy controller is developed for systems with unknown parameters,

$$u(t) = \frac{\sum_{i=1}^{q} \mu_i(x(t))(-F_i(t)x(t) + L_i(t)r(t))}{\sum_{i=1}^{q} \mu_i(x(t))}, \tag{9.49}$$

where $F_i(t)$ and $L_i(t)$ are the estimates of F_i^* and L_i^*, respectively, to be determined by an appropriate adaptation law.

After some manipulations, one has the error dynamics ($e(t) = x(t) - x_m(t)$),

$$\begin{aligned} \dot{e}(t) =& \frac{\sum_{i=1}^{q} \sum_{j=1}^{q} \mu_i(x(t))\mu_j(x(t))(A_m)_{ij}}{\sum_{i=1}^{q} \sum_{j=1}^{q} \mu_i(x(t))\mu_j(x(t))} e(t) \\ &+ \frac{\sum_{i=1}^{q} \sum_{j=1}^{q} \mu_i(x(t))\mu_j(x(t))B_i(-\tilde{F}_j(t)x(t) + \tilde{L}_j(t)r(t))}{\sum_{i=1}^{q} \sum_{j=1}^{q} \mu_i(x(t))\mu_j(x(t))}, \end{aligned} \tag{9.50}$$

where $\tilde{F}_j(t) = F_j(t) - F_j^*$, $\tilde{L}_j(t) = L_j(t) - L_j^*$, and B_i are unknown.

Assume that L_i^* is either positive-definite or negative-definite, and define $S_i^{-1} = \text{sgn}(l_i) L_i^*$, in which

$$\mathrm{sgn}(l_i) = \begin{cases} 1 & \text{if } L_i^* \text{ is positive definite} \\ -1 & \text{if } L_i^* \text{ is negative definite} \end{cases}$$

Taking $B_i = (B_m)_{ij} L_j^{*-1}$, (9.50) becomes

$$\dot{e}(t) = \frac{\sum_{i=1}^{q}\sum_{j=1}^{q} \mu_i(x(t))\mu_j(x(t))(A_m)_{ij}}{\sum_{i=1}^{q}\sum_{j=1}^{q} \mu_i(x(t))\mu_j(x(t))} e(t) + \frac{\sum_{i=1}^{q}\sum_{j=1}^{q} \mu_i(x(t))\mu_j(x(t))(B_m)_{ij} L_j^{*-1}(-\tilde{F}_j(t)x(t) + \tilde{L}_j(t)r(t))}{\sum_{i=1}^{q}\sum_{j=1}^{q} \mu_i(x(t))\mu_j(x(t))}, \tag{9.51}$$

In terms of the error dynamics (9.51), the adaptation law for the desired control parameters F_i^* and L_i^* is to be derived such that the closed-loop system (9.43) and (9.49) follows the SRM (9.44).

Consider a Lyapunov function candidate

$$V(e(t), \tilde{F}_i(t), \tilde{L}_i(t)) = e^T P e + \sum_{i=1}^{q} \mathrm{tr}(\tilde{F}_i^T S_i \tilde{F}_i + \tilde{L}_i^T S_i \tilde{L}_i), \tag{9.52}$$

where P is a positive-definite and symmetric matrix satisfying $(A_m)_{ij}^T P + P(A_m)_{ij} < 0$, and $\mathrm{tr}(\cdot)$ is the trace of a matrix.

After some tedious manipulations, one has

$$\dot{V} = -e(t)^T \frac{\sum_{i=1}^{q}\sum_{j=1}^{q} \mu_i(x(t))\mu_j(x(t))Q_{ij}}{\sum_{i=1}^{q}\sum_{j=1}^{q} \mu_i(x(t))\mu_j(x(t))} e(t) + 2\mathrm{tr}\left\{ -\frac{\sum_{i=1}^{q}\sum_{j=1}^{q} \mu_i(x(t))\mu_j(x(t))\tilde{F}_j(t)^T S_j (B_m)_{ij}^T \mathrm{sgn}(l_j)}{\sum_{i=1}^{q}\sum_{j=1}^{q} \mu_i(x(t))\mu_j(x(t))} P e(t) x(t)^T + \sum_{i=1}^{q} \tilde{F}_j(t)^T S_j \dot{\tilde{F}}_j(t) \right\}$$

$$+ 2\mathrm{tr}\left\{ \frac{\sum_{i=1}^{q}\sum_{j=1}^{q}\mu_i(x(t))\mu_j(x(t))\tilde{L}_j(t)^T S_j (B_m)_{ij}^T \mathrm{sgn}(l_j)}{\sum_{i=1}^{q}\sum_{j=1}^{q}\mu_i(x(t))\mu_j(x(t))} Pe(t)r(t)^T \right.$$
$$\left. + \sum_{i=1}^{q}\tilde{L}_j(t)^T S_j \dot{\tilde{L}}_j(t) \right\}. \tag{9.53}$$

Letting

$$\sum_{i=1}^{q} = \frac{\sum_{i=1}^{q}\sum_{j=1}^{q}\mu_i(x(t))\mu_j(x(t))\tilde{F}_i(t)^T S_i (B_m)_{ij}^T \mathrm{sgn}(l_i)}{\sum_{i=1}^{q}\sum_{j=1}^{q}\mu_i(x(t))\mu_j(x(t))} Pe(t)x(t)^T, \tag{9.54}$$

$$\sum_{i=1}^{q} = -\frac{\sum_{i=1}^{q}\sum_{j=1}^{q}\mu_i(x(t))\mu_j(x(t))\tilde{L}_i(t)^T S_i (B_m)_{ij}^T \mathrm{sgn}(l_i)}{\sum_{i=1}^{q}\sum_{j=1}^{q}\mu_i(x(t))\mu_j(x(t))} Pe(t)r(t)^T, \tag{9.55}$$

$\dot{V}$ can be negative, i.e.,

$$\dot{V} = -e(t)^T \frac{\sum_{i=1}^{q}\sum_{j=1}^{q}\mu_i(x(t))\mu_j(x(t))Q_{ij}}{\sum_{i=1}^{q}\sum_{j=1}^{q}\mu_i(x(t))\mu_j(x(t))} e(t) \le 0. \tag{9.56}$$

Therefore, the adaptation law is

$$\begin{aligned}\dot{\tilde{F}}_i(t) =& \dot{F}_i(t)\\ =& \left\{ \frac{\sum_{i=1}^{q}\mu_i(x(t))(B_m)_{ij}^T}{\sum_{i=1}^{q}\mu_i(x(t))} \right\} \left\{ \frac{\mu_j(x(t))}{\sum_{j=1}^{q}\mu_j(x(t))} \right\} \mathrm{sgn}(l_i)Pex(t)^T,\end{aligned} \tag{9.57}$$

$$\begin{aligned}\dot{\tilde{L}}_i(t) =& \dot{L}_i(t)\\ =& -\left\{ \frac{\sum_{i=1}^{q}\mu_i(x(t))(B_m)_{ij}^T}{\sum_{i=1}^{q}\mu_i(x(t))} \right\} \left\{ \frac{\mu_j(x(t))}{\sum_{j=1}^{q}\mu_j(x(t))} \right\} \mathrm{sgn}(l_i)Per(t)^T.\end{aligned} \tag{9.58}$$

The control law (9.49) together with the adaptation law (9.57) and (9.58) guarantee the boundedness of all signals in the closed-loop system. The properties of the adaptive fuzzy controller are given the following theorem.

Theorem 9.11. (Stability of the adaptive fuzzy controller for the TS fuzzy model [199]) *Consider the TS fuzzy model* (9.43) *and the reference fuzzy model* (9.44) *with the control law* (9.49) *and the adaptation law* (9.57) *and* (9.58). *Assume that the reference input* $r(t)$ *and the state* $x_m(t)$ *of the SRM are uniformly bounded, then the control law* (9.49) *and the adaptation law* (9.57), (9.58) *guarantee that*

(i) $F(t)$, $L(t)$, $e(t)$ *are bounded;*
(ii) $e \to 0$ *as* $t \to \infty$.

Proof. From (9.52) and (9.56), it follows that V is a Lyapunov function for the system (9.51), which implies the equilibrium $(F_ie, L_ie, e_e) = (F_i^*, L_i^*, 0)$ to be uniformly stable, which, in turn, implies that the trajectory $e(t), \tilde{F}(t), \tilde{L}(t)$ is bounded for all $t > 0$. According to that $e(t) = x(t) - x_m(t)$ and $x_m(t) \in L^\infty$ (L^∞ is a space of all essentially bounded functions), one has $x(t) \in L^\infty$. By (9.46) and $r(t) \in L^\infty$, one has $u(t) \in L^\infty$. Therefore, all variables in the closed-loop system are bounded.

Now, we prove that $e(t) \in L^\infty$. From (9.52) and (9.56), it is concluded that there exists a limit for V, that is,

$$\lim_{t\to\infty} V(e(t), \tilde{F}_i(t), \tilde{L}_i(t)) = V_\infty < \infty, \tag{9.59}$$

because V is bounded from below and is nonincreasing with time.

In terms of (9.56) and (9.59), one has

$$\begin{aligned}&\int_0^\infty e(t)^T \left(\frac{\sum \mu_i(x(t))\mu_j(x(t))Q_{ij}}{\sum \mu_i(x(t))\mu_j(x(t))}\right) e(t)d\tau \\ &= -\int_0^{infty} \dot{V} d\tau = (V_0 - V_\infty),\end{aligned} \tag{9.60}$$

where

$$V_0 = V(e(0), \tilde{F}_i(0), \tilde{L}_i(0)).$$

On the other hand, from $0 \leq \mu_i, \mu_j \leq 1$, and $\lambda_{\min}(Q_{ij}) \leq e(t)^T Q_{ij} e(t) \leq \lambda_{\max}(Q_{ij})$, which is due to $Q_{ij} = Q_{ij}^T > 0$.

One has

$$\begin{aligned}\{\lambda_{\min}(Q_{ij})\}_{\min}\|e(t)\|^2 &\leq e(t)^T \left(\frac{\sum \mu_i(x(t))\mu_j(x(t))Q_{ij}}{\sum \mu_i(x(t))\mu_j(x(t))}\right) e(t) \\ &\leq \{\lambda_{\max}(Q_{ij})_{\max}\|e(t)\|^2,\end{aligned} \tag{9.61}$$

where

$$\begin{aligned}\{\lambda_{\min}(Q_{ij})\}_{\min} &= \min\{\lambda_{\min}(Q_{11}), \dots, \lambda_{\min}(Q_{ll})\}, \\ \{\lambda_{\max}(Q_{ij})_{\max} &= \max\{\lambda_{\max}(Q_{11}), \dots, \lambda_{\max}(Q_{ll})\}.\end{aligned}$$

After inserting (9.61) into (9.60), and straightforward manipulation, one has

$$\begin{aligned}\frac{V_0 - V_\infty}{\{\lambda_{\min}(Q_{ij}\}_{\min}} &\leq \int_0^\infty \|e(t)\|^2 d\tau \\ &\leq \frac{(V_0 - V_\infty)}{t\{\lambda_{\max}(Q_{ij})\}_{\max}},\end{aligned} \tag{9.62}$$

which implies that $e(t) \in L^2$ (the space of square-integrable functions). Because $\tilde{e}(t), \tilde{K}_i(t), \tilde{L}_i(t), r(t) \in L^\infty$, it follows from (9.51) that $\dot{e} \in L^\infty$, which, together with $e(t) \in L^2$, implies that $e(t) \to 0$ as $t \to \infty$. This completes the proof.

Example 9.12. **Adaptive fuzzy control of the Lorenz system**
The TS fuzzy model of the Lorenz system was given in Section 6.4. The closed-loop eigenvalues for each subsystem of the SRM are set to be the same as $\lambda = [-1, -1-i, -1+i]$. It follows that

$$\begin{aligned}\dot{x}_m &= A_m x_m(t) + B_m r(t) \\ &= \begin{bmatrix} 0 & 1 & 0 \\ 0 & 0 & 1 \\ -2 & -4 & -3 \end{bmatrix} x_m(t) + \begin{bmatrix} 1 & 0 & 0 \\ 0 & 1 & 0 \\ 0 & 0 & 1 \end{bmatrix} r(t),\end{aligned} \tag{9.63}$$

here, $(A_m)_{ij} = A_m$ and $(B_m)_{ij} = B_m$ $(i, j = 1, 2)$.

The PDC fuzzy controller shares the same fuzzy sets in the premise as in Section 6.4 of the form:

$$\begin{aligned}&R^1 : \text{IF } x_1(t) \text{ is } \Gamma_1 \text{ THEN } u(t) = -F_1(t)x(t) + L_1(t)r(t), \\ &R^2 : \text{IF } x_1(t) \text{ is } \Gamma_2 \text{ THEN } u(t) = -F_2(t)x(t) + L_2(t)r(t).\end{aligned}$$

$F_i(t)$ and $L_i(t)$ $(i = 1, 2)$ are updated by an adaptation law so that the closed-loop system follows the SRM. The initial values of $F_i(t)$ and $L_i(t)$ $(i = 1, 2)$ are set from the nominal parameters of the system to be controlled, which are $a = 8$, $b = 5/3$, $c = 30$ and $d = 30$. Hence, the initial values of $F_i(t)$, $L_i(t)$ $(i = 1, 2)$ are given as follows:

$$\begin{aligned}&A_{10} - BF_{10} = A_{20} - BF_{20} = A_m, \quad BL_{10} = BL_{20} = B_m, \\ &F_{i0} = B(A_{i0} - A_m), \quad L_{i0} = B^{-1}B_m, \quad i = 1, 2, \\ &F_{10} = \begin{bmatrix} -10 & 9 & 0 \\ 28 & -1 & -31 \\ 2 & 34 & 0.3333 \end{bmatrix}, \quad F_{20} = \begin{bmatrix} -10 & 9 & 0 \\ 28 & -1 & 29 \\ 2 & -26 & 0.3333 \end{bmatrix}, \\ &L_{10} = L_{20} = \begin{bmatrix} 1 & 0 & 0 \\ 0 & 1 & 0 \\ 0 & 0 & 1 \end{bmatrix}.\end{aligned} \tag{9.64}$$

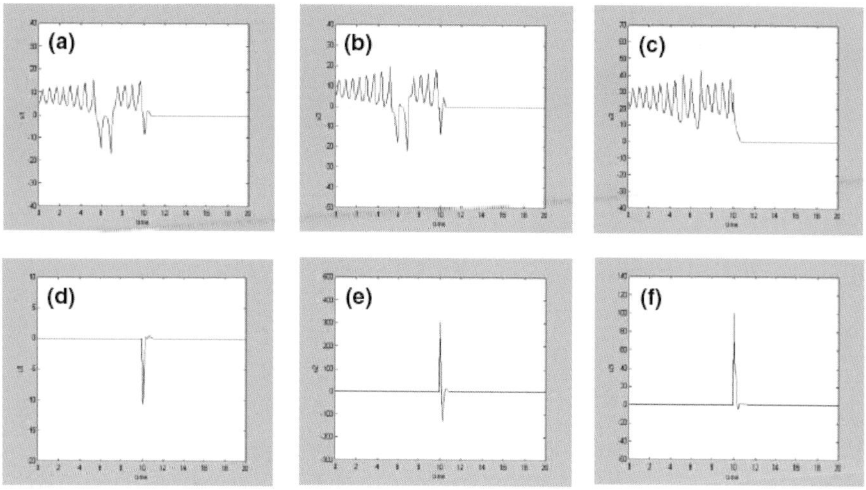

Fig. 9.13. Adaptive fuzzy control of the Lorenz system

Using (9.57) and (9.58), the following adaptation law for adjusting F_i and L_i can be derived:

$$\dot{F}_i(t) = \left\{ \frac{\mu_i(x(t))}{\sum_{i=1}^{2} \mu_i(x(t))} \right\} \text{sgn}(l_i) B_m^T P e(t) x(t)^T, \qquad i = 1, 2,$$

$$\dot{L}_i(t) = -\left\{ \frac{\mu_i(x(t))}{\sum_{i=1}^{2} \mu_i(x(t))} \right\} \text{sgn}(l_i) B_m^T P e(t) r(t)^T, \qquad i = 1, 2,$$

where

$$B_m^T = \begin{bmatrix} 1 & 0 & 0 \\ 0 & 1 & 0 \\ 0 & 0 & 1 \end{bmatrix}, \qquad P = \begin{bmatrix} 1.95 & 1.4 & 0.25 \\ 1.4 & 2.475 & 0.475 \\ 0.25 & 0.475 & 0.325 \end{bmatrix}.$$

The results of simulations are shown in Fig. 9.13, where the reference input $r(t) = 0$ and the control is activated at $t = 10$. It is seen that the adaptive fuzzy controller stabilizes the chaotic system to zero although the parameters of the chaotic system are not exactly known.

10

Synchronization of TS Fuzzy Systems

In this chapter, synchronization of TS fuzzy systems is discussed, where a fuzzy feedback law is adopted and realized via exact linearization (EL) techniques and by solving LMI problems. Two examples, synchronization of Chen's systems and hyperchaotic systems, are given for illustration.

10.1 Introduction

Since the discovery of synchronization of two weakly coupled pendulum clocks (hanging on the same beam) by Christian Huygens in 1665, various synchronizable systems have been found, studied and applied in the vast areas of physics, biology, chemistry, optics, and electrical and mechanical engineering [25, 201].

Synchronization is closely related to chaos. Synchrony can be commonly observed in coupled systems, arrays, or networks that can be periodic or chaotic [202]. Although two coupled chaotic systems intrinsically defy synchronization due to the extreme sensitivity to initial conditions, it is possible to coordinate their motions [204, 205].

Roughly speaking, synchronization refers to the tendency of two or more appropriately coupled chaotic systems to undergo resembling evolution in time, due to coupling or forcing. This ranges from complete agreement of trajectories to locking of phases [203, 206]. The basic idea in the seminal paper by Pecora and Carroll [204] was to take two identical three-dimensional dynamical systems described as $\dot{x} = f(x)$ with $x = (x, y, z)^T$ and $f(x)$ being a vector field, and to use one of them as so-called drive system $\dot{x}_d = f(x_d)$ to unidirectionally drive the second one, called response system $\dot{x}_r = f(x_r)$, by a suitable replacement of the dynamical variables in the response system. If $e(t) =: x_d - x_r \to 0$ as $t \to 0$, the dynamics of the response system approaches the time evolution of the drive system, and synchronization of these two systems is achieved.

Zhong Li: *Fuzzy Chaotic Systems*, StudFuzz **199**, 189–203 (2006)
www.springerlink.com

There are mainly two coupling fashions, unidirectional coupling and bidirectional coupling, which may lead to different synchronized states. For unidirectional synchronization, consider a drive (or master) system and a response (or slave) system, which are generally chaotic but not necessarily identical. One system evolves freely and drives the evolution of the other. As a result, the response system is slaved to follow the dynamics of the drive system. Typical examples are communication with chaos. While, for bidirectional synchronization, two systems are coupled with one another, and the coupling factor induces the rhythms onto a common synchronized manifold, thus inducing a mutual synchronization behavior [203]. Here, only unidirectional synchronization is discussed, and the adjective "unidirectional" is omitted below.

Synchronization of TS fuzzy models of chaotic systems has been extensively investigated [193], where a fuzzy feedback law was proposed for synchronization, which is realized via exact linearization techniques and by solving LMI problems. Although the exact linearization techniques ensure stability, the scheme is no longer suitable to chaotic communications due to the effects of signal masking and modulation. As synchronization issues are closely related to observer/controller design, generalized fuzzy response systems were proposed to solve the synchronization problem from the observer and controller points of view. This leads to a different concept for the design of LMI-based fuzzy chaotic synchronization and communication, which relaxes the EL condition [43, 207, 208, 209].

It is known from Chapter 6 that all well-known chaotic systems (either continuous or discrete) can be exactly represented as TS fuzzy models with only one premise variable. In particular, most systems may have common bias terms in their fuzzy models, for which a fuzzy driving signal can be adopted to achieve synchronization on two chaotic systems. When some LMI conditions are held, the design parameters exist and can be found. On the other hand, for fuzzy models without restricting common bias terms a typical (crisp) driving signal is chosen to be the same as the premise variable of the corresponding fuzzy chaotic model.

In this chapter, we first discuss the EL and relaxed EL conditions; then apply them to study synchronization issues between two chaotic TS fuzzy systems by solving LMI problems; finally, we show two examples, i.e., synchronization of Chen's systems and hyperchaotic systems, to illustrate the effectiveness of the synchronization methodology.

10.2 Exact Linearization

Consider the TS fuzzy model of the following form (refer to Chapter 6):

Plant Rule i:

IF $x_1(t)$ is Γ_1^i and $\cdots$ and $x_n(t)$ is Γ_n^i

THEN $sx(t) = A_i x(t) + B_i u(t)$, $i = 1, 2, \ldots, q$, (10.1) where Γ_j^i are fuzzy sets, $x(t) = [x_1(t), x_2(t), \ldots, x_n(t)]^T \in \Re^n$ is the state vector, $u(t) \in \Re^m$ is the control input vector, $A_i \in \Re^{n\times n}$ and $B_i \in \Re^{n\times m}$ are system matrix and input matrix, respectively, q is the number of rules of this TS fuzzy model, and $sx(t)$ denotes $\dot{x}(t)$ for continuous-time TS fuzzy systems (CFS) or $x(t+1)$ for discrete-time TS fuzzy systems (DFS).

The defuzzified output of this TS fuzzy system (10.1) is represented as follows:

$$sx(t) = \sum_{i=1}^{q} \mu_i(x(t))(A_i x(t) + B_i u(t)) , \tag{10.2}$$

where

$$\omega_i(x(t)) = \prod_{j=1}^{n} \Gamma_j^i(x_j(t)), \text{ and } \mu_i(x(t)) = \frac{\omega_i(x(t))}{\sum_{i=1}^{q} \omega_i(x(t))} ,$$

in which $\Gamma_j^i(x_j(t))$ is the grade of membership of $x_j(t)$ in Γ_j^i. Some basic properties of $\omega_i(t)$ are:

$$\omega_i(x(t)) \geq 0, \text{ and } \sum_{i=1}^{q} \omega_i(x(t)) > 0, \ i = 1, 2, \ldots, q .$$

To design a fuzzy controller from a TS fuzzy model, the parallel distributed compensation (PDC) is employed to determine its structure. Each control is constructed from the corresponding rule of a TS fuzzy model in the PDC. Thus, the designed fuzzy controller shares the same fuzzy sets with the fuzzy model in the premise parts.

Control Rule i:
IF $x(t)$ is Γ_1^i and $\cdots$ and $x_n(t)$ is Γ_n^i
THEN $u(t) = -F_i x(t), \quad i = 1, 2, \ldots, q ,$ (10.3)

The overall fuzzy controller is deduced as:

$$u(t) = -\frac{\sum_{i=1}^{q} \omega_i(x(t)) F_i x(t)}{\sum_{i=1}^{q} \omega_i(x(t))} = -\sum_{i=1}^{q} \mu_i(x(t)) F_i x(t). \tag{10.4}$$

For simplicity, assume that $B_i = B$ for $i = 1, 2, \ldots, q$ here and throughout in this chapter. Then, substituting (10.4) into (10.2) yields:

$$sx(t) = \sum_{i=1}^{q}\sum_{j=1}^{q} \mu_i(x(t))\mu_j(x(t))\{A_i - BF_j\}x(t), \tag{10.5}$$

The so-called exact linearization technique is to cancel the nonlinearity in chaotic systems via parallel distributed compensation (PDC). If EL is feasible, the resulting controller can be considered as a solution to the so-called global linearization and the feedback linearization problems [193].

Theorem 10.1. *The chaotic TS fuzzy system* (10.5) *is exactly linearized via the fuzzy controller* (10.4) *if there exist the feedback gains* F_i *such that*

$$\{(A_1 - BF_1) - (A_i - BF_i)\}^T \cdot \{(A_1 - BF_1) - (A_i - BF_i)\} = 0, \quad i = 2, 3, \ldots, q. \tag{10.6}$$

Then, the overall control system is linearized as $sx(t) = Gx(t)$*, in which* $G = A_1 - BF_1 = A_i - BF_i$.

Proof. It is straightforward to see that $G = A_1 - BF_1 = A_i - BF_i$ if the conditions (10.6) holds. It implies that the stability of the closed-loop system is reduced to stabilize $sx(t) = Gx(t)$.

The conditions (10.6) are applicable to both the CFS and the DFS. If B is a nonsingular matrix, the system can be exactly linearized using $F_i = B^{-1}(G - A_i)$. However, the assumption of a nonsingular matrix B is very strict. If B is not a nonsingular matrix, the conditions of Theorem 10.1 can still be relaxed by the following approximation technique.

$$X\{(A_1 - BF_1) - (A_i - BF_i)\}^T \cdot \{(A_1 - BF_1) - (A_i - BF_i)\}X < \beta S, \quad i = 2, 3, \ldots, q, \tag{10.7}$$

where X and S are positive-definite matrices with $S^T S < I$. The conditions (10.6) are likely to be satisfied if all entries in βS are near zero, i.e., $\beta S \approx 0$, in the above inequality. Using the Schur complement, one has

$$\begin{bmatrix} \beta S & X\{A_1 - BF_1) - (A_i - BF_i)\}^T \\ \{(A_1 - BF_1) - (A_i - BF_i)\}X & I \end{bmatrix} > 0, \quad i = 2, 3, \ldots, q.$$

Define $M_i = F_i X$ so that for $X > 0$ we have $F_i = M_i X^{-1}$. Substituting $F_i = M_i X^{-1}$ into the above inequalities yields:

$$\begin{bmatrix} \beta S & \{A_1 X - BM_1) - (A_i X - BM_i)\}^T \\ \{(A_1 X - BM_1) - (A_i X - BM_i)\} & I \end{bmatrix} > 0, \quad i = 2, 3, \ldots, q.$$

Note that G is not always a stable matrix even if the condition of Theorem 10.1 holds. In terms of the stability conditions in Section 9.4 and Theorem 10.1, the fuzzy controller designs can be solved for EL.

Stable Fuzzy Controller Design for EL – CFS:

$$\underset{X,S,M_1,M_2,\ldots,M_q}{\text{minimize}} \quad \beta$$
$$\text{subject to} \quad X > 0, \beta > 0, S > 0$$

$$\begin{bmatrix} I & S \\ S & I \end{bmatrix} > 0,$$

$$-A_iX + BM_i - XA_i^T + M_i^T B^T > 0, \quad i = 1, 2, \ldots, q,$$

$$\begin{bmatrix} \beta S & \{(A_1X - BM_1) - (A_iX - BM_i)\}^T \\ \{(A_1X - BM_1) - (A_iX - BM_i)\} & I \end{bmatrix} > 0,$$
$$i = 2, 3, \ldots, q,$$

where $X = P^{-1}$ and $M_i = F_iX$.
Stable Fuzzy Controller Design for EL – DFS:

$$\underset{X,S,M_1,M_2,\ldots,M_q}{\text{minimize}} \quad \beta$$
$$\text{subject to} \quad X > 0, \beta > 0, S > 0$$

$$\begin{bmatrix} I & S \\ S & I \end{bmatrix} > 0,$$

$$\begin{bmatrix} X & XA_i - M_i^T B^T \\ A_iX - BM_i & X \end{bmatrix} > 0, \quad i = 1, 2, \ldots, q,$$

$$\begin{bmatrix} \beta S & \{(A_1X - BM_1) - (A_iX - BM_i)\}^T \\ \{(A_1X - BM_1) - (A_iX - BM_i)\} & I \end{bmatrix} > 0,$$
$$i = 2, 3, \ldots, q,$$

where $X = P^{-1}$ and $M_i = F_iX$.

In the LMI, if all entries in $\beta \cdot S$ are close to zero, i.e., $\beta \cdot S \approx 0$, the above ELs for stable fuzzy controller designs are feasible. In this case, $G = A_i - BF_i$ for all i and G is a stable matrix.

10.3 Synchronization

To deal with synchronization of chaotic systems, one needs to design the control input so that the controlled system achieves asymptotic synchronization with the reference system, provided that the two systems start from different initial conditions. Here the reference system and controlled system have the same chaotic oscillator except that the controlled system has a control input (the controlled system can be viewed as an observer of the reference system). In this section, only full state feedback is considered. Two cases of the cancellation problem are discussed:

- Case 1: The cancellation problem is feasible, i.e., all entries in $\beta \cdot S$ are close to zero;
- Case 2: The cancellation problem is infeasible, i.e., all entries in $\beta \cdot S$ are not close to zero.

10.3.1 Case 1

Consider a reference fuzzy model which represents a reference chaotic system,

$$\begin{aligned}&\text{Reference Model Rule } i\text{:}\\ &\quad \text{IF } x_{R1}(t) \text{ is } \Gamma_1^i \text{ and } \cdots \text{ and } x_{Rn}(t) \text{ is } \Gamma_n^i\\ &\quad \text{THEN } sx_R(t) = A_i x_R(t),\ i = 1, 2, \ldots, q\,,\end{aligned} \tag{10.8}$$

where $sx_R(t)$ represents $\dot{x}_R(t)$ and $x_R(t+1)$ for CFS and DFS, respectively.

The defuzzified process is given as

$$sx_R(t) = \sum_{i=1}^{q} \mu_i(x_R(t)) A_i x_R(t), \tag{10.9}$$

Defining the error signal as $e(t) = x(t) - x_R(t)$, one has

$$se(t) = \sum_{i=1}^{q} \mu_i(x(t)) A_i x(t) - \sum_{i=1}^{q} \mu_i(x_R(t)) A_i x_R(t) + Bu(t), \tag{10.10}$$

where $se(t)$ represents $\dot{e}(t)$ and $e(t+1)$ for CFS and DFS, respectively.

The aim of the synchronization is to design the following fuzzy controller,

$$u(t) = -\sum_{i=1}^{q} \mu_i(x(t)) F_i x(t) + \sum_{i=1}^{q} \mu_i(x_R(t)) F_i x_R(t), \tag{10.11}$$

such that $\|e(t)\| \to 0$ as $t \to \infty$. Therefore, the design is to determine the feedback gains F_i. Substituting (10.11) into (10.10) gives:

$$
\begin{aligned}
se(t) = &\sum_{i=1}^{q} \mu_i(x(t))(A_i - BF_i)x(t) \\
&- \sum_{i=1}^{q} \mu_i(x_R(t))(A_i - BF_i)x_R(t).
\end{aligned} \tag{10.12}
$$

Then, if the condition of Theorem 10.1 holds, the error equation (10.12) can be linearized using the fuzzy control law (10.11). Thus, the linearized error system becomes $se(t) = Ge(t)$, where $G = A_i - BF_i$. It is noticed that G is not always a stable matrix even if the condition of Theorem 10.1 holds. Therefore, if there exist feedback gains F_i such that G is a stable matrix, then the error system (10.12) is linearized and stabilized. To design the fuzzy controller with the feedback gains F_i it remains to solve the LMI-based design problem.

10.3.2 Case 2

If the EL problem is infeasible, i.e., all entries in $\beta \cdot S$ are not close to zero, the error system cannot be linearized. The error system is rewritten as:

$$
\begin{aligned}
se(t) &= \sum_{i=1}^{q} \mu_i(x(t))A_i x(t) - \sum_{i=1}^{q} \mu_i(x_R(t))A_i x_R(t) + Bu(t) \\
&= \sum_{i=1}^{q} \mu_i(x(t))A_i e(t) \\
&\quad + \sum_{i=1}^{q} \{\mu_i(x(t)) - \mu_i(x_R(t))\} A_i x_R(t) + Bu(t).
\end{aligned} \tag{10.13}
$$

It is noted that $\sum_{i=1}^{q} \{\mu_i(x(t)) - \mu_i(x_R(t))\} A_i x_R(t) \approx 0$, if $\|e(t)\| \leq \delta$, in which δ is sufficiently small. As a result, system (10.13) is approximated as

$$
se(t) = \sum_{i=1}^{q} \mu_i(x(t))A_i e(t) + Bu(t). \tag{10.14}
$$

Consider the following fuzzy feedback law for the error system (10.14):

$$
u(t) = \begin{cases} \sum_{i=1}^{q} \mu_i(x(t))F_i e(t), & \|e(t)\| < \delta, \\ 0, & \text{otherwise.} \end{cases}
$$

Then, if there exist the feedback gains F_i satisfying the stability conditions of Theorem 9.4 or 9.5, the stability of the error system is guaranteed near the equilibrium points, i.e., $\|e(t)\| < \delta$. The feedback gains F_i can be found by solving the LMI. It should be noted that this approach guarantees local stability, only. This is the same idea as the OGY method [30]. Therefore, the convergence time to an equilibrium point is very long in general, but the control effort is small.

In the following, some simulation examples will be given to illustrate the effectiveness of the synchronization approaches.

10.4 Synchronization of Chen's Chaotic Systems

In the study of anti-controlling the Lorenz system, a new chaotic system, Chen's system, has been coined [210], which is described as

$$\begin{cases} \dot{x_1} = a(x_2 - x_1), \\ \dot{x_2} = (c-a)x_1 - x_1x_3 - cx_2, \\ \dot{x_3} = x_1x_2 - bx_3, \end{cases} \tag{10.15}$$

where a, b, c are parameters, and when $a = 35$, $b = 3$, $c = 28$, the system exhibits chaotic behavior. Assume that $x_1 \in [-d, d]$ and $d > 0$, then Chen's system can be exactly represented as a TS fuzzy model:

$$\begin{aligned} &\text{Rule 1: IF } x_1 \text{ is } M_1 \text{ THEN } \dot{x} = A_1 x(t), \\ &\text{Rule 2: IF } x_1 \text{ is } M_2 \text{ THEN } \dot{x} = A_2 x(t). \end{aligned} \tag{10.16}$$

where $x = [x_1, x_2, x_3]^T$, $A_1 = \begin{bmatrix} -a & a & 0 \\ c-a & c & -d \\ 0 & d & -b \end{bmatrix}$, $A_2 = \begin{bmatrix} -a & a & 0 \\ c-a & c & d \\ 0 & -d & -b \end{bmatrix}$, $M_1(x_1(t)) = \frac{1}{2}\left(1 + \frac{x_1}{d}\right)$, $M_2(x_1(t)) = \frac{1}{2}\left(1 - \frac{x_1}{d}\right)$, and $d = 30$.

The overall output of Chen's system is inferred as follows and its chaotic behavior is shown in Fig. 10.1,

$$\dot{x}(t) = \sum_{i=1}^{2} \mu_i(x_1(t)) A_i x(t). \tag{10.17}$$

where $\mu_i(x_1(t)) = M_i(x_1(t)) / \sum_{i=1}^{2} M_i(x_1(t))$.

Let (10.16) be the drive system, the fuzzy-controlled Chen's system can be the response system,

$$\begin{aligned} &\text{Rule 1: IF } x_{R1} \text{ is } M_1 \text{ THEN } \dot{x}_R = A_1 x_R(t), \\ &\text{Rule 2: IF } x_{R1} \text{ is } M_2 \text{ THEN } \dot{x}_R = A_2 x_R(t). \end{aligned} \tag{10.18}$$

where $x_R = [x_{R1}, x_{R2}, \ldots, x_{Rn}]^T$.

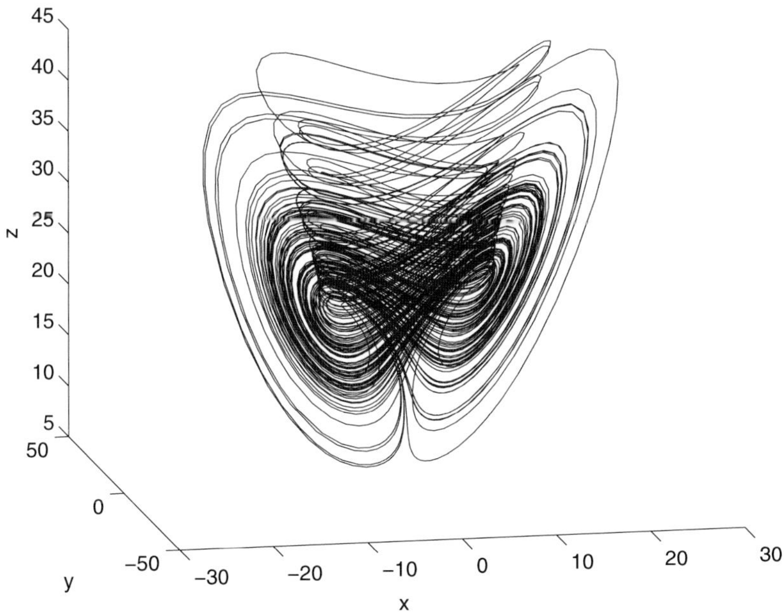

Fig. 10.1. Phase portrait of Chen's fuzzy system

The overall output of (10.18) is given as

$$\dot{x}_R(t) = \sum_{i=1}^{2} \mu_i(x_{R1}(t))A_i x_R(t) + Bu(t).$$

Assume that $e(t) = x_R(t) - x(t)$, one has

$$\dot{e}(t) = \sum_{i=1}^{2} \mu_i(x_{R1})A_i x_R(t) - \sum_{i=1}^{2} \mu_i(x_1)A_i x(t) + Bu(t). \tag{10.19}$$

The fuzzy controller can be designed to realize the synchronization as following [211]:

$$u(t) = -\sum_{i=1}^{2} \mu_i(x_{R1})F_i x_R(t) + \sum_{i=1}^{2} \mu_i(x_1)F_i x(t). \tag{10.20}$$

Substituting (10.20) into (10.19) results in the error system,

$$\dot{e}(t) = \sum_{i=1}^{2} \mu_i(x_{R1})(A_i - BF_i)x_R(t) - \sum_{i=1}^{2} \mu_i(x_1)(A_i - BF_i)x(t). \tag{10.21}$$

In order to synchronize the drive and response systems, the error system (10.21) can exactly be linearized according to Theorem 10.1. In terms of the stability conditions, the feedback gains can be obtained by solving the LMIs to synchronize the drive and response systems. Let B be an identity matrix, using Matlab to solve the LMI problem yields the feedback gains of the error system (10.21),

$$F_1 = \begin{bmatrix} -34.5000 & 13.8737 & -0.1086 \\ 14.1263 & 28.5000 & -31.1732 \\ 0.1086 & 31.1732 & -2.5000 \end{bmatrix},$$

$$F_2 = \begin{bmatrix} -34.5000 & 13.8737 & -0.1086 \\ 14.1263 & 28.5000 & 28.8268 \\ 0.1086 & -28.8268 & -2.5000 \end{bmatrix}. \tag{10.22}$$

The control result of the error system is shown in Fig. 10.2.

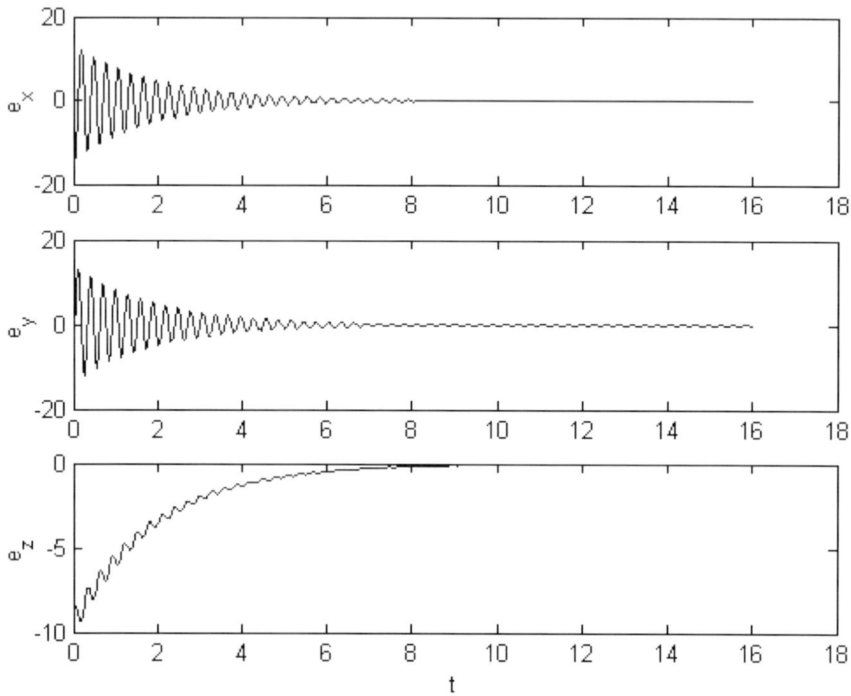

Fig. 10.2. Control result with feedback gains (10.22)

It should be noted that the fuzzy controller with the feedback gains (10.22) uses all state variables. If the synchronization is used in secure communication, it is not safe to transmit all the states for synchronizing the chaotic systems. In order to improve the controller, the feedback gains take the form

$$F_1 = F_2 = \begin{bmatrix} 0 & 0 & 0 \\ 0 & k & 0 \\ 0 & 0 & 0 \end{bmatrix}.$$

First, let $k = 0$, $\beta = 0.0001$. Solving the LMIs with $k = k + 1$ until the LMIs are feasible, one can derive that $k = 23$ by trial-and-error. Figure 10.3 shows the control results with the fuzzy controller via only state variable y.

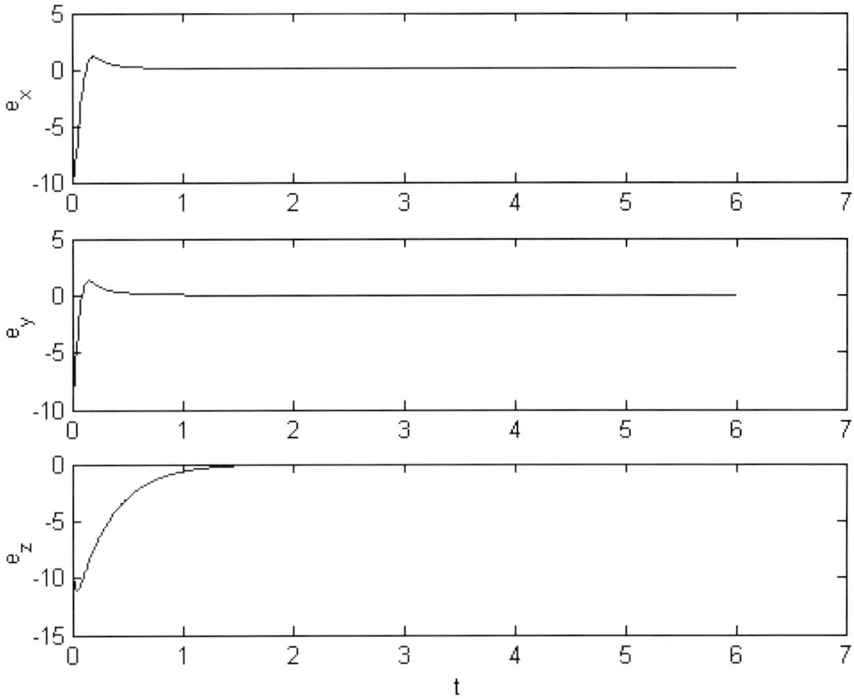

Fig. 10.3. Control result via single variable

10.5 Synchronization of Hyperchaotic Systems

Consider the forth order Rössler system [212],

$$\begin{cases} \dot{x_1} = -x_2 - x_3, \\ \dot{x_2} = x_1 + \alpha x_2 + w, \\ \dot{x_3} = x_3 x_3 + \beta, \\ \dot{w} = -0.5x_3 + 0.05w, \end{cases} \tag{10.23}$$

where x_1, x_2, x_3, w are state variables, α, β are the parameters, and when $\alpha = 0.25$, $\beta = 3$, this system exhibits hyperchaotic behavior. Assume that

$x_1 \in [c-d, c+d]$ and $d > 0$, the forth order Rössler system can exactly be modeled as a TS fuzzy system as follows,

$$\begin{aligned} &\text{Rule 1: IF } x_1(t) \text{ is } M_1 \text{ THEN } \dot{x}(t) = A_1 x(t) + b_1, \\ &\text{Rule 2: IF } x_1(t) \text{ is } M_2 \text{ THEN } \dot{x}(t) = A_2 x(t) + b_2, \end{aligned} \tag{10.24}$$

where $x = [x_1, x_2, x_3, w]^T$, $A_1 = \begin{bmatrix} 0 & -1 & -1 & 0 \\ 1 & \alpha & 0 & 1 \\ 0 & 0 & d & 0 \\ 0 & 0 & -0.5 & 0.05 \end{bmatrix}$, $A_2 = \begin{bmatrix} 0 & -1 & -1 & 0 \\ 1 & \alpha & 0 & 1 \\ 0 & 0 & -d & 0 \\ 0 & 0 & -0.5 & 0.05 \end{bmatrix}$,

$b_1 = b_2 = \begin{bmatrix} 0 \\ 0 \\ \beta \\ 0 \end{bmatrix}$, $M_1(x_1) = \frac{1}{2}\left(1 + \frac{x_1}{d}\right)$, $M_2 = \frac{1}{2}\left(1 - \frac{x_1}{d}\right)$.

Choose the constants $c = -32$ and $d = 50$ in terms of the region of interest. The overall output of the forth order Rössler TS fuzzy system is inferred as follows and the hyperchaotic behavior is shown in Fig. 10.4,

$$\dot{x}(t) = \sum_{i=1}^{2} \mu_i(x_1(t))\{A_i x(t) + b_i\}. \tag{10.25}$$

To achieve synchronization of hyperchaotic systems based on the TS fuzzy model, consider the following fuzzy control system as response system, given the fuzzy hyperchaotic drive system (10.24),

$$\begin{aligned} &\text{Rule 1: IF } x_{R1}(t) \text{ is } M_1 \text{ THEN } \dot{x}_R(t) = A_1 x_R(t) + b_1 + Bu(t), \\ &\text{Rule 2: IF } x_{R1}(t) \text{ is } M_2 \text{ THEN } \dot{x}_R(t) = A_2 x_R(t) + b_2 + Bu(t), \end{aligned} \tag{10.26}$$

The defuzzification output is as follows,

$$\dot{x}_R(t) = \sum_{i=1}^{2} \mu_i(x_1(t))\{A_i x_R(t) + b_i\} + Bu(t). \tag{10.27}$$

It is noted that the fuzzy drive system (10.24) and the fuzzy response system (10.26) represent the same hyperchaotic oscillator except that the controlled system has a control input $u(t)$. Defining the error signal as $e(t) = x_R(t) - x(t)$, one has

$$\begin{aligned} \dot{e}(t) = &\sum_{i=1}^{2} \mu_i(x_{R1}(t))\{A_i x_R(t) + b_i\} \\ &- \sum_{i=1}^{2} \mu_i(x_1(t))\{A_i x_1(t) + b_i\} + Bu(t). \end{aligned} \tag{10.28}$$

The aim of synchronization is to design the following fuzzy controllers,

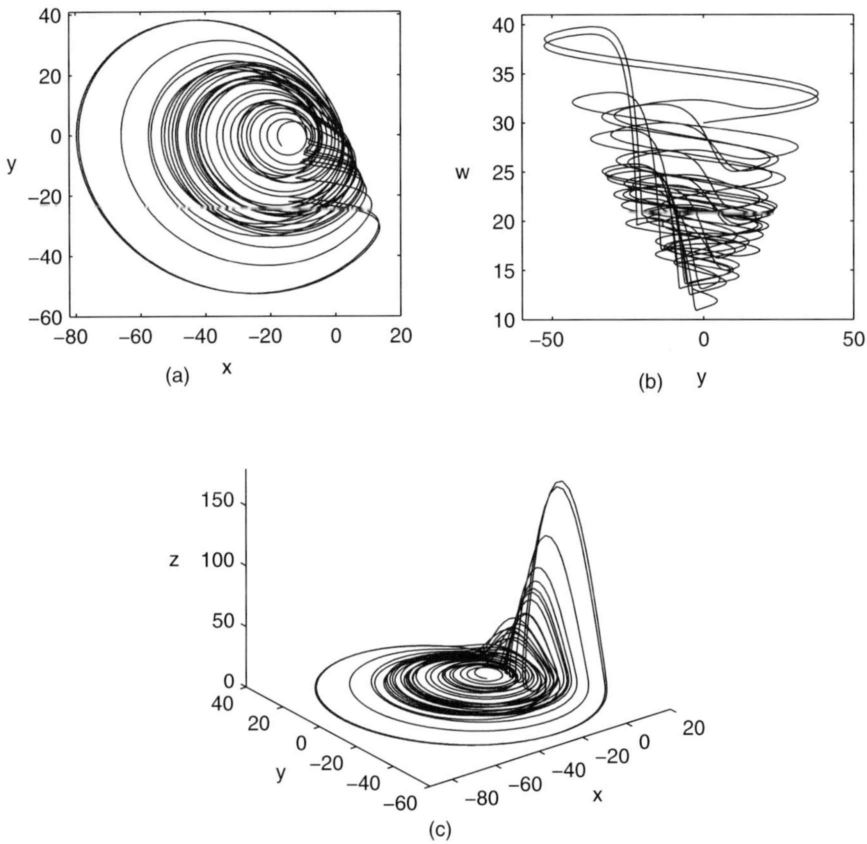

Fig. 10.4. Hyperchaotic behavior of TS fuzzy-model-based fourth order Rössler system: (a) $x-y$ plane; (b) $y-w$ plane; (c) $x-y-z$ space

$$u(t) = -\sum_{i=1}^{2} \mu_i(x_{R1}(t))F_i x_R(t) + \sum_{i=1}^{2} \mu_i(x_1(t))F_i x(t), \tag{10.29}$$

such that $\|e(t)\| \to 0$ as $t \to 0$. To determine the feedback gains F_i, substituting (10.29) into (10.28) yields

$$\begin{aligned}\dot{e}(t) =& \sum_{i=1}^{2} \mu_i(x_{R1}(t))(A_i - BF_i)x_R(t) \\ & - \sum_{i=1}^{2} \mu_i(x_1(t))(A_i - BF_i)x(t).\end{aligned} \tag{10.30}$$

In order to synchronize the drive and response systems, using Theorem 10.1, the error system (10.30) can exactly be linearized as $\dot{e}(t) = Ge(t)$

via the fuzzy controller, where $G = A_1 - BF_1 = A_2 - BF_2$. In terms of the stability conditions, the feedback gains can be obtained by solving the LMIs to synchronize the drive and response systems. Let B be an identity matrix, using Matlab to solve the LMI problem yields the feedback gains of the error system (10.30),

$$F_1 = \begin{bmatrix} 0.5000 & 0.0000 & -0.5000 & -0.0000 \\ 0.0000 & 0.7500 & 0.0000 & 0.5000 \\ -0.5000 & 0.0000 & 50.5000 & -0.2500 \\ -0.0000 & 0.5000 & -0.2500 & 0.5500 \end{bmatrix},$$

$$F_2 = \begin{bmatrix} 0.5000 & 0.0000 & -0.5000 & -0.0000 \\ 0.0000 & 0.7500 & 0.0000 & 0.5000 \\ -0.5000 & 0.0000 & -49.5000 & -0.2500 \\ -0.0000 & 0.5000 & -0.2500 & 0.5500 \end{bmatrix}. \tag{10.31}$$

The synchronization error is shown in Fig. 10.5(a). For the same reason as mentioned in the last section, the feedback gains F_i can have this form:

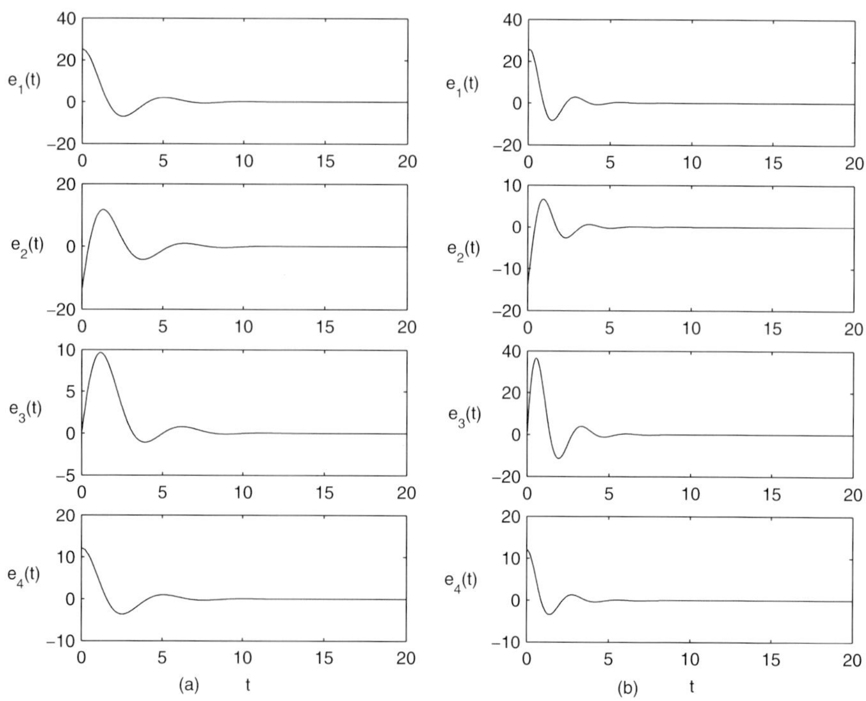

Fig. 10.5. Hyperchaotic synchronization error of the fourth order Rössler system with (a) fuzzy controller (10.31); (b) fuzzy controller (10.32)

$$F_i = \begin{bmatrix} 0 & 0 & 0 & 0 \\ 0 & 0 & 0 & 0 \\ a_i & b_i & c_i & d_i \\ 0 & 0 & 0 & 0 \end{bmatrix}.$$

The following feedback gains can similarly be derived to linearize and stabilize the error system.

$$F_1 = \begin{bmatrix} 0 & 0 & 0 & 0 \\ 0 & 0 & 0 & 0 \\ -3.5712 & -2.3372 & 52.5806 & -6.6192 \\ 0 & 0 & 0 & 0 \end{bmatrix},$$
$$F_2 = \begin{bmatrix} 0 & 0 & 0 & 0 \\ 0 & 0 & 0 & 0 \\ -3.5712 & -2.3372 & -47.4194 & -6.6192 \\ 0 & 0 & 0 & 0 \end{bmatrix}. \tag{10.32}$$

The synchronization error is shown in Fig. 10.5(b).

11

Chaotifying TS Fuzzy Systems

In this chapter, chaotifying both discrete-time and continuous-time Takagi-Sugeno (TS) fuzzy systems is introduced. To chaotify discrete-time TS fuzzy systems, the parallel distributed compensation (PDC) method is employed to determine the structure of a fuzzy controller so as to make all Lyapunov exponents of the controlled TS fuzzy system strictly positive. But for continuous-time ones, the chaotification approach is based on fuzzy feedback linearization and a suitable approximate relationship between a time-delay differential equation and a discrete map. The time-delay feedback controller, chosen among several candidates, is a simple sinusoidal function of the delay states of the system, which can have an arbitrarily small amplitude. These anti-control approaches are all proved to be mathematically rigorous in the sense of Li and Yorke. Some examples are given to illustrate the effectiveness of the proposed anti-control methods.

11.1 Introduction

Nowadays, it is well known that most conventional control methods and many special techniques can be used for chaos control [213]. No matter the purpose is to reduce harmful or undesirable chaos or to introduce useful or beneficial chaos, numerous control methodologies have been proposed, developed, tested and applied. Similar to conventional systems control, the concept of "controlling chaos" is first to mean ordering or suppressing chaos in the sense of stabilizing chaotic system responses. In this pursuit, numerical and experimental simulations have convincingly demonstrated that chaotic systems respond well to these control strategies. These methods of ordering chaos include the now-familiar OGY method [78, 214], feedback controls [215, 216], and fuzzy control [198, 217, 193, 33], to list just a few.

However, controlling chaos encompasses also many nontraditional tasks, particularly those of enhancing or generating chaos when it is beneficial. Thus, the process of chaos control is now understood as a transition between chaos

Zhong Li: *Fuzzy Chaotic Systems*, StudFuzz **199**, 205–238 (2006)
www.springerlink.com

and order, depending on the application of interest. The task of purposely creating chaos, sometimes called chaotification or anti-control of chaos, has attracted increasing attention in recent years due to its great potential in nontraditional applications such as those found within the contexts of physical, chemical, mechanical, electrical, optical, and particularly biological and medical systems [218, 219, 220]. Recently, there have been some successful reports on anti-controlling chaos [219, 220, 221]. Although these reports are essentially experimental or semianalytical, in the sense that no explicit and quantitative computational formulae are provided with rigorous mathematical justification, especially for the continuous-time case, they are nevertheless interesting and promising.

One simple, yet mathematically rigorous control method from the engineering feedback control approach was developed [222, 223, 224], where a linear state-feedback controller with an uniformly bounded control-gain sequence can be designed to make all Lyapunov exponents of the controlled system strictly positive and arbitrarily assigned. Moreover, such a controller can be designed for an arbitrarily given, n-dimensional dynamical system that could originally be nonchaotic or even asymptotically stable. The goal of chaotification is finally achieved with a simple modulus operation or a sawtooth (or even a sine) function. The design criterion is to use the definition of chaos given by Devaney [20] or Li-Yorke [19], while for the n-dimensional case the Marotto theorem [130] was used for a proof. For the continuous-time case, a general approach to make an arbitrarily given autonomous system chaotic has also been proposed recently [226, 227, 228, 229]. Here, the main tool to use is time-delay feedback perturbation on a system parameter or as an exogenous input [227].

This chapter studies anti-control of chaos for both continuous-time and discrete-time TS fuzzy systems. The anti-control approaches are extensions of formerly developed methods for general systems to TS fuzzy systems, and are proved to be mathematically rigorous in the sense of Li and Yorke. Examples are given finally for illustration.

11.2 Chaotifying Discrete-time TS Fuzzy Systems

11.2.1 Discrete-time TS Fuzzy System via Mod-Operation

Rewrite the discrete-time TS fuzzy system (6.5) to the following general version:

$$\begin{aligned}
&\text{Discrete-time TS fuzzy model:}\\
&\quad\text{Plant Rule } i\text{:}\\
&\qquad\text{IF } x_1(k) \text{ is } \Gamma_1^i \text{ and } \ldots \text{ and } x_n(k) \text{ is } \Gamma_n^i\\
&\qquad\text{THEN } x(k+1) = A_i x(k) + B_i u(k),
\end{aligned} \tag{11.1}$$

where

$$x^T(k) = [x_1(k), x_2(k), \ldots, x_n(k)],$$
$$u^T(k) = [u_1(k), u_2(k), \ldots, u_m(k)],$$

$i = 1, 2, \ldots, q$, in which q is the number of IF-THEN rules, Γ_j^i are fuzzy sets and the equation $x(k+1) = A_i x(k) + B_i u(k)$ is the output of the ith IF-THEN rule.

Assume that A_i and B_i, $i = 1, 2, \ldots, q$, are uniformly bounded, i.e., there are constants N and Q such that

$$\sup_{1 \le i \le q} \|A_i\| \le N < \infty \text{ and } \sup_{1 \le i \le q} \|B_i\| \le Q < \infty$$

where $\|\cdot\|$ denotes the spectral norm of a finite-dimensional matrix, i.e., the largest singular value of the matrix.

Now, given one pair of $(x(k), u(k))$, the final output of the fuzzy system at step $k+1$ is inferred as follows:

$$x(k+1) = \frac{1}{\sum_{i=1}^{q} \omega_i(k)} \sum_{i=1}^{q} \omega_i(k)\{A_i x(k) + B_i u(k)\}, \tag{11.2}$$

where

$$\omega_i(k) = \prod_{j=1}^{n} \Gamma_j^i(x_j(k)).$$

in which $\Gamma_j^i(x_j(k))$ is the degree of membership of $x_j(k)$ in Γ_j^i, with

$$\begin{cases} \sum_{i=1}^{q} \omega_i(k) > 0, \\ \omega_i(k) \ge 0, \end{cases} \quad i = 1, 2, \ldots, q.$$

By introducing $\mu_i(k) = \frac{\omega_i(k)}{\sum_{i=1}^{q} \omega_i(k)}$ instead of $\omega_i(k)$, (11.2) is rewritten as

$$x(k+1) = \sum_{i=1}^{q} \mu_i(k)\{A_i x(k) + B_i u(k)\}. \tag{11.3}$$

Note that

$$\begin{cases} \sum_{i=1}^{q} \mu_i(k) = 1, \\ \mu_i(k) \ge 0, \end{cases} \quad i = 1, 2, \ldots, q, \tag{11.4}$$

in which $\{\mu_i(k)\}_{i=1}^q$ can be regarded as the normalized weight of the IF-THEN rules.

In terms of the Marotto Theorem 4.3, the definition of a chaotic TS fuzzy model can be given as follows.

Definition 11.1 (Chaotic TS Fuzzy Model [39]). *The Takagi-Sugeno (TS) fuzzy model (11.3) is said to be chaotic in the sense of Li and Yorke if it has a snap-back repeller.*

11.2.2 Anti-controller Design

Applying the PDC controller (9.3) to (11.3) yields the controlled system

$$x(t+1) = \sum_{i=1}^{q}\sum_{j=1}^{q} \mu_i(t)h_j(t)\{A_i + B_iF_j\}x(t). \tag{11.5}$$

For simplicity, assume that $B_i = B$, $i = 1, 2, \ldots, q$, so that (11.5) can be rewritten as

$$x(k+1) = \sum_{i=1}^{q} \mu_i(k)\{A_i + BF_i\}x(k) = \sum_{i=1}^{q} \mu_i(k)G_{ii}x(k). \tag{11.6}$$

In terms of Theorem 10.1, if $G = A_1 + BF_1 = A_i + BF_i = G_{ii}$, the overall controlled system can be linearized as

$$x(k+1) = Gx(k) \tag{11.7}$$

where $G = G_{ii}$, $i = 2, 3, \ldots, q$.

The jth Lyapunov exponent of the orbit of the controlled system (11.7), starting from the given x_0, is defined by

$$\lambda_j = \lim_{k\to\infty} \frac{1}{k} \ln\left|\mu_j(G^k)\right|, j = 1, 2, \ldots, n, \tag{11.8}$$

where $\mu_j(G^k)$ is the jth singular value of matrix G^k.

In the controlled system (11.7), one is able to design the constant gain matrices $\{F_i\}_{i=1}^q$, given in (9.2), such that the Lyapunov exponents of the controlled system orbit $\{x_k\}_{k=0}^\infty$ can be arbitrarily assigned:

$$\lambda_j(x_0) = \sigma_j, \quad j = 1, 2, \ldots, n, \tag{11.9}$$

where $\{\sigma_j\}_{j=1}^n$ are arbitrarily chosen to be positive, zero, or negative (but all finite).

A convenient choice for the matrix G in (11.7) is, simply,

$$G = diag\{e^{\sigma_1}, e^{\sigma_2}, \ldots, e^{\sigma_n}\}.$$

It is clear that the eigenvalues of G are all larger than 1 if $\sigma_j > 0$, $j = 1, 2, \ldots, n$. The desired matrices F_i, $i = 1, 2, \ldots, q$, can then be obtained.

11.2.3 Verification of the Anti-control Design

Theorem 11.2. *[39] The resulting overall controlled system (11.7), along with the mod-operation,*

$$x(k+1) = Gx(k) \text{ (mod-1)}, \tag{11.10}$$

where $G = diag\{e^{\sigma_1}, e^{\sigma_2}, \ldots, e^{\sigma_n}\}$ and $\sigma_i > 0$, $i = 1, 2, \ldots, n$, is chaotic in the sense of Li and Yorke.

Proof: The controlled system (11.7) is

$$x(k+1) = Gx(k) \text{ (mod-1)} \equiv g(x(k)). \tag{11.11}$$

It is now to prove that the fixed point $x^* = 0$ of (11.7) is a snap-back repeller. To do so, define two n-dimensional vectors, $b = [1, 1, \ldots, 1]^T$ and

$$x^0 = G^{-2}b = \begin{bmatrix} e^{-2\sigma_1} & 0 \cdots & 0 \\ 0 & 0 \cdots & 0 \\ \vdots & \vdots \ddots & \vdots \\ 0 & 0 \cdots & e^{-2\sigma_n} \end{bmatrix} b \neq 0. \tag{11.12}$$

Since $\sigma_i > 0$, $i = 1, 2, \ldots, n$, $\|x^0\|_\infty < 1$. Clearly, after two-step iterations on (11.10) with $x^1 = g(x^0) = G^{-1}b$, one has

$$g^2(x^0) = g(x^1) = 0 = x^*. \tag{11.13}$$

Let r be a given constant satisfying $\|x^0\|_\infty \leq r \leq 1$. For any $x \in B_r \equiv \{x \in \Re^n | \; \|x\|_\infty \leq r\}$, it is also clear that all the eignevalues of G exceed unity. Therefore, the fixed point $x^* = 0$ of (11.7) is a snap-back repeller. By the Marotto theorem, the controlled system (11.6)–(11.7) is chaotic in the sense of Li and Yorke. Thus, it completes the proof.

11.2.4 A Simulation Example

Consider a nonchaotic discrete-time TS fuzzy model given as follows:

$$\text{Rule 1: IF } x(k) \text{ is } \Gamma_1, \text{ THEN } \begin{bmatrix} x(k+1) \\ y(k+1) \end{bmatrix} = A_1 \begin{bmatrix} x(k) \\ y(k) \end{bmatrix} + Bu(k),$$

$$\text{Rule 2: IF } x(k) \text{ is } \Gamma_2, \text{ THEN } \begin{bmatrix} x(k+1) \\ y(k+1) \end{bmatrix} = A_2 \begin{bmatrix} x(k) \\ y(k) \end{bmatrix} + Bu(k),$$

where

$$A_1 = \begin{bmatrix} d & 0.3 \\ 1 & 0 \end{bmatrix}, \quad A_2 = \begin{bmatrix} -d & 0.3 \\ 1 & 0 \end{bmatrix},$$

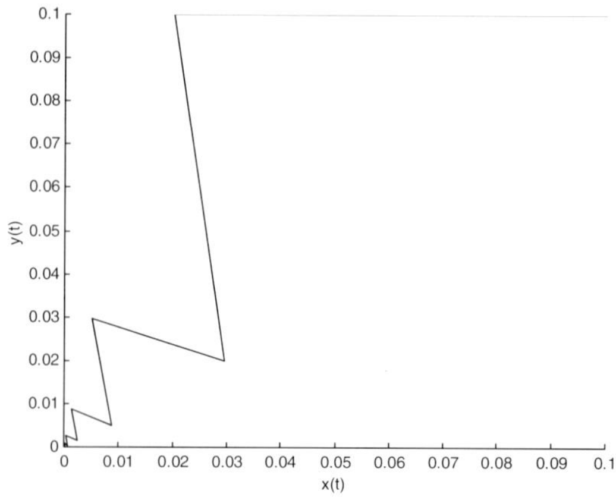

Fig. 11.1. System orbit without control input

$x(k) \in [-d, d]$ and $d > 0$, with membership functions

$$\Gamma_1 = \frac{1}{2}\left(1 - \frac{x(k)}{d}\right) \text{ and } \Gamma_2 = \frac{1}{2}\left(1 + \frac{x(k)}{d}\right).$$

Without control, i.e., $u \equiv 0$, the system is stable as shown in Fig. 11.1.

Here, we choose two desired Lyapunov exponents, $\sigma_1 = \ln(1.9) = 0.6418539$ and $\sigma_2 = \ln(2.0) = 0.6931472$.

For simplicity, assume that $B = \begin{bmatrix} 1 & 0 \\ 0 & 1 \end{bmatrix}$ and $d = 2$. The design of the feedback controller is completed by following the procedure described above. The controlled system output is shown in Figs. 11.2–11.3. The output trajectory is displayed in the phase plane after some mod-2 operations (they are obviously equivalent to mod-1 operations for anti-control), which has the above-indicated Lyapunov exponents $\lambda_1 = \sigma_1$ and $\lambda_2 = \sigma_2$.

11.3 Chaotifying Discrete-time TS Fuzzy Systems via a Sinusoidal Function

For simplicity, further assume that $B_i = I$ in the discrete-time TS fuzzy system (11.1), so that (11.3) is in the form

$$\begin{aligned} x(k+1) &= \sum_{i=1}^{q} \mu_i(k)\{A_i x(k) + u(k)\} \\ &= \left[\sum_{i=1}^{q} \mu_i(k) A_i\right] x(k) + u(k). \end{aligned} \tag{11.14}$$

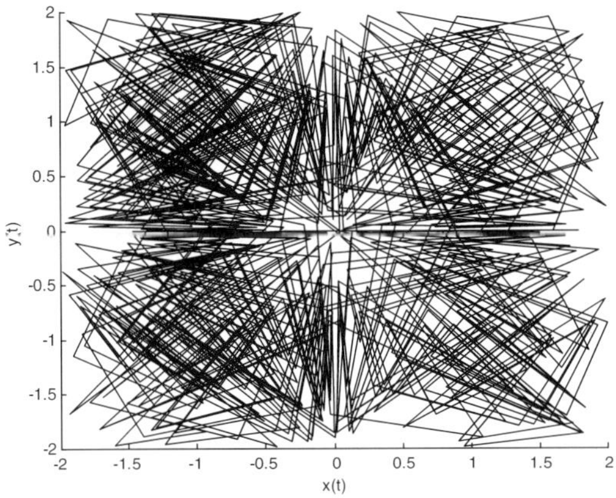

Fig. 11.2. Chaotic orbit of the anti-controlled system

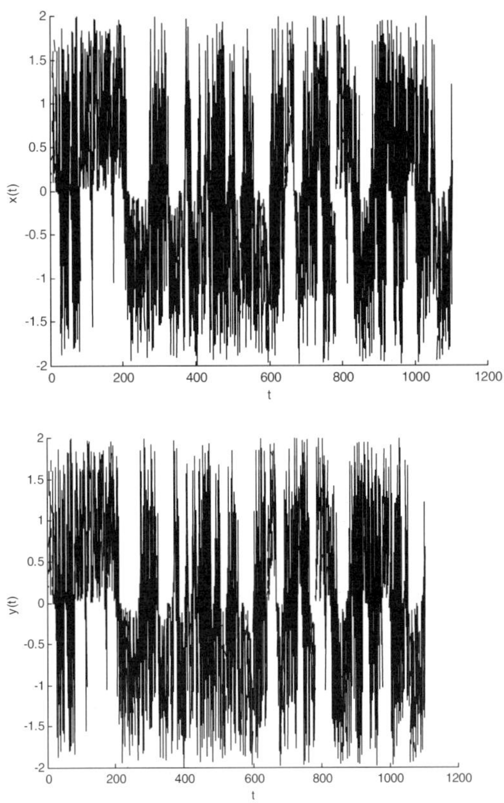

Fig. 11.3. Phase portraits of (a) $t - x(k)$; (b) $t - y(k)$

The chaotification problem is to design a control input sequence, $\{u(k)\}_{k=0}^{\infty}$, with an arbitrarily small magnitude, $\sigma > 0$, namely,

$$\|u(k)\|_{\infty} \leq \sigma, \text{ for all } i = 1, 2, \ldots, q, \tag{11.15}$$

such that the controlled system (11.14) becomes chaotic.

For this purpose, among several possible candidates, a simple sinusoidal function is used to construct the control input as follow [40]:

$$\begin{aligned} u(k) =& \Phi(\beta x(k)) \\ \equiv& \left[\varphi(\beta x_1(k)), \varphi(\beta x_2(k)), \ldots, \varphi(\beta x_n(k))\right]^T, \end{aligned} \tag{11.16}$$

where $x(k) = \left[x_1(k), x_2(k), \ldots, x_n(k)\right]^T$, β is a constant, and $\varphi : \Re \rightarrow \Re$ is a continuous sinusoidal function defined by (see Fig. 11.4)

$$\varphi(x) = \sigma \sin\left(\frac{\pi}{\sigma} x\right). \tag{11.17}$$

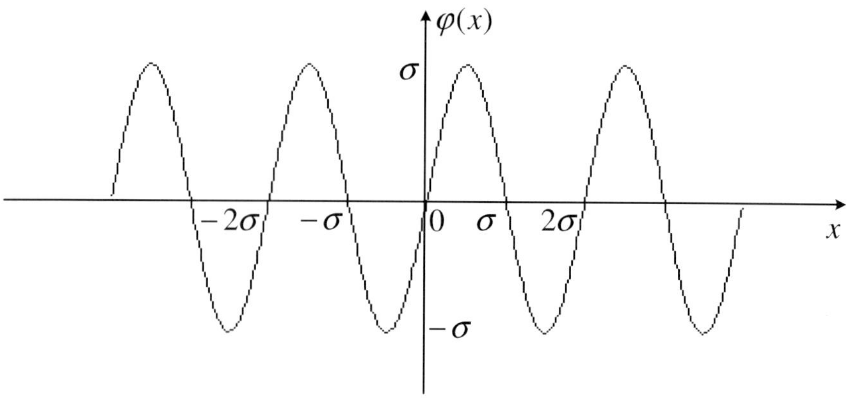

Fig. 11.4. Sinusoidal function used for chaotification

Obviously, $|\varphi(x)| \leq \sigma$ for all $x \in \Re$, so that $\|u(k)\| \leq \sigma$, where σ can be arbitrarily small, as required by condition (11.15). Here, the sinusoidal function actually serves as a smooth version of the modulus operation.

Lemma 11.3 (Boundedness). *The state vector of the controlled system (11.14), under the control of the controller (11.16) and (11.17), is uniformly bounded by a constant,* $\sigma(1-\alpha)^{-1}$.

Proof: The solution of the controlled system (11.14) can be written as

$$x(k) = \left[\sum_{i=1}^{q} \mu_i(k) A_i\right]^k x(0) + \sum_{j=1}^{k-1} \left[\sum_{i=1}^{q} \mu_i(k) A_i\right]^{k-1-j} u(j).$$

Since $\max_{1\le i\le q}\{\|A_i\|\} = \alpha < 1$ and $\|u(k)\| \le \infty$, one has a decreasing sequence $\{x(k)\}_{k=0}^{\infty}$ with

$$
\begin{aligned}
\lim_{k\to\infty} \|x(k)\| \le &\lim_{k\to\infty} \left\{ \left[\left\|\sum_{i=1}^{q} \mu_i(k) A_i\right\|\right]^k \|x(0)\| + \sum_{j=1}^{k-1} \left[\left\|\sum_{i=1}^{q} \mu_i(k) A_i\right\|\right]^{k-1-j} \|u(j)\| \right\} \\
&\le \lim_{k\to\infty} \alpha^k \|x(0)\| + \lim_{k\to\infty} \sum_{j=1}^{k-1} \alpha^{k-1-j} \|u(j)\| \\
&\le \sigma \lim_{k\to\infty} \sum_{j=1}^{k-1} \alpha^{k-1-j} \\
&= \frac{1}{1-\alpha}\sigma.
\end{aligned}
$$

This means that the sinusoidal function folds an expanding trajectory back toward the origin when the trajectory becomes too large in magnitude, thus bounding the controlled system trajectory globally. On the other hand, it will be shown in the next section that if β is chosen to be large enough, the controller designed above can lead all eigenvalues of the controlled system's Jacobian, at every time step, to exceed unity in absolute value. Consequently, it can be proven that all the Lyapunov exponents of the controlled system are strictly positive, so that the system trajectory is locally expanding in all directions. The combination of these two effects, stretching and folding, will then yield complex chaotic dynamics within the bounded region of the controlled system trajectories.

In this case, the fuzzy control rule structure (9.2) has the following form:

$$
\begin{aligned}
&\text{Control Rule } i: \\
&\text{IF } x_1(k) \text{ is } \Gamma_1^i \text{ and } \ldots \text{ and } x_n(k) \text{ is } \Gamma_n^i \\
&\text{THEN } u(k) = \Phi(\beta x(k)), \\
&i = 1, 2, \ldots, q.
\end{aligned}
\tag{11.18}
$$

The fuzzy control rules have linear state feedback laws in the consequent parts. The overall fuzzy controller is represented by

$$
u(k) = \frac{1}{\sum_{i=1}^{q} \omega_i(k)} \sum_{i=1}^{q} \omega_i(k) \Phi(\beta x(k)) = \Phi(\beta x(k)). \tag{11.19}
$$

In terms of the Marotto theorem, the theoretical result of the above controller design is summarized as follows [40].

Theorem 11.4. *Suppose that $\mu_i(k)$, $i = 1, 2, \ldots, q$, are continuously differentiable in the neighborhood of the fixed point, $x^* = 0$, of the controlled system (11.14). Then, there exists a positive constant $\bar{\beta}$ such that if $\beta > \bar{\beta}$ then the controlled TS fuzzy system (11.14) and (11.16) is chaotic in the sense of Li and Yorke.*

Proof: The controlled system (11.14) and (11.16) is

$$
\begin{aligned}
x(k+1) &= \sum_{i=1}^{q} \mu_i(k)\{A_i x(k) + u(k)\} \\
&= \left[\sum_{i=1}^{q} \mu_i(k) A_i\right] x(k) + \Phi(\beta x(k)) \equiv g(x(k)).
\end{aligned} \tag{11.20}
$$

Obviously, $x^* = 0$ is a fixed point of (11.20), which is now proven to be a snap-back repeller. Differentiating (11.20) at this fixed point yields

$$
\|g'(0)\| = \left\| \left[\sum_{i=1}^{q} \mu_i(k) A_i\right]_0 + \pi\beta I \right\|. \tag{11.21}
$$

If $\beta > \frac{(1+\alpha)}{\pi}$, then $\|g'(0)\| > 1$. By the continuity of $g'(x)$ in the neighborhood of the fixed point, there exists a small positive constant, r, such that when $x \in B(x^*; r)$, $\|g'(x)\| > 1$.

Therefore, the Gerschgorin theorem [232] implies that all eigenvalues of $g'(x)$ exceed unity in absolute value for all $x \in B(x_0; r)$.

Next, it is shown that there exists a point, $x_0 \in B(x^*; r)$, such that $g^2(x_0) = 0 = x^*$ and $(g^2(x_0))' \neq 0$.

Indeed, it is easy to see that if $\beta > \frac{3\alpha}{2}$, then there exist $x_1 = \left[\frac{\sigma}{2\beta}, \ldots, \frac{\sigma}{2\beta}\right]^T$ and $x_2 = \left[\frac{3\sigma}{2\beta}, \ldots, \frac{3\sigma}{2\beta}\right]^T$, such that

$$
g(x_1) > 0 \text{ and } g(x_2) < 0.
$$

Therefore, by the mean value theorem in calculus, there exists a point, $x_1 < x_1^* < x_2$, such that $g(x_1^*) = 0$.

Let $\tilde{x} = [r, r, \ldots, r]^T$. It is clear that there exists a constant $\beta_1 > 0$ such that if $\beta > \beta_1$ then

$$
g(0) = 0 < x_1^* \text{ and } g(\tilde{x}) > x_1^*.
$$

Using the mean value theorem again, one concludes that there exists a point, $x_0 \in B(x^*; r)$, such that $g(x_0) = x_1^*$. Therefore,

$$
g^2(x_0) = g(x_1^*) = 0.
$$

On the other hand, there exists a constant $\beta_2 > 0$ such that $g'(x_1^*) < 0$, for $\cos(\frac{\pi}{\sigma\beta x_1^*}) < 0$. Therefore,

$$(g^2)'(x_0) = g'(x_1^*)g'(x_0) \neq 0.$$

To conclude, if $\beta > \bar{\beta} \equiv \max\{(1+\alpha)/\pi, (3\sigma)/2, \beta_1, \beta_2\}$, then $x_0 = 0$ is a snap-back repeller of the map g defined in (11.20), so the controlled system (11.14) and (11.16) is chaotic in the sense of Li and Yorke.

Example 11.5. To visualize the theoretical analysis and design, the same example as given in Subsection 11.2.4 is used for illustration.

The controlled TS fuzzy system is described as follows,

$$\begin{aligned} x(k+1) &= \sum_{i=1}^{q} \mu_i(k)\{A_i x(k) + u(k)\} = \sum_{i=1}^{q} \mu_i(k) A_i x(k) + u(k) \\ &= \sum_{i=1}^{q} \mu_i(k) A_i x(k) + \sigma \sin\left(\frac{\pi}{\sigma}\beta x(k)\right). \end{aligned}$$

In the simulation, the magnitude of the control input is arbitrarily chosen to be $\sigma = 0.1$. Thus, $\|u(k)\| < \infty$, and β can also be regarded as a control parameter. Without control, the TS fuzzy model is stable, as shown in Fig. 11.1. For β taking the values 0.25, 0.4, 0.45, 0.5 and 1.3, the phase portraits, time evolutions, and bifurcation diagrams are shown in Fig. 11.5–11.11, respectively. These numerical results validate the theoretical analysis and the design of the chaos generator.

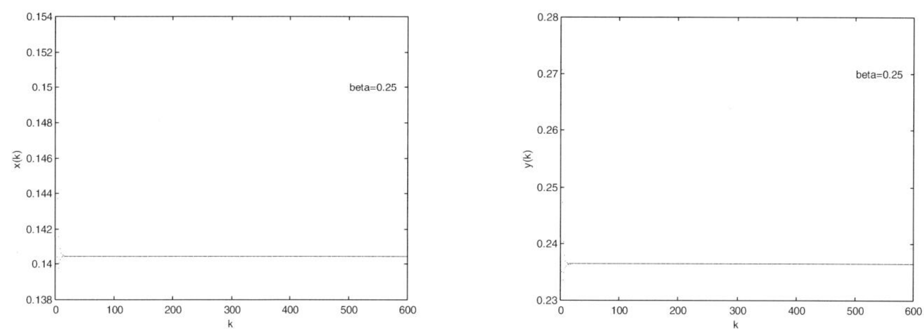

Fig. 11.5. Periodic orbits at $\beta = 0.25$

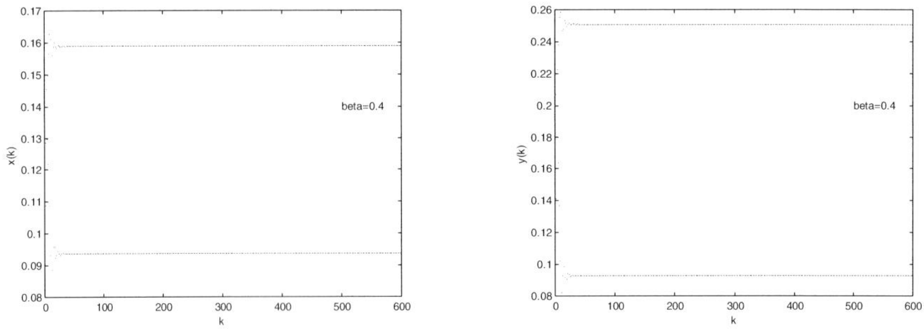

Fig. 11.6. Period-doubling bifurcation at $\beta = 0.4$

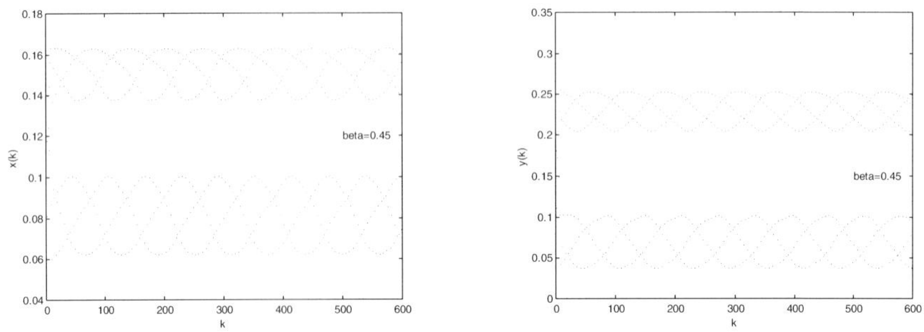

Fig. 11.7. Special period-doubling bifurcation at $\beta = 0.45$

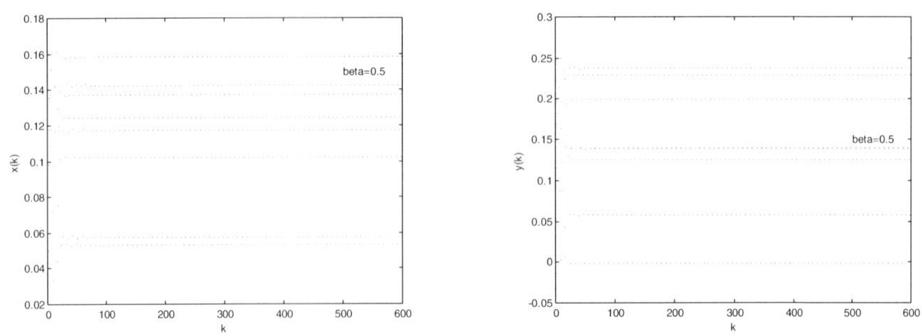

Fig. 11.8. Period-8 bifurcation at $\beta = 0.5$

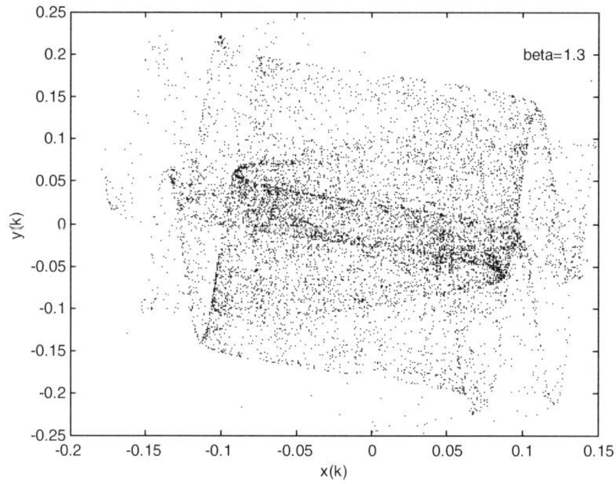

Fig. 11.9. Phase portrait at $\beta = 1.3$

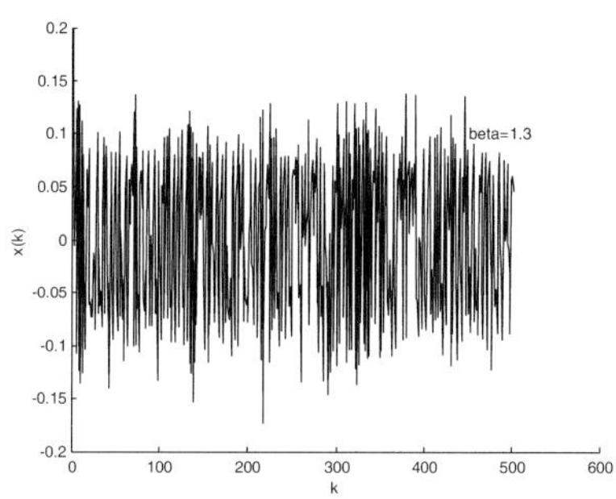

(a)$k - x(k)$

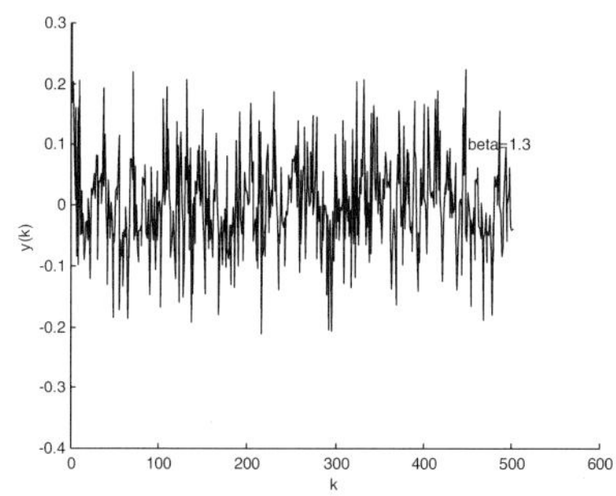

(b)$k - y(k)$

Fig. 11.10. Time evolution at $\beta = 1.3$

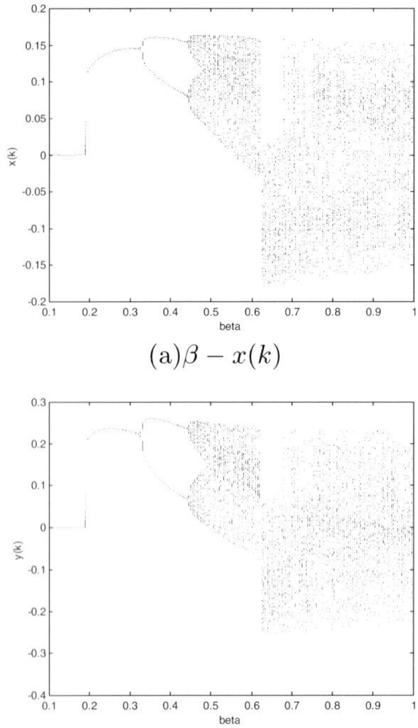

(a)$\beta - x(k)$

(b)$\beta - y(k)$

Fig. 11.11. Bifurcation diagrams

11.4 Chaotifying Continuous-time TS Fuzzy Systems via Discretization

In order to make a nonchaotic or even stable continuous-time TS fuzzy system chaotic, based on the above proposed anti-control approaches a natural and straightforward approach is to convert it to a discrete-time version, i.e., to discretize the continuous-time TS fuzzy system. Conversion of a continuous-time controller to an equivalent digital controller is known as digital redesign, which is to be introduced in Chapter 12. The discretization theorem was given in Chapter 6 as Theorem 6.1.

For the continuous-time TS fuzzy system (6.2), if the subsystems are stable in each local subspace, and $\mu_i(x(t))$, $i = 1, 2, \ldots, q$, are continuously differentiable in a neighborhood of the origin, then local stability of the overall system is guaranteed by the following theorem.

Theorem 11.6 (Local Stability Theorem). *In (6.2), if A_i, $i = 1, 2, \ldots, q$, are all $n \times n$ Hurwitz stable matrices, then the uncontrolled system (6.3) (with*

$u(t) = 0$) and uncontrolled discretized system (6.5) (with $u(k) = 0$) are both stable in the neighborhood of the origin.

Proof: It is first noted that A_i, $i = 1, 2, \ldots, q$, are all $n \times n$ Hurwitz stable matrices, i.e., all of their eigenvalues have negative real parts.

1) For the uncontrolled system (6.3), the origin $x_0 = 0$ is obviously the fixed point of the uncontrolled system (6.3), and its Jacobian at the origin is $\sum_{i=1}^{q} \mu_i(x_0) A_i$. The characteristic equation is

$$\begin{aligned} \left| \lambda I - \sum_{i=1}^{q} \mu_i(x_0) A_i \right| &= \left| \sum_{i=1}^{q} \mu_i(x_0)(\lambda I - A_i) \right| \\ &= \sum_{i=1}^{q} \mu_i(x_0) \prod_{j=1}^{n} (\lambda - \lambda_{ij}) \\ &= 0, \end{aligned}$$

where λ_{ij}, $j = 1, 2, \ldots, n$, are the eigenvalues of A_i, $i = 1, 2, \ldots, q$, which all have negative real parts. If $\lambda \geq 0$, then $|\lambda I - \sum_{i=1}^{q} h_i A_i| > 0$. So $|\lambda I - \sum_{i=1}^{q} h_i A_i| = 0$ holds only when $\lambda < 0$. This means that $\sum_{i=1}^{q} \mu_i(x_0) A_i$ is a Hurwitz stable matrix, hence the uncontrolled system (6.3) is stable in a neighborhood of the origin.

2) Consider the uncontrolled discretized system (6.5). Since A_i, $i = 1, 2, \ldots, q$, are Hurwitz stable matrices, $G_i = \exp(A_i T_s)$ are Schur stable matrices, i.e., $\rho(G_i) < 1$. Therefore, there exists a certain norm, $\|\cdot\|$, such that $\|G_i\| < 1$. From $\mu_i(x(t)) \geq 0$ and $\sum_{i=1}^{q} \mu_i(x(t)) = 1$ and the convexity of the matrix norm, it follows that

$$\begin{aligned} \left\| \sum_{i=1}^{q} h_i G_i \right\| &\leq \sum_{i=1}^{q} \mu_i \|G_i\| \\ &\leq \left[\sum_{i=1}^{q} \mu_i \right] \max\{\|G_i\|\} \\ &= \max\{\|G_i\|\} \\ &= \alpha < 1. \end{aligned}$$

Thus,

$$\|x(k+1)\| \leq \left\| \sum_{i=1}^{q} h_i G_i \right\| \|x(k)\| \leq \alpha \|x(k)\|.$$

By the contraction mapping theorem, it is concluded that the uncontrolled discretized system (6.5) is stable in the neighborhood of the origin.

After converting the continuous-time TS fuzzy system to a discrete-time counterpart, an anti-controller can be designed for the discretized TS fuzzy system to make it chaotic. The anti-controller's design and the verification of chaos in the controlled TS fuzzy system are similar to that discussed in Sections 11.2 and 11.3, so they are omitted.

Example 11.7. Consider a continuous-time TS fuzzy system, which is the TS fuzzy model of the Lorenz equation, given by

$$\text{Rule 1: IF } x_1(t) \text{ is } \Gamma_1, \text{ THEN } \frac{d}{dt}\begin{bmatrix} x_1(t) \\ x_2(t) \\ x_3(t) \end{bmatrix} = A_1 \begin{bmatrix} x_1(t) \\ x_2(t) \\ x_3(t) \end{bmatrix}$$

$$\text{Rule 2: IF } x_1(t) \text{ is } \Gamma_2, \text{ THEN } \frac{d}{dt}\begin{bmatrix} x_1(t) \\ x_2(t) \\ x_3(t) \end{bmatrix} = A_2 \begin{bmatrix} x_1(t) \\ x_2(t) \\ x_3(t) \end{bmatrix}$$

where

$$A_1 = \begin{bmatrix} -d & d & 0 \\ r & -1 & -x_{1\,\mathrm{min}} \\ 0 & x_{1\,\mathrm{min}} & -b \end{bmatrix} \text{ and } A_2 = \begin{bmatrix} -d & d & 0 \\ r & -1 & -x_{1\,\mathrm{max}} \\ 0 & x_{1\,\mathrm{max}} & -b \end{bmatrix}$$

and the membership functions are

$$\Gamma_1 = \frac{-x_1 + x_{1\,\mathrm{max}}}{x_{1\,\mathrm{max}} - x_{1\,\mathrm{min}}} \text{ and } \Gamma_2 = \frac{x_1 - x_{1\,\mathrm{min}}}{x_{1\,\mathrm{max}} - x_{1\,\mathrm{min}}},$$

where Γ_i, $i = 1, 2$, are positive-semidefinite for all $x_1 \in [x_{1\,\mathrm{min}}, x_{1\,\mathrm{max}}]$, and d, r, and b are parameters.

With the parameters $(d, r, b) = (10, 28, 8/3)$ and initial values $(0.1, 0.1, 0.1)$, the trajectory of the TS fuzzy system of the Lorenz system is shown in Fig. 6.2.

In terms of Theorem 6.1, the discretized TS fuzzy model of the original continuous-time TS fuzzy Lorenz system is obtained as follows:

$$\text{Rule 1: IF } x_1(k) \text{ is } \Gamma_1, \text{ THEN } \begin{bmatrix} x_1(k+1) \\ x_2(k+1) \\ x_3(k+1) \end{bmatrix} = G_1 \begin{bmatrix} x_1(k) \\ x_2(k) \\ x_3(k) \end{bmatrix}$$

$$\text{Rule 2: IF } x_1(k) \text{ is } \Gamma_2, \text{ THEN } \begin{bmatrix} x_1(k+1) \\ x_2(k+1) \\ x_3(k+1) \end{bmatrix} = G_2 \begin{bmatrix} x_1(k) \\ x_2(k) \\ x_3(k) \end{bmatrix}$$

where

$$G_1 = \begin{bmatrix} 1 - dT_s & dT_s & 0 \\ rT_s & 1 - T_s & -x_{1\,\mathrm{min}}T_s \\ 0 & x_{1\,\mathrm{min}}T_s & 1 - bT_s \end{bmatrix} \text{ and } G_2 = \begin{bmatrix} 1 - dT_s & dT_s & 0 \\ rT_s & 1 - T_s & -x_{1\,\mathrm{max}}T_s \\ 0 & x_{1\,\mathrm{max}}T_s & 1 - bT_s \end{bmatrix}.$$

Figure 6.5 shows the trajectory of the discrete-time version of the continuous-time TS fuzzy Lorenz model, with $T_s = 0.004$. It can be seen that the overall shape of the trajectory is very similar to that in Fig. 11.11 (a).

When the parameters are chosen as $d = 200$, $r = 40$, and $b = 200$, the eigenvalues of A_i are -132.1267, -68.8733 and -200.0000, so they are Hurwitz stable matrices. Hence, by Theorem 11.6, the continuous-time and the corresponding discretized TS fuzzy Lorenz system are both stable. The anti-controller is designed as described in (11.16)–(11.17). The controlled TS fuzzy Lorenz system is described as

$$\begin{aligned} x(k+1) &= \sum_{i=1}^{q} \mu_i \{G_i x(k) + u(k)\} \\ &= \sum_{i=1}^{q} h_i G_i x(k) + u(k) \\ &= \sum_{i=1}^{q} h_i G_i x(k) + \sigma \sin\left(\frac{\pi}{\sigma} \beta x(k)\right). \end{aligned}$$

In the simulation, the magnitude of the control input is arbitrarily chosen to be $\sigma = 0.1$. Thus, $\|u(k)\| < \sigma$, and β can also be regarded as a control parameter. For β taking the values 0.6 and 2.1, the time evolutions, phase portraits, and bifurcation diagrams are shown in Fig. 11.12–11.15, respectively. These numerical results validate the theoretical analysis and the design of the anti-controller developed above.

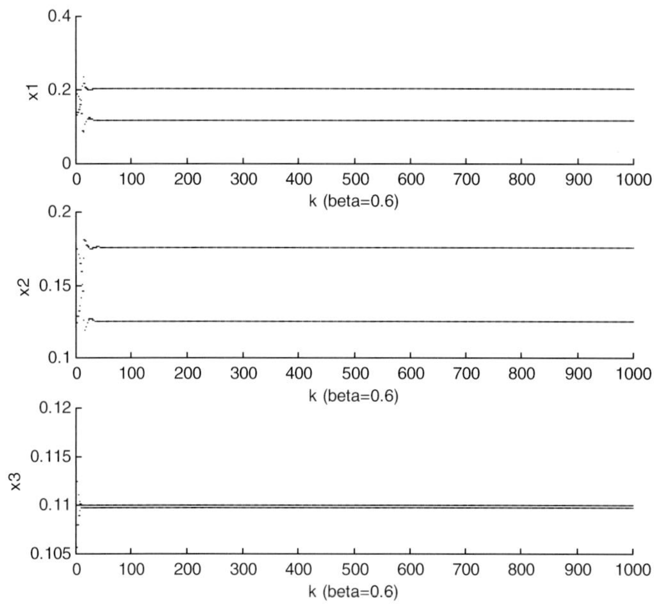

Fig. 11.12. Period-doubling bifurcations at $\beta = 0.6$

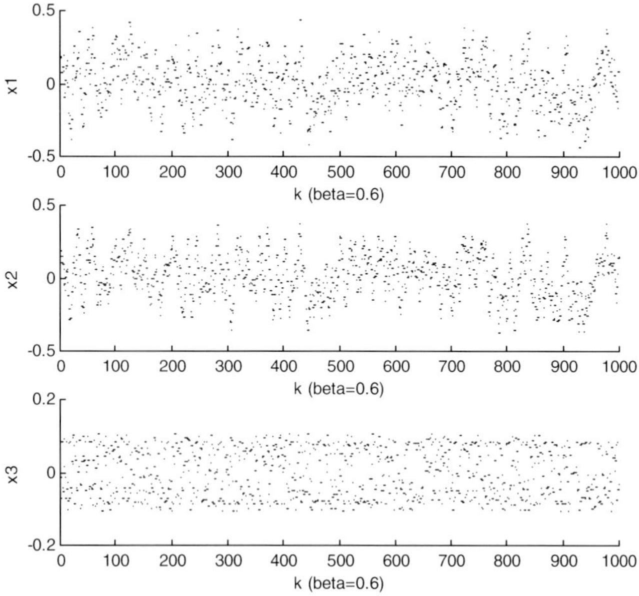

Fig. 11.13. Chaotic time evolutions

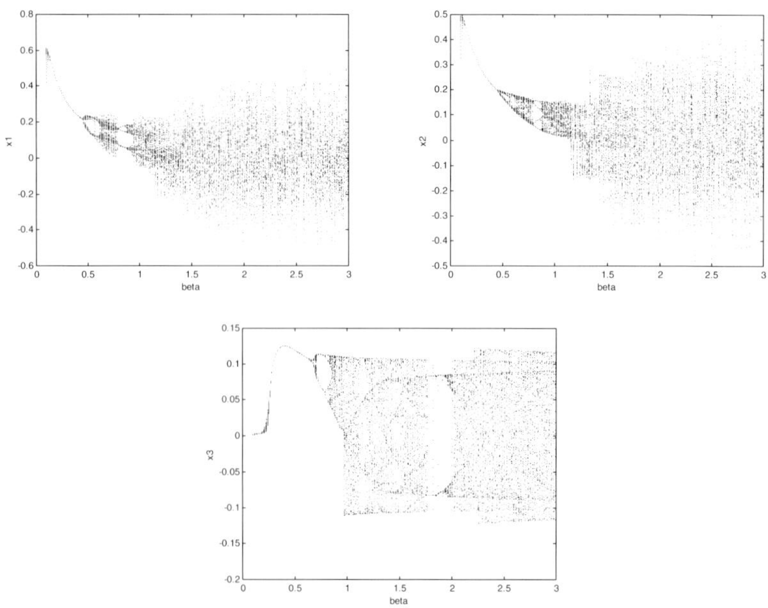

Fig. 11.14. Bifurcation diagrams

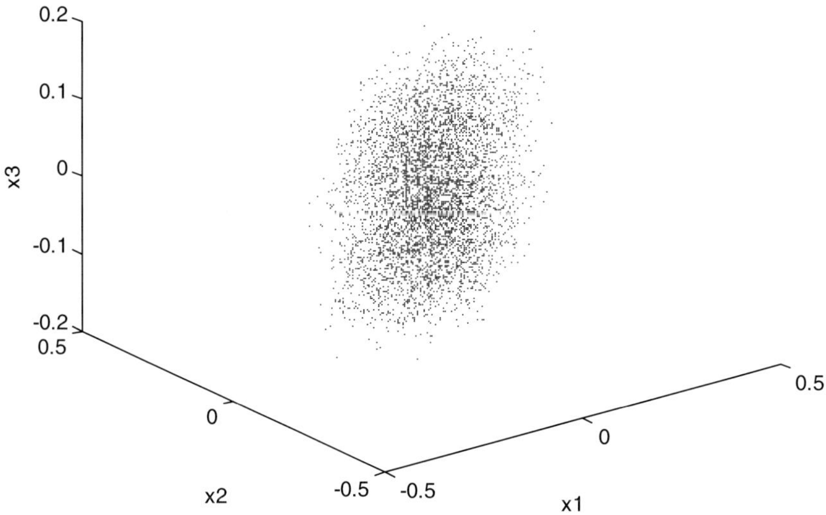

Fig. 11.15. Chaotic phase portrait at $\beta = 2.1$

11.5 Chaotifying Continuous-Time TS Fuzzy Systems via Time-delay Feedback

A new technique was proposed to chaotify continuous-time TS fuzzy systems, which is general and yet by nature very different from the aforementioned approaches [42]. It can make an arbitrarily given continuous-time TS fuzzy system chaotic via fuzzy feedback linearization, differential geometric control theory, and time-delay feedback perturbations [237, 238, 239]. In this approach, asymptotic analysis is used to establish an approximate relationship between a time-delay differential equation and a discrete chaotic map, so that a time-delay feedback control term can be constructed to make the controlled TS fuzzy system chaotic, where the generated chaos is in the sense of Li and Yorke [19]. Systems with time-delay are inherently infinite-dimensional. Therefore, it is possible to produce complicated dynamics such as bifurcation and chaos, even in a very simple first-order system [240].

Specifically, consider a general *single-input* TS fuzzy system in the following form:

$$\begin{aligned} &\text{Plant Rule } i: \\ &\quad \text{IF } x_1(t) \text{ is } \Gamma_1^i \ldots \text{and } x_n(t) \text{ is } \Gamma_n^i, \\ &\quad \text{THEN } \dot{x}(t) = A_i x(t) + B_i u(t), \end{aligned} \tag{11.22}$$

where $x(t) \in \Re^n$, $u(t) \in \Re$, $A_i \in \Re^{n\times n}$, $B_i \in \Re^n$, $i = 1, 2, \cdots, q$, in which q is the number of IF-THEN rules, Γ_j^i are fuzzy sets, and the equation $\dot{x}(t) = A_i x(t) + B_i u(t)$ is the output from the ith IF-THEN rule.

Similarly, by using $\mu_i = \omega_i / \sum_{j=1}^{r} \omega_j$ instead of ω_i, for a given pair of $(x(t), u(t))$, the final output of the fuzzy system is inferred by

$$\begin{aligned}
\dot{x}(t) =& \frac{1}{\sum_{i=1}^{r} \omega_i} \sum_{i=1}^{r} \omega_i \{\mathbf{A}_i \mathbf{x}(t) + \mathbf{B}_i u(t)\} \\
=& \sum_{i=1}^{q} \mu_i \{A_i x(t) + B_i u(t)\} \\
=& \left[\sum_{i=1}^{q} \mu_i A_i\right] x(t) + \left[\sum_{i=1}^{q} \mu_i B_i\right] u(t) \\
=& \bar{A} x(t) + \bar{B} u(t),
\end{aligned} \tag{11.23}$$

where $\bar{A} \equiv \sum_{i=1}^{q} \mu_i A_i$, and $\bar{B} \equiv \sum_{i=1}^{q} \mu_i B_i$.

Definition 11.8 (Local and Global Controllability). *If (A_i, B_i), $i = 1, 2, \ldots, q$, are controllable pairs, then the fuzzy model (11.23) is called **locally controllable**. If $(\bar{A}, \bar{B})$ is a controllable pair, then the fuzzy model (11.23) is called **globally controllable**.*

For a special kind of locally controllable TS fuzzy systems and a more general type of TS fuzzy systems, fuzzy feedback linearization, nonlinear control theory, and the PDC with time-delay can be employed to make them chaotic.

11.5.1 PDC Controller for Locally Controllable TS Fuzzy Systems

The PDC controller discussed in Section 9.3 is employed here to determine the structure of a fuzzy controller from a given TS fuzzy model [198, 241, 242, 243, 244]. The PDC provides the following fuzzy-control rule structure from the fuzzy model (11.23):

Control Rule i :

IF $x_1(t)$ is Γ_1^i … and $x_n(t)$ is Γ_n^i,

THEN $u(t) = -F_i x(t) + \nu, \qquad i = 1, 2, \cdots, q,$ (11.24)

where F_i are feedback control gains, ν is a time-delay feedback term of the form

$$\nu(t - \tau) = \omega(h(x(t - \tau))),$$

in which h is a scalar function remaining to be further determined.

The fuzzy control rules have linear state feedback laws and a scalar function in the consequent parts. The overall fuzzy controller is represented by

$$
\begin{aligned}
u(t) &= -\frac{1}{\sum_{i=1}^{q} \omega_i} \sum_{i=1}^{q} \omega_i \left(F_i x(t) + \nu \right) \\
&= -\sum_{i=1}^{q} \mu_i F_i x(t) + \nu.
\end{aligned} \tag{11.25}
$$

Chaotifying the TS fuzzy system (11.23) is generally a very difficult task. However, in some special cases, this can be done. Consider the case of a common B in (11.23), i.e., $B_1 = B_2 = \ldots = B_q = B$.

Definition 11.9. *The TS fuzzy system (11.23) is called* exactly linearizable *via the feedback controller (11.25) if there exist feedback gains F_i such that*

$$A_i - BF_i = G, \tag{11.26}$$

for $i = 2, 3, \ldots, q$.

Notice that having a common G might not always be possible even if (A_i, B_i), $i = 1, 2, \ldots, q$, are controllable.

If the locally controllable TS fuzzy system (11.23) is exactly linearizable, one can choose F_i such that $A_i - BF_i = G$ for $i = 1, 2, \cdots, q$, where G is a Hurwitz stable matrix with desired eigenvalues, and (G, B) is also controllable. Thus, the global fuzzy model (11.23) is reduced to the following form:

$$\dot{x}(t) = Gx(t) + B\nu. \tag{11.27}$$

The feedback control law (11.25) can then be designed by using conventional linear system theory [237] so that the eigenvalues of $A_i - BF_i$ are the specified ones. The feedback gain F_i can be obtained by using Ackermann's formula [237], in the case of a single-input system, as follows:

$$F_i = [0\,0 \ldots 0\,1] \left[B \,|\, A_i B \,|\, \cdots \,|\, A_i^{n-1} B \right]^{-1} \phi(A_i), \tag{11.28}$$

where $\phi(A_i) = A_i^n + a_1 A_i^{n-1} + \ldots + a_{n-1} A_i + a_n I$, $i = 1, 2, \ldots, q$, are Hurwitz stable polynomials, and a_1, a_2, $\cdots$, a_n are coefficients of the characteristic polynomial:

$$
\begin{aligned}
|\lambda I - G| &= (\lambda - \mu_1)(\lambda - \mu_2) \ldots (\lambda - \mu_n) \\
&= \lambda^n + a_1 \lambda^{n-1} + \ldots + a_{n-1} \lambda + a_n.
\end{aligned} \tag{11.29}
$$

Define a transformation matrix T by

$$T = MW, \tag{11.30}$$

where M is the controllability matrix of the form

$$M = \left[B \,|\, GB \,|\, \ldots \,|\, G^{n-1}B\right]$$

and

$$W = \begin{bmatrix} a_{n-1} & a_{n-2} & \cdots & a_1 & 1 \\ a_{n-2} & a_{n-3} & \cdots & 1 & 0 \\ \vdots & \vdots & \ddots & \vdots & \vdots \\ a_1 & 1 & \cdots & 0 & 0 \\ 1 & 0 & \cdots & 0 & 0 \end{bmatrix}.$$

Also define a new state vector $\hat{x}$ by

$$x = T\hat{x}.$$

The rank of the controllability matrix M is n, because (G, B) is controllable. Then, the inverse of matrix T exists and (11.28) can be modified to

$$\dot{\hat{x}} = T^{-1}GT\hat{x} + T^{-1}B\nu, \tag{11.31}$$

where

$$T^{-1}GT = \begin{bmatrix} 0 & 1 & 0 & \cdots & 0 \\ 0 & 0 & 1 & \cdots & 0 \\ \vdots & \vdots & \vdots & \ddots & \vdots \\ 0 & 0 & 0 & \cdots & 1 \\ -\alpha_n & -\alpha_{n-1} & -\alpha_{n-2} & \cdots & -\alpha_1 \end{bmatrix}$$

and

$$T^{-1}B = \begin{bmatrix} 0 \\ 0 \\ \vdots \\ 0 \\ 1 \end{bmatrix},$$

with

$$\nu = \omega(\hat{x}_1(t-\tau)) = \omega((T^{-1}x)_1(t-\tau)), \tag{11.32}$$

in which $\hat{x}_1$ means the first component of vector $\hat{x}$. This will be further determined in the following.

11.5.2 Controller Design for General TS Fuzzy Systems

Two different approaches are employed to study the problem of chaotification for a more general TS fuzzy system.

Approximate Linearization Approach

Obviously, the origin is an equilibrium of the uncontrolled system (11.23) with $u = 0$ therein, and by Theorem 11.6 it is asymptotically stable within its sufficiently small neighborhood if the subsystem (11.22) is stable.

In a small neighborhood of the origin, system (11.23) can be represented by its linearization, evaluated at the origin, as

$$\dot{x}(t) = \bar{A}_0 x(t) + \bar{B}_0 u(t). \tag{11.33}$$

Suppose $(\bar{A}_0, \bar{B}_0)$ is controllable, a feedback control law similar to the above-proposed formula (11.30) can be used to convert it to the controllable canonical form (11.31).

Globally Exact Linearization Approach

To achieve the intended chaotification, differential-geometric control theory can be applied [238, 245, 246]. Assume that $y = h(x)$ is a scalar output of system (11.23), where h is a smooth function satisfying $h(x) = 0$. Combined with (11.23), it can be described as a SISO affine nonlinear system as follows:

$$\begin{cases} \dot{x}(t) = \bar{A}x(t) + \bar{B}u(t) := f(x) + g(x) \\ \quad y = h(x). \end{cases} \tag{11.34}$$

Let $L_f^i h(x)$ denote the ith *Lie derivative* of the smooth function $h(x)$ with respect to a vector field $f(x)$ and $ad_f^i g(x)$ the ith *Lie bracket* of the two smooth vector fields $f(x)$ and $g(x)$.

Definition 11.10. *The SISO affine system (11.34) is said to have a relative degree r at x^* if there exists a neighborhood U of x^* such that*

$$L_g L_f^k h(x^*) = 0, \forall\, x \in U,\ 0 \le k < r - 1;$$
$$L_g L_f^{r-1} h(x) \neq 0.$$

Definition 11.11 (Involutive Distributions). *A distribution $\Delta = span\{f_1, \ldots, f_m\}$ is called involutive if for any two vector fields τ_1, $\tau_2 \in \Delta$, their Lie bracket $ad_{\tau_1}\tau_2 \in \Delta$.*

The following result is well known [238].

Lemma 11.12. *The SISO system (11.34) has relative degree n at x^* if and only if there exists a neighborhood U of x^* such that*

(i) rank $\left[g(x),\, ad_f g(x),\, \ldots,\, ad_f^{n-1} g(x)\right] = n$ *for all* $x \in U$,

(ii) span $\left\{g(x),\, ad_f g(x),\, \ldots,\, ad_f^{n-2} g(x)\right\}$ *is involutive in* U.

Definition 11.13. *The SISO system (11.34) is called globally feedback-linearizable if it has relative degree n at x^*.*

In this case, the output $y = h(x)$ is a solution to the following partial differential equation:

$$\frac{\partial h(x)}{\partial x}\left[g(x),\ ad_f g(x),\ \ldots,\ ad_f^{n-2} g(x)\right] = 0.$$

If the output $y = h(x)$ results in a relative degree of n, then one may use

$$y = \Phi(x) = \begin{pmatrix} h(x) \\ L_f h(x) \\ L_f^2 h(x) \\ \vdots \\ L_f^{n-1} h(x) \end{pmatrix} \tag{11.35}$$

and the control law

$$u(x) = \frac{1}{L_g L_f^{n-1} h(x)}(-L_f^n h(x) + \nu) \tag{11.36}$$

to yield the linear system

$$\dot{\mathbf{y}} = \begin{bmatrix} 0 & 1 & 0 & \cdots & 0 \\ 0 & 0 & 1 & \cdots & 0 \\ \vdots & \vdots & \vdots & \ddots & \vdots \\ 0 & 0 & 0 & \cdots & 0 \\ 0 & 0 & 0 & \cdots & 0 \end{bmatrix} \mathbf{y} + \begin{bmatrix} 0 \\ 0 \\ \vdots \\ 0 \\ 1 \end{bmatrix} u, \tag{11.37}$$

where $\mathbf{y} = \left[y, \dot{y}, \cdots, y^{(n-1)}\right]^T$, and

$$\begin{aligned} u =& \frac{1}{L_g L_f^{n-1}}\left(-L_f^n h(x) + \omega(y(t-\tau))\right) \\ =& \frac{1}{L_g L_f^{n-1}}\left(-L_f^n h(x) + \omega(h(x(t-\tau)))\right), \end{aligned} \tag{11.38}$$

in which the time-delay term also remains to be further determined in the following.

11.5.3 Verification of Chaos

The controlled TS fuzzy system (11.33) with time-delay term (11.37) can be recast in the following n-dimensional state-space form:

$$\dot{x}(t) = Gx(t) + B\nu, \tag{11.39}$$

where G and B are in the controllable canonical form, namely,

$$G = \begin{bmatrix} 0 & 1 & \cdots & 0 \\ \vdots & \vdots & \ddots & \vdots \\ 0 & 0 & \cdots & 1 \\ -\alpha_n & -\alpha_{n-1} & \cdots & -\alpha_1 \end{bmatrix} \text{ and } B = \begin{bmatrix} 0 \\ 0 \\ \vdots \\ 0 \\ 1 \end{bmatrix}.$$

Since G is a Hurwitz stable matrix and $\nu(t)$ is uniformly bounded, the solution of (11.39) is bounded for any bounded initial condition and can be computed iteratively on each τ-time interval $(m\tau,\, (m+1)\tau]$ for $m = 0, 1, \ldots$.

Denote $x(t) \equiv x(m\tau + \hat{t}) \equiv x(m, \hat{t})$ for $t \equiv m\tau + \hat{t}$, $\hat{t} \in (0, \tau]$. It follows that

$$x(t) = x(m, \hat{t}) = e^{G\hat{t}} x(m-1, \hat{t}) + \int_0^{\hat{t}} e^{G(\hat{t}-t')} B\omega(x_1(m-1, t'))dt'. \quad (11.40)$$

Lemma 11.14. *[247, 227] Let $\delta(t-t_0)$ be the scalar-valued Dirac distribution centered at $t_0 \geq 0$, and let $d\xi(t, t_0) = e^{G(t_0-t)}dt$ be a matrix-valued measure defined on $[0, \tau]$. If it is imposed that $d\xi(t, t_0) = C(t, t_0)\delta(t - t_0)dt$, then $C(t, t_0) \approx -G^{-1}e^{G(t_0-t)}$ for a sufficiently large $\tau > 0$. Moreover, $C(t, t_0) \to -G^{-1}$ as $t \to t_0$.*

Lemma 11.15. *For a sufficiently large τ and a large $\hat{t} \in (t_0, \tau]$,*

$$x_1(m, \hat{t}) \approx \omega(x_1(m-1, \hat{t})) \text{ and } x_i(m, \hat{t}) \approx 0, \quad (11.41)$$

for $m = 0, 1, \ldots$ and $i = 2, \ldots, n$.

Proof. Note that for any given bounded initial condition, $x(t)$ is uniformly bounded and $e^{G\hat{t}}x(m-1, \hat{t})$ tends to 0 rapidly as $\hat{t} \to \infty$. Therefore, it follows from (11.34) that

$$\begin{aligned} x(t) &\approx \int_0^{\hat{t}} e^{G(\hat{t}-t')} B\omega(x_1(m-1, t'))dt' \\ &\approx \int_0^{\hat{t}} C(t', \hat{t})\delta(t' - \hat{t}) B\omega(x_1(m-1, t'))dt' \\ &\approx C(t', \hat{t}) B\omega(x_1(m-1, \hat{t})) \\ &\approx -G^{-1}\omega(x_1(m-1, t')). \end{aligned}$$

Since

$$G^{-1} = \begin{bmatrix} -\frac{\alpha_1}{\alpha_0} & -\frac{\alpha_2}{\alpha_0} & \cdots & -\frac{\alpha_{n-1}}{\alpha_0} & -\frac{1}{\alpha_0} \\ 1 & 0 & \cdots & 0 & 0 \\ 0 & 1 & \ddots & 0 & 0 \\ 0 & 0 & \cdots & 0 & 0 \\ 0 & 0 & \cdots & 1 & 0 \end{bmatrix},$$

one has

$$x(m,\hat{t}) \approx \left(\alpha_0^{-1}\omega(x_1(m-1,\hat{t})),\, 0,\, \cdots,\, 0\right)^T.$$

The proof is thus completed.

Lemma 11.15 establishes an asymptotically approximate relationship between the time-delay Eq. (11.39) and the difference Eq. (11.41). Although there is an essential difference between the dynamics of a time-delay equation and that of its associated difference equation [248], it is reasonable to expect that the first state component of the time-delay Eq. (11.39) is chaotic if $\omega(\cdot)$ is a bounded chaotic map and the delay time is sufficiently large. This implies that one can use time-delay feedback to drive system (11.39) to be chaotic.

Theorem 11.16. *If $\omega(\cdot)$ is a bounded chaotic map and if the delay time is sufficiently large, then the TS fuzzy system (11.23) controlled by the time-delayed feedback control law is chaotic.*

Proof. It follows directly from Lemma 11.15.

There are many well-known chaotic maps, such as the Logistic map, Hénon map or baker's map, which can be used to construct the time-delay feedback $\omega(\cdot)$, thus making the controlled TS fuzzy system (11.23) chaotic. One simple choice is

$$\nu(t) = \omega(x_1(t-\tau)) = \sigma \sin\left(\frac{\pi}{\sigma}\beta x_1(t-\tau)\right), \tag{11.42}$$

which is shown in Fig. 11.4.

Obviously, $|\, \nu(t) \,| \leq \sigma$ for all $x_1 \in \Re$, which can be arbitrarily small.

As mentioned above, if the map (11.42) is chaotic, then one can expect that the time-delay PDC will make the TS fuzzy system (11.23) chaotic, provided that the delay time τ is sufficiently large. Mathematically, it is shown that the map (11.42) is indeed chaotic in the sense of Li and Yorke by arguments similar to that given in [19, 20]. The bifurcation diagram of map (11.42) is shown in Fig. 11.16 with $\sigma = 0.1$, which reveals the chaotic nature of the map (11.42).

11.5.4 Simulation Examples

To visualize the theoretical analysis and design, two examples are included here for illustration.

Example 11.17. Consider the fuzzy system

$$\begin{aligned} &\text{Rule 1: IF } x_2(t) \text{ is } \Gamma_1, \text{ THEN } \dot{x} = A_1 x + Bu, \\ &\text{Rule 2: IF } x_2(t) \text{ is } \Gamma_2, \text{ THEN } \dot{x} = A_2 x + Bu, \end{aligned}$$

where

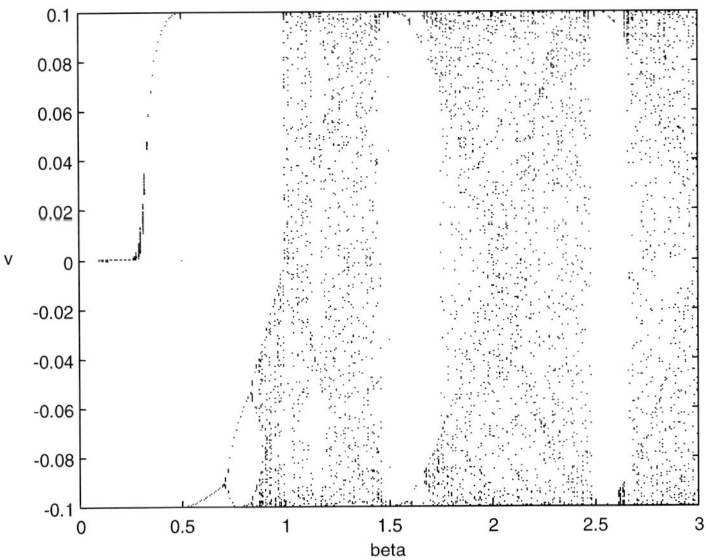

Fig. 11.16. Bifurcation diagram of map (11.42) with $\sigma = 0.1$

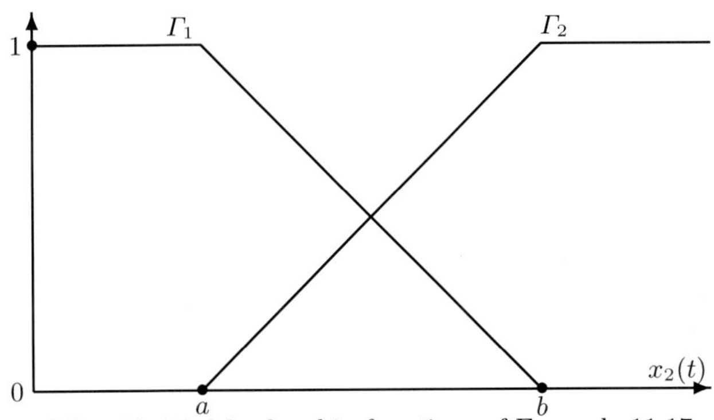

Fig. 11.17. Membership functions of Example 11.17

$$A_1 = \begin{bmatrix} 1 & -0.5 \\ 1 & 0 \end{bmatrix}, \; A_2 = \begin{bmatrix} -1 & -0.5 \\ 1 & 0 \end{bmatrix}, \; \text{and } B = \begin{bmatrix} 1 \\ 0 \end{bmatrix}.$$

Figure 11.17 shows the membership functions of Γ_1 and Γ_2.

Since A_1 and A_2 are stable, the linear subsystems are stable, and the overall system is stable in a neighborhood of the origin by Theorem 11.6. Employing the feedback controller (11.25) and choosing the closed-loop eigenvalues as $[-0.35, -0.5]$, we obtain

$$F_1 = [1.85, -0.325], \; F_2 = [-0.15, -0.325],$$

and

$$A_1 - BF_1 = A_2 - BF_2 = G = \begin{bmatrix} -1.85 & -0.175 \\ 1 & 0 \end{bmatrix}.$$

The closed-loop system becomes

$$\dot{x} = Gx + B\nu$$

.

Define a transformation matrix, $T = MW = \begin{bmatrix} 0 & 1 \\ 1 & 0 \end{bmatrix}$, and a new state vector $\hat{x}$ by $x = T\hat{x}$, where $M = \begin{bmatrix} 1 & -0.85 \\ 0 & 1 \end{bmatrix}$ and $W = \begin{bmatrix} 0.85 & 1 \\ 1 & 0 \end{bmatrix}$. One has

$$\dot{\hat{x}} = \begin{bmatrix} 0 & 1 \\ -0.175 & -0.85 \end{bmatrix} \hat{x} + \begin{bmatrix} 0 \\ 1 \end{bmatrix} \nu,$$

where

$$\nu = \sigma \sin\left(\frac{\pi}{\sigma}\beta\hat{x}_1(t-\tau)\right) = \sigma \sin\left(\frac{\pi}{\sigma}\beta(T^{-1}x)_1(t-\tau)\right) = \sigma \sin\left(\frac{\pi}{\sigma}\beta x_2(t-\tau)\right).$$

When $\sigma = 1$, $\beta = 141$ and $\tau = 1$, and the control starts at $t = 3$, the time response and chaotic attractor of the controlled TS fuzzy system are obtained as shown in Fig. 11.18 and 11.19, respectively.

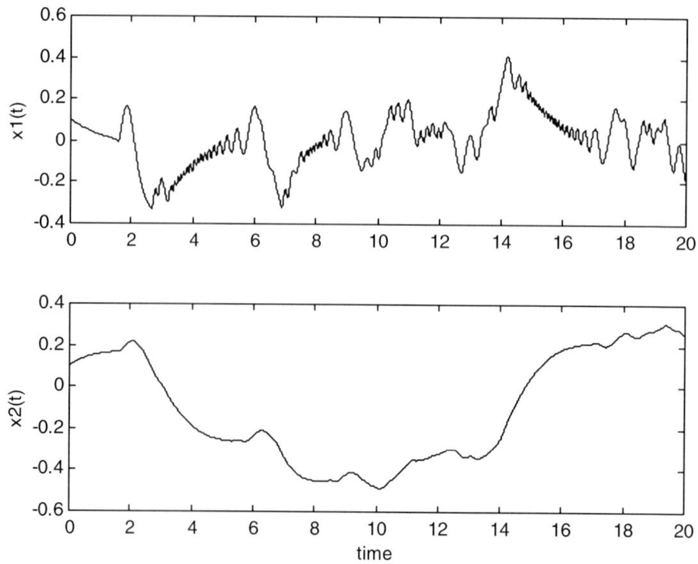

Fig. 11.18. Time response of the controlled TS fuzzy system (the first component is chaotic)

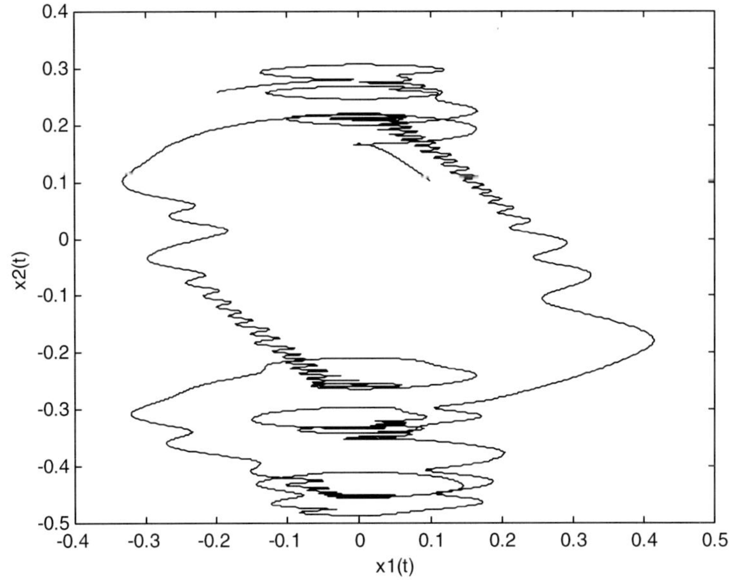

Fig. 11.19. Chaotic attractor of the controlled TS fuzzy system (phase portrait)

Example 11.18. Consider a continuous-time TS fuzzy system, the fuzzy model of Chen's system [236], given by

$$\text{Rule 1: IF } x_1(t) \text{ is } \Gamma_1, \text{ THEN } \frac{d}{dt}\begin{bmatrix} x_1(t) \\ x_2(t) \\ x_3(t) \end{bmatrix} = A_1 \begin{bmatrix} x_1(t) \\ x_2(t) \\ x_3(t) \end{bmatrix} + Bu$$

$$\text{Rule 2: IF } x_1(t) \text{ is } \Gamma_2, \text{ THEN } \frac{d}{dt}\begin{bmatrix} x_1(t) \\ x_2(t) \\ x_3(t) \end{bmatrix} = A_2 \begin{bmatrix} x_1(t) \\ x_2(t) \\ x_3(t) \end{bmatrix} + Bu,$$

where

$$A_1 = \begin{bmatrix} -a & a & 0 \\ c-a & c & -x_{1\,\min} \\ 0 & x_{1\,\min} & -b \end{bmatrix}, \quad A_2 = \begin{bmatrix} -a & a & 0 \\ c-a & c & -x_{1\,\max} \\ 0 & x_{1\,\max} & -b \end{bmatrix},$$

with membership functions

$$\Gamma_1 = \frac{-x_1 + x_{1\,\max}}{x_{1\,\max} - x_{1\,\min}}, \text{ and } \Gamma_2 = \frac{x_1 - x_{1\,\min}}{x_{1\,\max} - x_{1\,\min}},$$

where Γ_i, $i = 1, 2$, are positive-semidefinite for all $x_1 \in [x_{1\,\min}, x_{1\,\max}] = [-30, 30]$, and a, b, and c are parameters.

We fixed $a = 35$, $b = 3$, and let c vary. For $c < 12.8$, the origin of the uncontrolled subsystems is a globally exponentially stable equilibrium point. For $17.5061 \approx c_{H_1} < c < 20$, the uncontrolled Chen's system has two locally exponentially stable equilibrium points given by the following formula: $P_{\pm} = \left(\pm\sqrt{b(a-2c)},\, \pm\sqrt{b(a-2c)},\, a-2c\right)$, and the system is chaotic only if $c > c_{H_2} \approx 37.9868$ or $12.8 < c < c_{H_1} \approx 17.5061$.

A. Approximate Linearization Approach

In the simulation, we took $c = 10$, so the origin of the uncontrolled overall system is locally stable. Its linearization is

$$\dot{x} = \begin{bmatrix} -35 & 35 & 0 \\ -25 & 10 & 0 \\ 0 & 0 & -3 \end{bmatrix} x + \begin{bmatrix} 1 \\ 1 \\ 1 \end{bmatrix} u.$$

The feedback gain $F = [-20.2,\ -1.8,\ 0]$ can be chosen to make the system have the desired eigenvalues -1, -2, and -3. The system becomes

$$\dot{x} = \begin{bmatrix} -14.8 & 36.8 & 0 \\ -4.8 & 11.8 & 0 \\ 20.2 & 1.8 & -3 \end{bmatrix} x + \begin{bmatrix} 1 \\ 1 \\ 1 \end{bmatrix} \nu.$$

Define a transformation matrix T by

$$T = \begin{bmatrix} -54.69 & 22.6 & 1 \\ -18.69 & 7.6 & 1 \\ 411.51 & 19.6 & 1 \end{bmatrix},$$

and define a new state vector $\hat{x}$ by

$$x = T\hat{x}$$

.

The controller is designed as

$$\begin{aligned} u =& Fx + \sigma \sin\left(\frac{\pi}{\sigma}\beta(T^{1}x)_1(t-\tau)\right) \\ =& Fx + \sigma \sin\left(\frac{\pi}{\sigma}\beta(-0.00170x_1(t-\tau) - 0.0004x_2(t-\tau) + 0.0022x_3(t-\tau))\right). \end{aligned}$$

In the simulation, we fixed $\tau = 1$. Figure 11.20 shows the chaotic attractor of the controlled Chen's TS fuzzy system, with $\sigma = 1$ and $\beta = 31$.

B. Globally Exact Linearization Approach

In the simulation, assume that $c = 19$. In this setting, the uncontrolled TS fuzzy model of Chen's system has two locally exponentially stable equilibrium points. Here, we use c as a control parameter, and denote $c = c_0 + \delta c$. The controlled Chen's TS fuzzy system becomes

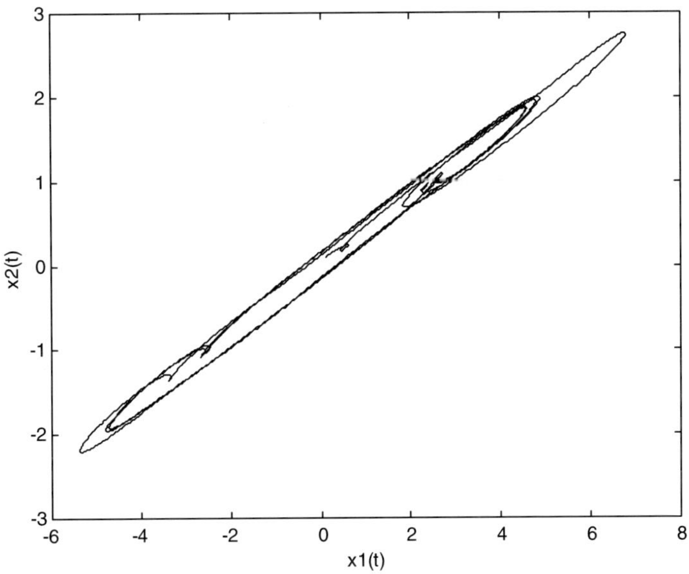

Fig. 11.20. Chaotic attractor of the controlled Chen's TS fuzzy system

$$\dot{x} = \begin{bmatrix} a(x_2 - x_1) \\ (c-a)x_1 - x_1x_3 - cx_2 \\ x_1x_2 - bx_3 \end{bmatrix} + \begin{bmatrix} 0 \\ x_1 + x_2 \\ 0 \end{bmatrix} \delta c$$

with

$$ad_f g = \begin{bmatrix} -a(x_1 + x_2) \\ ax_2 - 2ax_1 - x_1x_3 \\ -x_1(x_1 + x_3) \end{bmatrix}.$$

We still fixed $a = 35$, $b = 3$ and $c = 19$, but let $0 < \delta c < 1.48$. It can be verified that conditions (i) and (ii) in Lemma 11.12 are satisfied for all $\mathbf{x} \neq 0$, and so the relative degree of the system at the nontrivial equilibrium points is three. It follows from (11.35) that

$$\frac{\partial h}{\partial x} g(x) = \frac{\partial h}{\partial x_2}(x_1 + x_2) = 0,$$
$$\frac{\partial h}{\partial x} ad_f g(x) = 0,$$

so that

$$\frac{\partial h}{\partial x_2} = 0, \; a\frac{\partial h}{\partial x_1} + x_1\frac{\partial h}{\partial x_3} = 0.$$

The solution is given by

$$y = h(x) = 0.5x_1^2 - ax_3.$$

Therefore, we may take

$$\delta c = \sigma \sin\left(\frac{\pi}{\sigma}\beta\left(0.5x_1^2(t-\tau) - ax_3(t-\tau)\right)\right).$$

In the simulation, we fixed $\tau = 1$. For $\sigma = 1$ and $\beta = 31$, the controlled system has two separated chaotic attractors; each is near to one of the two originally stable fixed points, as shown by Fig. 11.21 and 11.22, respectively.

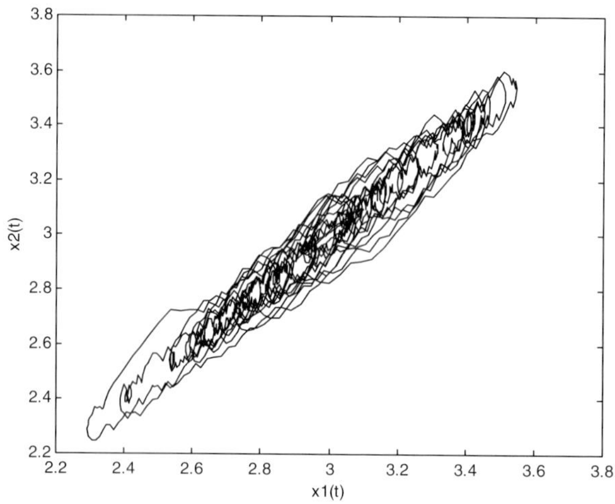

Fig. 11.21. One separated chaotic attractor of the controlled Chen's TS fuzzy system

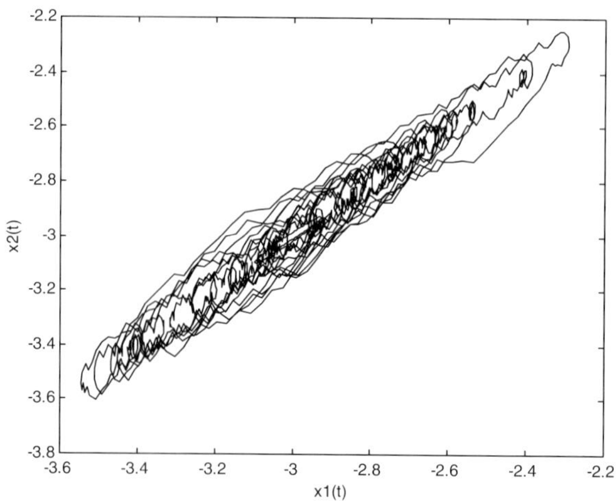

Fig. 11.22. Another separated chaotic attractor of the controlled Chen's TS fuzzy system

For $\sigma = 5$, Fig. 11.23 shows that the two separated attractors merge into one chaotic attractor with β unchanged, and its corresponding three-dimensional phase portrait is shown in Fig. 11.24.

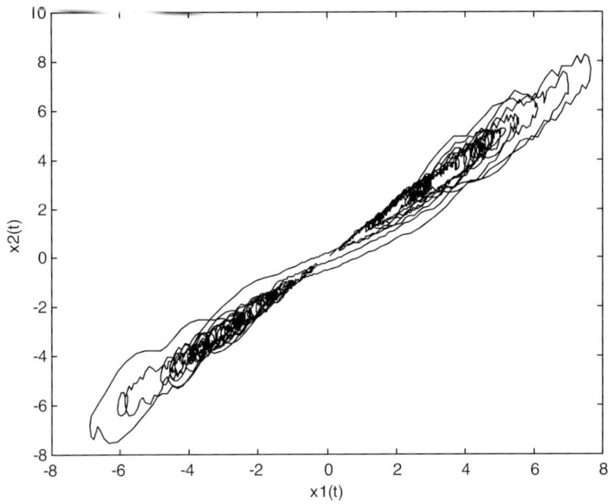

Fig. 11.23. Projection of the chaotic attractor of the controlled Chen's TS fuzzy system on $x_1 - x_2$ plane

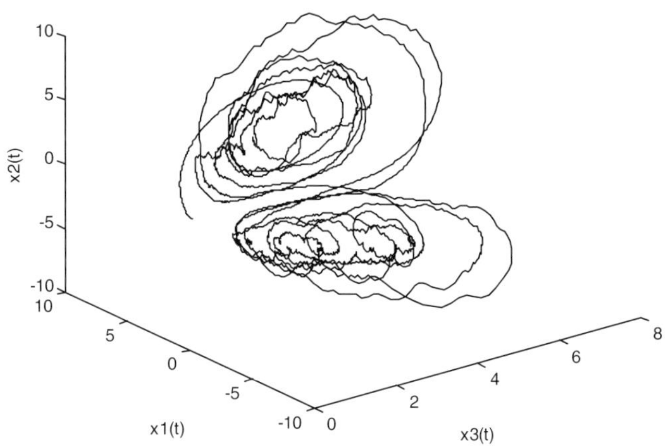

Fig. 11.24. The 3-D chaotic attractor of the controlled Chen's TS fuzzy system

Finally, it is remarked that although the fundamental idea of the anti-control algorithm used in this section to chaotify continuous-time TS fuzzy systems is correct and insightful, the time-delay differential equation is only approximated by a related discrete map. Inspired by this fundamental idea, [228] improved the technical contents by deriving a similar yet rigorous design method for chaotification, which may be modified to apply it in the TS fuzzy systems setting.

12

Intelligent Digital Redesign for TS Fuzzy Systems

This chapter introduces digital control of chaotic systems represented by TS fuzzy systems using intelligent digital redesign (IDR) techniques. The term, intelligent digital redesign, involves converting an existing analog TS fuzzy-model-based controller into an equivalent digital counterpart in the sense of state-matching. The IDR problem is viewed as a minimization problem of norm distances between nonlinearly interpolated linear operators to be matched. The main features of this method are that its constructive condition, with global rather than local state-matching for given chaotic systems, is formulated in terms of linear matrix inequalities (LMI). The stability property is preserved by the proposed IDR algorithm. A set-point regulation example of a chaotic system is demonstrated to visualize the feasibility of the developed methodology, which implies safe digital implementation of chaos control systems.

12.1 Introduction

Dynamical behavior of most physical systems is characterized by a set of differential equations in continuous-time setting. Therefore, it is advantageous to design a controller in the continuous-time framework, called an analog control design [249, 129].

It is now known that controlling chaos in continuous-time setting is usually implemented by analog circuits. However, the use of digital devices in the control of complex dynamical systems is becoming more popular due to modern high-speed computers and microelectronics, which results in hybrid control systems, i.e., continuous-time systems under the control of digital controllers. Therefore, digital implementation of a controller is indeed very desirable especially when the designed controller uses some sophisticated and advanced control algorithms that require a considerable amount of computing efforts. Unfortunately, an analog plant is often controlled by feeding sampled outputs back with analog-to-digital and digital-to-analog interfacing devices, in

Zhong Li: *Fuzzy Chaotic Systems*, StudFuzz **199**, 239–253 (2006)
www.springerlink.com

which continuous-time and discrete-time signals coexist. This renders traditional analysis tools for homogeneous signal systems not directly applicable.

The former main-stream in designing a suitable digital controller was to discretize the analog plant first and, then, determine a digital controller for the discretized plant, which is called the *direct digital design approach* [250]. However, there are some difficulties in determining the continuous-time specifications of a control system, because digital control laws often ignore the system's inter-sampling behavior. To take advantage of digital technology in control engineering as well as to surmount the theoretical obstacles, various digital control techniques have been proposed. Among them, yet another efficient approach is the so-called digital redesign (DR) [251, 252, 253, 254, 255, 256, 250, 257, 258, 259, 260, 261]. It consists in converting a well-designed analog control into an equivalent digital one maintaining the property of the original analog control system in the sense of state-matching, by which the benefits of both analog control and advanced digital technology can be achieved.

The DR techniques were first studied in [233] and, then, developed by many others [251, 252, 253, 254, 255, 256, 250, 257, 258, 259, 260, 261]. It should be noted that although the DR techniques are attractive for digital implementation of advanced analog control, these schemes basically work for linear systems, only, but are not applicable to nonlinear systems, especially, chaotic systems. Therefore, it is necessary to develop some intelligent digital redesign (IDR) methodologies for complex chaotic systems, the first attempt of which was made by Joo et al. [230]. They synergetically merged both Takagi-Sugeno (TS) fuzzy-model-based control and the DR technique for a class of complex chaotic systems. Then, Chang et al. extended it to uncertain chaotic TS fuzzy systems [262].

Although state-of-the-art IDR techniques allow the control engineer to apply the classical DR techniques to nonlinear systems represented by TS fuzzy systems, some problems still remain to be addressed. These IDRs are required to satisfy a local state-matching criterion for each sub-closed-loop system – the control loop closed by the ith plant and the ith control rule, but not a global one. As stability is concerned, it has implicitly been assumed that it is retained by closely mathing the analogously controlled stable state with the digitally controlled state. Involving an explicit stability condition into the IDR algorithm has not been treated, which leaves a theoretical challenge to be tackled.

In this chapter, an intelligent digital redesign approach for TS fuzzy systems is introduced, which guarantees the global state-matching condition and stability of the digital control systems. Some sufficient conditions will be derived in terms of linear matrix inequalities for the global intelligent digital redesign of continuous-time TS fuzzy-model-based controllers. In particular, the global state-matching condition is represented as a convex minimization problem of the norm distance between nonlinearly interpolated linear operators to be matched. Thus, it can be cast into the LMI framework. Two numerical examples are provided to illustrate the effectiveness of the proposed methodology.

12.2 Digital Fuzzy Systems and Their Discretization

This section discusses a desired digital fuzzy control system and its discretization.

Recall Chapter 6, and denote the continuous-time TS fuzzy system as

$$\begin{aligned} R^i &: \text{IF } x_{c1}(t) \text{ is about } \Gamma_1^i \text{ and } \cdots \text{ and } x_{cn}(t) \text{ is about } \Gamma_n^i \\ &\quad \text{THEN } \dot{x}_c(t) = A_i x_c(t) + B_i u_c(t) \end{aligned} \tag{12.1}$$

where $x_c(t) = [x_{c1}, \ldots, x_{cn}]^T$ are the premise variables, R^i denotes the ith fuzzy inference rule, Γ_j^i, $i = 1, 2, \ldots, q$, $j = 1, 2, \ldots, n$, is the fuzzy set of the jth premise variable in the ith fuzzy inference rule.

Using center-average defuzzification, product inference, and singleton fuzzifier, the global dynamics of the TS fuzzy system (12.1) are described as

$$\dot{x}_c(t) = \sum_{i=1}^{q} \mu_i(x_c(t))(A_i x_c(t) + B_i u_c(t)), \tag{12.2}$$

where

$$\omega_i(x_c(t)) = \prod_{j=1}^{n} \Gamma_j^i(x_{cj}(t)), \text{ and } \mu_i(x_c(t)) = \frac{\omega_i(x_c(t))}{\sum_{i=1}^{q} \omega_i(x_c(t))}$$

in which $\Gamma_j^i(x_{cj}(t))$ is the grade of membership of $x_{cj}(t)$ in Γ_j^i. Some basic properties of $\mu_i(t)$ are:

$$\mu_i(x_c(t)) \geq 0, \text{ and } \sum_{i=1}^{q} \mu_i(x_c(t)) = 1, \ i = 1, 2, \ldots, q.$$

It is assumed that a well-designed continuous-time state-feedback TS fuzzy-model-based control law is predesigned, which is to be used in redesigning the digital control law. The controller is of the form

$$\begin{aligned} R^i &: \text{IF } x_{c1}(t) \text{ is about } \Gamma_1^i \text{ and } \cdots \text{ and } x_{cn}(t) \text{ is about } \Gamma_n^i \\ &\quad \text{THEN } u_c(t) = F_c^i x_c(t). \end{aligned} \tag{12.3}$$

The defuzzified output of the controller is given by

$$u_c(t) = \sum_{i=1}^{q} \mu_i(x_c(t)) F_c^i x_c(t). \tag{12.4}$$

Substituting (12.4) into (12.2) gives the global closed-loop continuous-time TS fuzzy system,

$$\dot{x}_c(t) = \sum_{i=1}^{q}\sum_{j=1}^{q} \mu_i(x_c(t))\mu_j(x_c(t))(A_i + B_i F_c^j)x_c(t). \tag{12.5}$$

By sharing the same premise parts as in (12.1), a desired digital fuzzy control system is represented by

$$\dot{x}_d(t) = \sum_{i=1}^{q} \mu_i(x_d(t))(A_i x_d(t) + B_i u_d(t)) \tag{12.6}$$

where $u_d(t) = u_d(kT) \in \Re^m$ is a piecewise constant control to be determined within the time interval $[kT, kT+T)$, $k \in \mathcal{Z}^+$, and $T > 0$ is the sampling period.

The IDR problem is usually set up in discrete-time setting, to find some relevant digital control satisfying the state-matching criterion at each sampling time instant. Thus, it is more convenient to timely discretize (12.6) to derive the IDR condition.

The discretization theorem has been stated as Theorem 6.1, and the discretized form of (12.6) is

$$x_d(kT+T) = \sum_{i=1}^{q} \mu_i(x_d(kT))(G_i x_d(kT) + H_i u_d(kT)) \tag{12.7}$$

where $G_i = e^{A_i T}$ and $H_i = (G_i - I)A_i^{-1}B$.

For (12.6), the intelligently redesigned digital fuzzy control takes the following form:

$$\begin{aligned} &R^i : \text{IF } x_{d1}(kT) \text{ is about } \Gamma_1^i \text{ and } \cdots \text{ and } x_{dn}(kT) \text{ is about } \Gamma_n^i, \\ &\qquad \text{THEN } u_d(t) = F_d^i x_d(kT) \\ &\quad i = 1, 2, \ldots, q, \end{aligned}$$

for $t \in [kT, kT+T)$, and the overall control is given by

$$u_d(t) = \sum_{i=1}^{q} \mu_i(x_d(kT)) F_d^i x_d(kT) \tag{12.8}$$

during the sampling time interval.

Substituting (12.8) into (12.7) yields the discretized closed-loop system

$$x_d(kT+T) = \sum_{i=1}^{q}\sum_{j=1}^{q} \mu_i(x_d(kT))\mu_j(x_d(kT))(G_i + H_i F_d^j)x_d(kT) \tag{12.9}$$

Corollary 12.1. *The dynamical behavior of* (12.5) *can also be approximately discretized as*

$$x_c(kT+T) = \sum_{i=1}^{r}\sum_{j=1}^{r} \mu_i(x_c(kT))\mu_j(x_c(kT))\phi_i x_c(kT) \tag{12.10}$$

where $\phi_i = e^{(A_i+BF_c^i)T}$.

Proof. It can be proven in a straightforward way by the discretization Theorem 6.1.

12.3 Global State-matching Intelligent Digital Redesign

For practical digital implementation of the predesigned continuous-time fuzzy-model-based controller, one may desire to convert it into a digital controller maintaining the control performance in the sense of state-matching.

The goal is to develop an IDR technique for TS fuzzy systems so that the global dynamical behavior of (12.6) with the intelligently redesigned digital fuzzy control can be retained as that of the closed-loop TS fuzzy system with the existing analog fuzzy control, and that stability of the digitally controlled TS fuzzy system is preserved, too. To this end, the IDR problem is formulated as:

Problem 12.2 (Global intelligent digital redesign problem). Given well-designed analog control gain matrices F_c^i for a continuous-time TS fuzzy-model-based controller, *redesign* the gain matrices F_d^i, $i = 1, \ldots, q$, for a digital fuzzy-model-based control law (12.8) such that the following design objectives are satisfied:

1. The state $x_d(kT)$ of the digitally controlled continuous-time TS fuzzy system (12.9) matches the state $x_c(kT)$ of the discrete-time representation (12.10) of the analogously controlled system (12.5) at $t = kT$ as closely as possible.
2. The digitally controlled TS fuzzy system (12.9) is globally asymptotically stable.

To achieve the first objective of Problem 12.2, viz., comparing (12.9) with (12.10) in order to realize $x_c(kT+T) = x_d(kT+T)$ under the assumption of $x_c(kT) = x_d(kT)$, it is necessary to determine F_d^i such that the following matrix equality constraint holds

$$\Phi_i = G_i + H_i F_d^j. \tag{12.11}$$

Then, the state $x_d(t)$ will closely match the state $x_c(t)$ globally at each sampling time instance, $t = kT$, $k = 1, 2, \ldots$, provided that their initial conditions are the same, i.e., $x_c(0) = x_d(0) = x_0$.

It should be remarked that F_d^i in (12.11) can be solved if the dimension of the state vector is not larger than that of the control input vector and H_i, $i = 1, 2, \ldots, q$, is nonsingular, which is, however, unusual even if in linear time-invariant (LTI) systems. In the LTI systems setting, various techniques for finding approximate solutions to (12.11) have been developed to overcome the difficulties [250, 260, 254, 263, 264]. One of these techniques has been used in the TS fuzzy system setting [230, 262], which, however, allows local state-matching, only. It is further remarked that equalities (12.11) are hardly fulfilled in many cases of TS fuzzy-model-based control, since each F_d^i must satisfy q different matrix equality constraints (12.11). In addition, it is hard to carry out stability analysis of sampled-data TS fuzzy systems due to the hybridism of the system states.

In order to overcome these difficulties, Problem 12.2 should be relaxed, and F_d^i can be found using a suboptimal way instead, i.e., a numerical optimization technique. Here, the aim is to find F_d^i such that the norm distance between Φ_{ij} and $G_i + H_i F_d^i$ is minimized. The stability condition for the discretized version of the controlled sampled-data TS fuzzy system is commutative [265, 266]. Problem 12.2 can be relaxed as:

Problem 12.3 (γ-suboptimal global intelligent digital redesign problem). Given a well-designed gain matrix F_c^i for the stabilizing continuous-time TS fuzzy-model-based controller (12.4), find gain matrices F_d^i, $i = 1, 2, \ldots, q$, for the digital fuzzy-model-based control law (12.8) such that the following constraints are satisfied:

1. Minimize γ subject to $\|\Phi_i - G_i - H_i F_d^j\| < \gamma$ in the sense of the induced 2-norm measure, where $i, j = 1, 2, \ldots, q$.
2. The discretized closed-loop system (12.9) is globally asymptotically stable in the sense of the Lyapunov stability criterion.

Note that Problem 12.3 is a constrained convex optimization problem, hence, can be solved numerically by formulation in terms of LMIs, which can be concluded from Theorem 12.4.

Theorem 12.4 (γ-suboptimal global intelligent digital redesign). *If there exist a symmetric positive-definite matrix Q, a symmetric positive-semidefinite matrix O, matrices K_i, and a possibly small positive scalar γ such that the following minimization problem has solutions*

$$\underset{Q,O,K_i}{\text{minimize}} \quad \gamma \quad \textit{subject to}$$

$$\begin{bmatrix} -\gamma Q & \star \\ \Phi_{ij}Q - G_iQ - H_iK_j & -\gamma I \end{bmatrix} < 0, \quad i,j = 1,2,\ldots,q. \tag{12.12}$$

$$\begin{bmatrix} -Q + (q-1)O & \star \\ G_iQ + H_iK_i & -Q \end{bmatrix} < 0, \quad i = 1,2,\ldots,q. \tag{12.13}$$

$$\begin{bmatrix} -Q - O & \star \\ \left(\frac{G_iQ+H_iK_j+G_jQ+H_jK_i}{2}\right) & -Q \end{bmatrix} < 0, \tag{12.14}$$

$$i = 1,2,\ldots,j, \quad j = i+1,\ldots,q,$$

then the state $x_d(t)$ of the discrete-time TS fuzzy system (12.6) *controlled by the intelligently redesigned digital fuzzy-model-based controller* (12.8) *closely matches the state $x_c(t)$ of the continuous-time stable TS fuzzy system* (12.5)*. Furthermore, the discretized sampled-data TS fuzzy system* (12.9) *is globally asymptotically stabilizable in the sense of the Lyapunov stability criterion, where $\star$ denotes the transposed element in symmetric positions.*

Proof. Introducing a free matrix variable X, one has

$$\begin{aligned} \|\Phi_i - G_i - H_iF_d^j\|_2 &< \widehat{\gamma} \\ &= \widehat{\gamma}\frac{1}{\|X^TX\|_2}\cdot\|X^TX\|_2 \\ &= \gamma\|X^TX\|_2 \end{aligned} \tag{12.15}$$

where $\gamma = \widehat{\gamma}/\|X^TX\|_2$ is a positive scalar and X has full column rank. Accordingly, X^TX is symmetric and positive-definite. Without loss of generality, one can choose P as X^TX, which is reasonable, since P is definitely bounded from (12.12), and symmetric and positive-definte.

From the definition of the induced 2-norm, $\|\cdot\|_2 = \sqrt{\lambda_{\max}((\cdot)^T(\cdot))}$, the following inequality holds

$$(\Phi_i - G_i - H_iF_d^j)^T(\phi_i - G_i - H_iF_d^j) < \gamma^2 P. \tag{12.16}$$

Using the Schur complement, (12.16) can be represented with LMIs of the form

$$\begin{bmatrix} -\gamma P & \star \\ \Phi_i - G_i - H_iF_d^j & -\gamma I \end{bmatrix} < 0. \tag{12.17}$$

Further, applying the Congruence transformation to (12.17) with $\text{diag}[P^{-1}, I]$, and denoting $F_i = F_d^iP^{-1}$ yields (12.12).

Next, the LMIs (12.13) and (12.14) directly follow from the standard Lyapunov stability criterion for the discrete-time TS fuzzy system and, in turn, applying Schur complement and Congruence transformation. The details are shown in the Appendix of this chapter. This completes the proof of the theorem.

It is remarked that the LMI (12.14) in Theorem 12.4, for the pair (i, j) such that the union of the fuzzy sets Γ_l^i and Γ_l^j is a null set, for any l, $l = 1, 2, \ldots, n$, on the compact set U, and for all $t \geq 0$, does not have to be solved in redesigning the digital gain matrices. Further, it should be noted that different from the local intelligent digital redesign [230], the newly proposed intelligent digital redesign approach tries to match the states of the global dynamical systems, not local ones, and incorporates the stability conditions of the discretized closed-loop system with the digitally redesign fuzzy-model-based controller.

12.4 Digital Redesign for Duffing-like Chaotic Oscillator

Recall Chapter 6, the Duffing-like chaotic oscillator [159]:

$$\ddot{y}(t) - ay(t) + by(t)|y(t)| = -\epsilon(\zeta\dot{y}(t) - c\sin(\omega t)), \tag{12.18}$$

can be represented as a TS fuzzy model of the form

$$\begin{aligned} R^1 &: \text{IF } x_{c1}(t) \text{ is about } \Gamma_1^1, \text{ THEN } \dot{x}_c(t) = A_1 x_c(t) \\ R^2 &: \text{IF } x_{c1}(t) \text{ is about } \Gamma_1^2, \text{ THEN } \dot{x}_c(t) = A_2 x_c(t) \end{aligned}$$

where

$$A_1 = \begin{bmatrix} 0 & 1 & 0 & 0 \\ a & -\epsilon\zeta & \epsilon c & 0 \\ 0 & 0 & 0 & \omega \\ 0 & 0 & -\omega & 0 \end{bmatrix}, \quad A_2 = \begin{bmatrix} 0 & 1 & 0 & 0 \\ a - bM & -\epsilon\zeta & \epsilon c & 0 \\ 0 & 0 & 0 & \omega \\ 0 & 0 & -\omega & 0 \end{bmatrix},$$

where $x_c(t) = [y(t), \dot{y}(t), \sin(\omega t), \cos(\omega t)]^T$. For the design of a suitable fuzzy-model-based controller, the input matrices are assumed to be

$$B_1 = B_2 = \begin{bmatrix} 0 \\ 1 \\ 1 \\ 0 \end{bmatrix},$$

which preserve the controllability of the system.

In terms of Theorem 3 in [153], the well-designed gain matrices for the continuous-time fuzzy-model-based controller are derived as

$$\begin{aligned} F_c^1 =& [-9.1263, -2.6634, -1.3199, 0.0400], \\ F_c^2 =& [-8.3869, -3.5096, -0.7655, -0.3501]. \end{aligned}$$

Applying Theorem 12.4 yields the digitally redesigned fuzzy-model-based control gain matrices for the sampling period $T = 0.3$ sec as

$$F_d^1 = [-4.4147, -2 - 0323, -1.2822, -0.3181],$$
$$F_d^2 = [-3.9342, -2.2794, -1.0726, -0.4246].$$

The initial values $x_c(0) = x_d(0) = x_0 = [1, 0, 0, 1]^T$ are taken and the simulation time is 6 sec. The following performance measure [254] is employed to qualitatively compare the performance of this intelligent digital redesign technique with other methods, for instance, the one proposed in [230],

$$\mathcal{P} = \sum_{i=1}^{4} \left(\int_0^6 |x_{ci}(t) - x_{di}(t)| dt \right). \tag{12.19}$$

Figure 12.1 shows the time response diagram, where the input is activated at $t = 1.2$ sec. It is shown that after control is activated, all trajectories converge to the origin. However, Fig. 12.2 illustrates the excellence of the proposed method. It is emphasized that the trajectory of the system controlled by the proposed method closely matches that of the original one, while the local intelligent digital redesign method yields poor state-matching. The performance measures of two methods regarding state-matching clearly validate the the superiority of the proposed method.

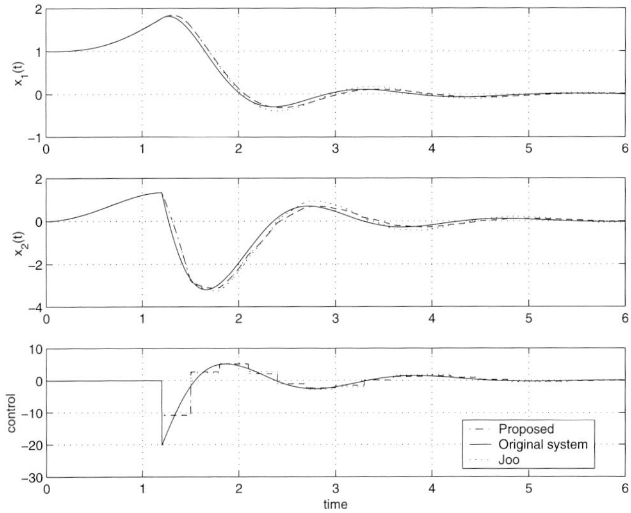

Fig. 12.1. Time responses of the controlled Duffing-like chaotic oscillator (control is fired at $t = 1.2$ sec and $T = 0.3$ sec); solid line: continuous-time control, dotted line: digital control proposed in [230], dash-dotted line: proposed digital control

To illustrate the superiority of the proposed method to the other in terms of stabilizability and control performance, a longer sampling period, $T = 0.6$ sec, is taken. According to Theorem 12.4, the gain matrices for $T = 0.6$ sec are derived as

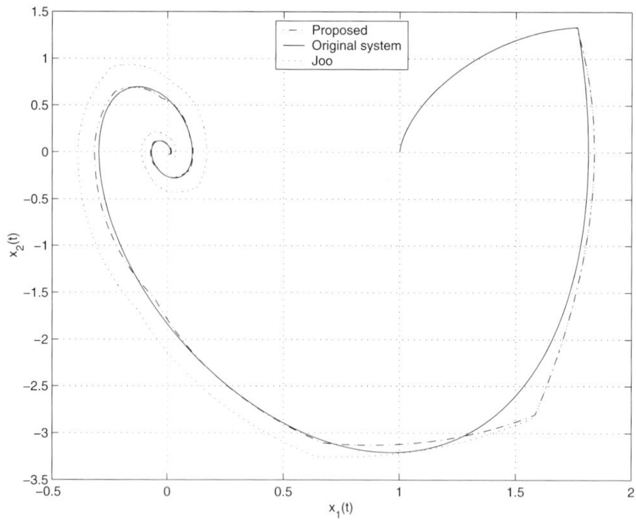

Fig. 12.2. Trajectories of the controlled Duffing-like chaotic oscillator (control is fired at $t = 1.2$ sec and $T = 0.3$ sec); solid line: continuous-time control, dotted line: digital control proposed in [230], dash-dotted line: proposed digital control

$$
\begin{aligned}
F_d^1 =& [-2.0022, -1.4249, -1.1689, -0.5963], \\
F_d^2 =& [-1.3277, -1.2906, -1.1066, -0.5905].
\end{aligned}
$$

Figures 12.5 and 12.6 show the time response diagram and the trajectories of the two digitally controlled systems. It is shown that the proposed method not only drives the trajectories to the origin, but also matches more closely the trajectory of the original system. As compared, the other method gives a deteriorated state-matching performance.

Other simulations for various sampling periods are depicted in Figs. 12.3, 12.4, 12.7 and 12.8, and the performance measures are shown in Table 12.1. It is further emphasized that the proposed method guarantees the stability of the controlled system, while the other one may fail to stabilize the system, especially for relatively long sampling periods, which is another advantage of the proposed method. This is because the proposed method incorporates the stability criterion in the redesign condition, whereas the other approach does not.

12.5 Appendix

Proof. The closed-loop system (12.9) can be rewritten as

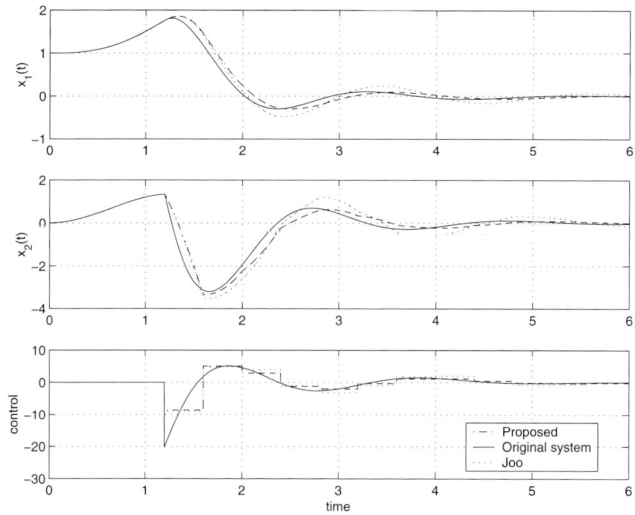

Fig. 12.3. Time responses of the controlled Duffing-like chaotic oscillator (control is fired at $t = 1.2$ sec and $T = 0.4$ sec); solid line: continuous-time control, dotted line: digital control proposed in [230], dash-dotted line: proposed digital control

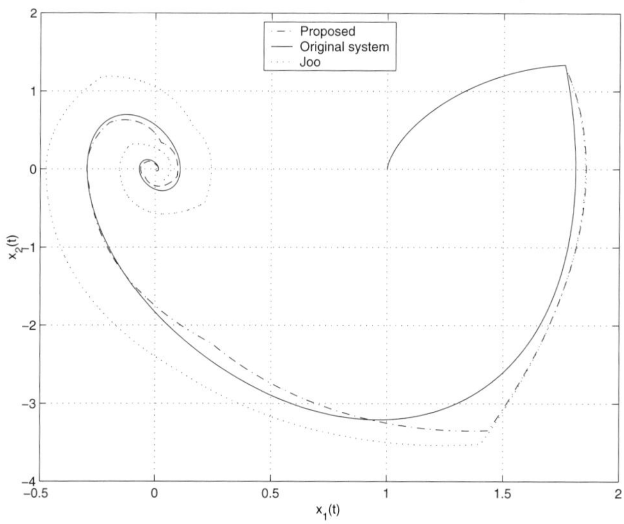

Fig. 12.4. Trajectories of the controlled Duffing-like chaotic oscillator (control is fired at $t = 1.2$ sec and $T = 0.4$ sec); solid line: continuous-time control, dotted line: digital control proposed in [230], dash-dotted line: proposed digital control

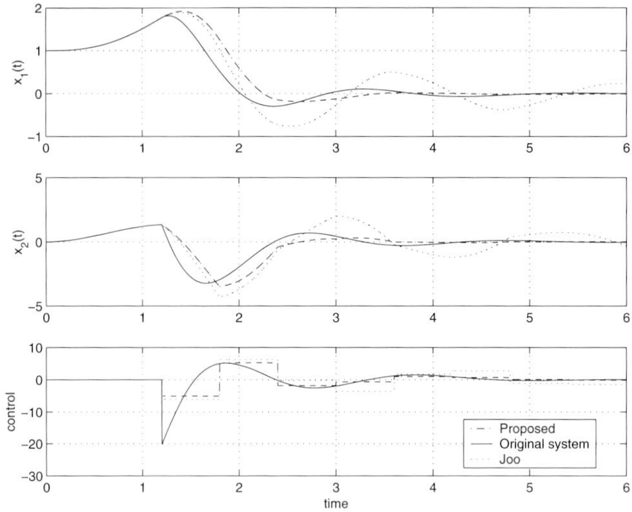

Fig. 12.5. Time responses of the controlled Duffing-like chaotic oscillator (control is fired at $t = 1.2$ sec and $T = 0.6$ sec); solid line: continuous-time control, dotted line: digital control proposed in [230], dash-dotted line: proposed digital control

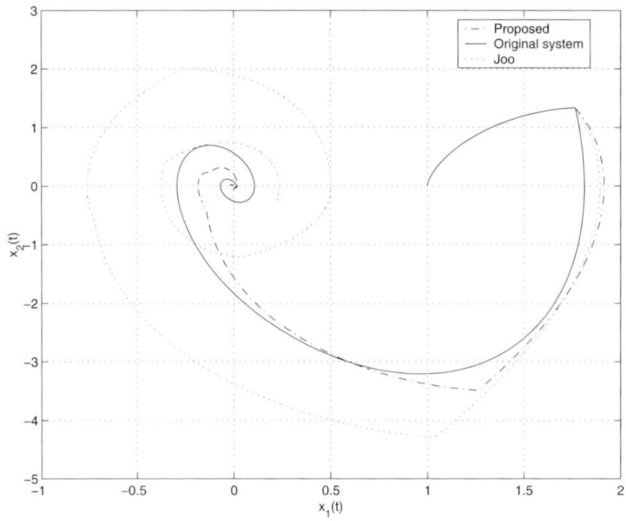

Fig. 12.6. Trajectories of the controlled Duffing-like chaotic oscillator (control is fired at $t = 1.2$ sec and $T = 0.6$ sec); solid line: continuous-time control, dotted line: digital control proposed in [230], dash-dotted line: proposed digital control

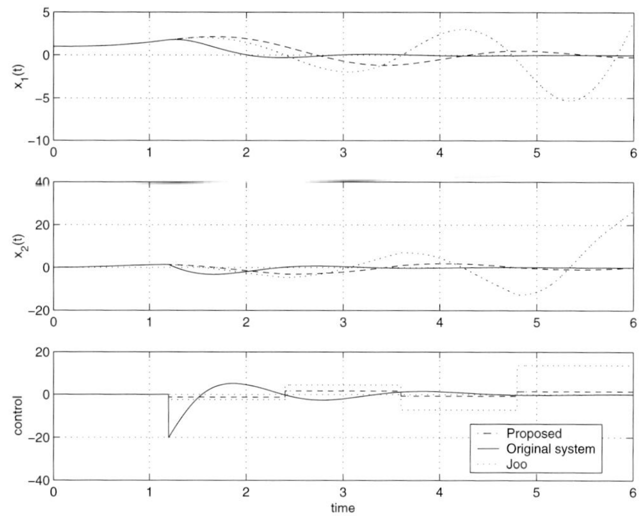

Fig. 12.7. Time responses of the controlled Duffing-like chaotic oscillator (control is fired at $t = 1.2$ sec and $T = 1.2$ sec); solid line: continuous-time control, dotted line: digital control proposed in [230], dash-dotted line: proposed digital control

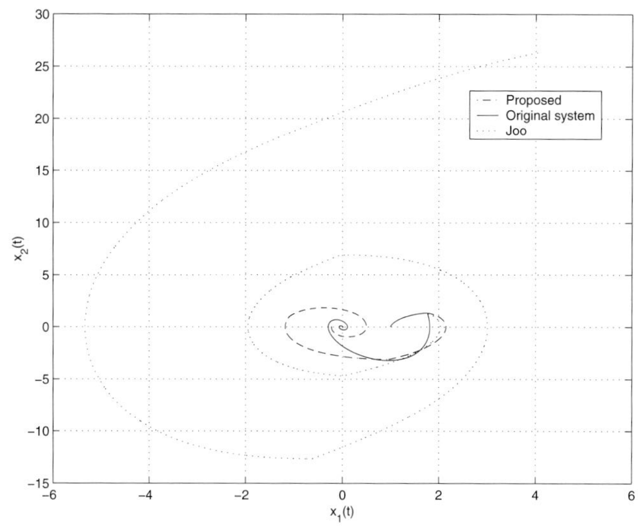

Fig. 12.8. Trajectories of the controlled Duffing-like chaotic oscillator (control is fired at $t = 1.2$ sec and $T = 1.2$ sec; solid line: continuous-time control, dotted line: digital control proposed in [230], dash-dotted line: proposed digital control

Table 12.1. Comparison of the proposed method's performance $\mathcal{P}$ with the other method one's for various sampling periods T

Method	Sampling period T (s)			
	0.3	0.4	0.6	1.2
Joo [230]	2.4273	4.6280	12.3751	unstable
Proposed	2.0798	3.3829	6.3144	24.1874

$$
\begin{aligned}
x_d(kT+T) &= \sum_{i=1}^{q} \theta_i^2(z(kT)) \left(G_i + H_i K_d^i\right) x_d(kT) \\
&\quad + 2\sum_{i<j}^{q} \theta_i(z(kT))\theta_j(z(kT)) \left(\frac{G_i + H_i K_d^j + G_j + H_j K_d^i}{2}\right) x_d(kT)
\end{aligned} \tag{12.20}
$$

Now consider the Lyapunov functional candidate

$$
V(x_d(kT)) = x_d(kT)^T P x_d(kT) \tag{12.21}
$$

where P is a time-invariant, symmetric and positive-definite matrix. Clearly, $V(0) = 0, V(x_d(kT)) > 0$, for all $k > 0$, and radially unbounded in any neighborhood of $x_d(kT) = 0$. By applying Corollary 12.1 in [153], the rate of increase of $V(x_d(kT))$ is

$$
\begin{aligned}
&\Delta V(x_d(kT)) \\
&\quad = V(x_d(kT+T)) - V(x_d(kT)) \\
&\quad = x_d(kT+T)^T P x_d(kT+T) - x_d(kT)^T P x_d(kT) \\
&\quad \le \frac{1}{4}\sum_{i=1}^{q}\sum_{j=1}^{q} \theta_i(x_d(kT))\theta_j(x_d(kT)) x_d(kT)^T \\
&\quad\quad \times ((G_i + H_i K_d^j + G_j + H_j K_d^i)^T P \\
&\quad\quad \times (G_i + H_i K_d^j + G_j + H_j K_d^i) - 4P) x_d(kT) \\
&\quad \le \sum_{i=1}^{q} \theta_i^2(x_d(kT)) x_d(kT)^T ((G_i + H_i K_d^i)^T P \\
&\quad\quad \times (G_i + H_i K_d^i) - P + (q-1)W) x_d(kT) \\
&\quad\quad + 2\sum_{i<j}^{q} \theta_i(x_d(kT))\theta_j(x_d(kT)) x_d(kT)^T \left(\left(\frac{G_i + H_i K_d^j + G_j + H_j K_d^i}{2}\right)^T \right. \\
&\quad\quad \left. \times P \left(\frac{G_i + H_i K_d^j + G_j + H_j K_d^i}{2}\right) - P - W\right) x_d(kT)
\end{aligned} \tag{12.22}
$$

where W is a positive-semidefinite matrix. Therefore, if the following two inequalities are negative-definite, then the controlled system is globally asymptotically stable.

$$(G_i + H_i K_d^i)^T P (G_i + H_i K_d^i) - P + (q-1)W < 0, \qquad i = 1, 2, \ldots, q. \tag{12.23}$$

$$\left(\frac{G_i + H_i K_d^j + G_j + H_j K_d^i}{2} \right)^T P \left(\frac{G_i + H_i K_d^j + G_j + H_j K_d^i}{2} \right) - P - W < 0,$$
$$i = 1, \ldots, j., j = i+1, \ldots, q. \tag{12.24}$$

Applying Schur complement and Congruence transformation with diag $\left[P^{-1}, I\right]$ to (12.23) and (12.24), and letting $Q = P^{-1}, O = P^{-1}WP^{-1}$ and $F_i = K_d^i P^{-1}$ yield the LMIs (12.13) and (12.14).

13

Spatiotemporal Chaos and Synchronization in Complex Fuzzy Systems

In this chapter, spatiotemporal chaos in arrays of coupled fuzzy-logic-based chaotic oscillators is observed, and the effects of network topology on synchronization through defining a synchronization index are investigated in the framework of complex networks.

13.1 Introduction

It is already known that fuzzy systems can exhibit chaotic phenomena. This chapter aims to show some other interesting but more complicated chaotic phenomena in complex fuzzy systems built by coupled fuzzy logic chaotic units [267, 268, 269, 129].

The artificial reproduction of some exotic phenomena such as spatiotemporal chaos and synchronization provides a means to study and understand collective dynamics that are commonly observed in nature [271]. In particular, the synchronization of oscillators in a complex system is a classical topic related to different research fields: from circuit theory to biological and social systems, which has attracted ever increasing attention in the fascinating field of complex systems and networks.

The behavior of arrays of large numbers of oscillators by considering both the effects due to the dynamics of the single units and those related to network topology so that the global system achieves synchronization has been studied in [272, 273].

This chapter aims to illustrate a class of complex systems, arrays of coupled fuzzy chaotic oscillators, and then to characterize the synchronization of these systems, both in a qualitative and a quantitative way.

This study of complex systems has been focused on the relationships among the features of the fundamental cells, the architecture of the network under consideration and the spatiotemporal dynamics achieved [272]. Therefore, their characterization can be investigated from two different points of

Zhong Li: *Fuzzy Chaotic Systems*, StudFuzz **199**, 255–273 (2006)
www.springerlink.com

view, viz., as single units and as macro-system. At the single-unit level, attention is focused on the effects of nonlinear dynamics of each unit due to its own parameters, without taking into account the neighborhood interferences on the collective behavior. While, at the macro-system level, the aim is to correlate the global dynamics in terms of the complex system's topology through the number of connections and their spatial distribution.

A generalization of the synchronization principles for an array of fuzzy-logic-based chaotic dynamical systems is suggested and evaluated as alternative approach to build locally connected fuzzy complex systems by varying both the rules driving the cells and the network architecture. A discrete fuzzy oscillator showing a chaotic behavior has been chosen as fundamental unit. This fuzzy system is obtained through a linguistic description of the stretching and folding features for an assigned value of the Lyapunov exponent [37]. At macro-level, initially, the considered fuzzy cells have been connected with a regular topology. Different structures have been implemented using different parameter values as the Lyapunov exponent and the number of connections, and a qualitative comparison of their dynamic features have been performed. The definition of a synchronization index gives the opportunity to quantify the synchronization properties of the considered collective behavior. The introduction of this index speeds up the process of pattern comparison. Therefore, it is possible to consider a wide number of topologies and architectures for a more exhaustive and detailed study. Further, the investigation at the macro-level has been carried out by comparing the global dynamics obtained through different topologies: regular, small-world and random [274]. This investigation opens a new way to see the complex real-world phenomena that are often not easy to describe using mathematical models.

13.2 Complex Fuzzy Systems

To characterize the spatiotemporal dynamics of a nonlinear fuzzy one-dimensional array, it is necessary to determine the dynamics of its elementary cell and network structure. A discrete chaotic fuzzy oscillator is employed to serve as the fundamental element of this network, and its chaotic behavior is characterized by varying the Lyapunov exponent in a suitable range. The macro-system is built up by connecting a large number of identical cells in a regular configuration. The study of the global dynamics of this fuzzy system starts from the characterization of the spatiotemporal patterns through visual inspection and frequency analysis.

13.2.1 Single-unit: Chaotic Fuzzy Oscillator

Recall Chapter 5 to summarize the fuzzy modeling approach in the following. A chaotic fuzzy oscillator can be characterized using two variables x and d, which represent the nominal value of the state and the uncertainty on the center value, respectively, generate the desired evolution and perform the stretching and folding features. The discrete dynamics of this fuzzy oscillator can be considered as a two dimensional chaotic map with the following structure [269, 129]:

$$\begin{cases} x(k+1) = \Psi(x(k), d(k)), \\ d(k+1) = \Phi(x(k), d(k)), \end{cases} \tag{13.1}$$

where Ψ and Φ are the fuzzy inference functions described through a set of fuzzy rules for each variable as shown in Table 13.1.

Table 13.1. Rules of the fuzzy inference system

$x(k)/d(k)$	$zero\,(Z)$	$small\,(S)$	$large\,(L)$	$very\ large\ (VL)$
$large.left\,(LL)$	SL/Z	SL/M	SL/VL	SR/L
$small.left\,(SL)$	LL/Z	LL/M	LL/VL	LL/S
$small.right\,(SR)$	LR/Z	LR/M	LR/VL	LR/S
$left.right\,(LR)$	SR/Z	SR/M	SR/VL	SL/L

The range of the variables is chosen according to a suitable embedding zone performing the main oscillatory dynamics leading to an uncertain evolution. The designed fuzzy sets are shown in Fig. 13.1. The relation between specific chaotic dynamics and the fuzzy sets is formalized by (13.2), where the center of each membership function is evaluated starting from a value of the Lyapunov exponent l.

$$\begin{aligned} \frac{C_{medium}}{C_{small}} &= e^{l}, \\ \frac{C_{verylarge}}{C_{large}} &= e^{l}. \end{aligned} \tag{13.2}$$

The chaotic time series generated by such a fuzzy oscillator with $l = 0.1$ is shown in Fig. 13.2.

13.2.2 Macro-system: Chain of Fuzzy Oscillators

The macro-system is constructed by an array of one-dimensional fuzzy oscillators with a regular distribution of the connections (C). This means that each fundamental fuzzy unit has $2 \cdot C$ neighbors, half on its right (C) and half on its left (C). The equations of the array's single element are rewritten according to the number of oscillators (N), and the diffusion coefficient (D) that weights the information exchange, as described in Eq. (13.3).

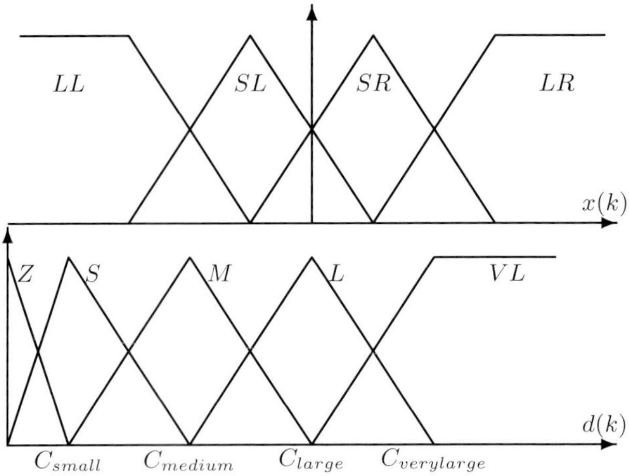

Fig. 13.1. Fuzzy sets for $x(k)$ (upper) and $d(k)$ (lower)

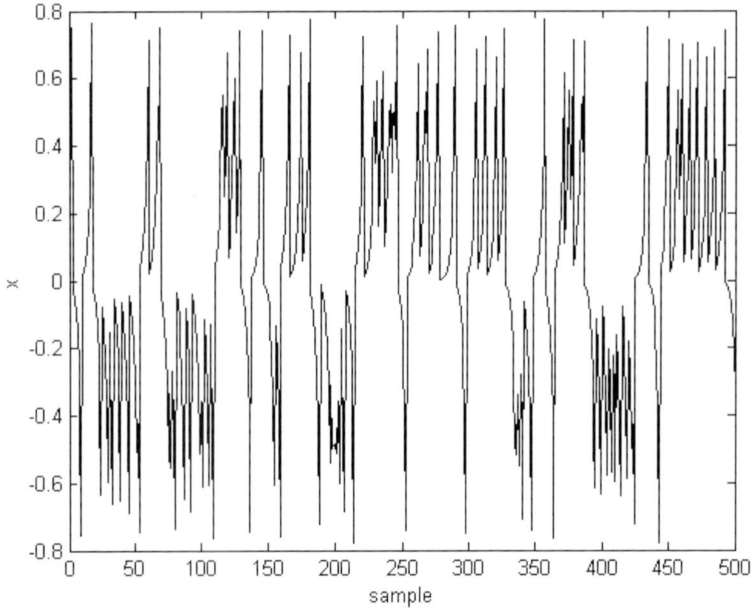

Fig. 13.2. Time series of the fuzzy chaotic oscillator with $l = 0.1$

$$
\begin{cases}
x_i(k+1) = \Psi\left(x_i(k) + D\left(-2Cx_i(k) + \sum_{j=-C, j\neq 0}^{C} x_{i+j}(k)\right), d_i(k)\right) \\
d_i(k+1) = \Phi(x_i(k), d_i(k)), \qquad i = 1, \ldots, N.
\end{cases}
\tag{13.3}
$$

The state value of each fuzzy oscillator depends on its own state value at the previous sample time instant, and on the contributions coming from the state values of the other fuzzy oscillators connected through a bi-directional information exchange. In this way, the one-dimensional array can be designed by coupling $N = 200$ discrete fuzzy chaotic oscillators with $l = 0.1$ and a constant diffusion coefficient $D = 0.005$, and the single unit starts its evolution from random initial conditions.

By varying the number of connections for each unit $(2 \cdot C)$ in the fuzzy chain, four different global dynamics can be observed. The pattern formations related to these four different collective behaviors are displayed in Fig. 13.3. Each spatiotemporal map is built for a particular value of C, where the index i associated to each cell is plotted versus sampling time, meanwhile the state value x_i of each cell is represented through a 64-color scale, with blue for low value and red for high values.

It is illustrated in Fig. 13.3 that four behaviors can be observed by increasing the numbers of connections (C), i.e., from the initial spatiotemporal chaos at $C = 4$ (Fig. 13.3a), then a regular synchronized behavior at $C = 16$ (Fig. 13.3b), through a transition phase at $C = 25$ (Fig. 13.3c), to finally the chaotic synchronized behavior at $C = 46$ (Fig. 13.3d).

To understand the different spatiotemporal dynamics, the time series of the nominal value x_i of two cells are plotted in Fig. 13.4 and their Fourier spectra in Fig. 13.5, where the solid line is for x_{100} and the dotted line is for x_{101}.

It is shown that in the spatiotemporal chaos at $C = 4$ each chaotic oscillator evolves with its own dynamics due to the different initial conditions (Fig. 13.4a), and that the absence of a regular oscillatory behavior is identifiable in the spread Fourier spectra (Fig. 13.5a).

It is also shown that in the regular synchronized behavior at $C = 16$ the regular oscillations of the two cells are visible (Fig. 13.4b), and that their values of picks in a particular time frequency can be distinctly observed (Fig. 13.5b).

It is further shown that in the transition phase at $C = 25$ the time evolutions of the two cells become almost the same in long time intervals (Fig. 13.4c), and that the spread Fourier spectra indicate the disappearance of the regular oscillatory behavior identifiable in the narrow pick (Fig. 13.5c).

Finally, it is shown that in the chaotic synchronized behavior at $C = 46$ a perfect synchronization of the two cells is observed (Fig. 13.4d), and that the chaotic feature of the signals is exhibited with broad band spectra, which are perfectly superimposed (Fig. 13.5d).

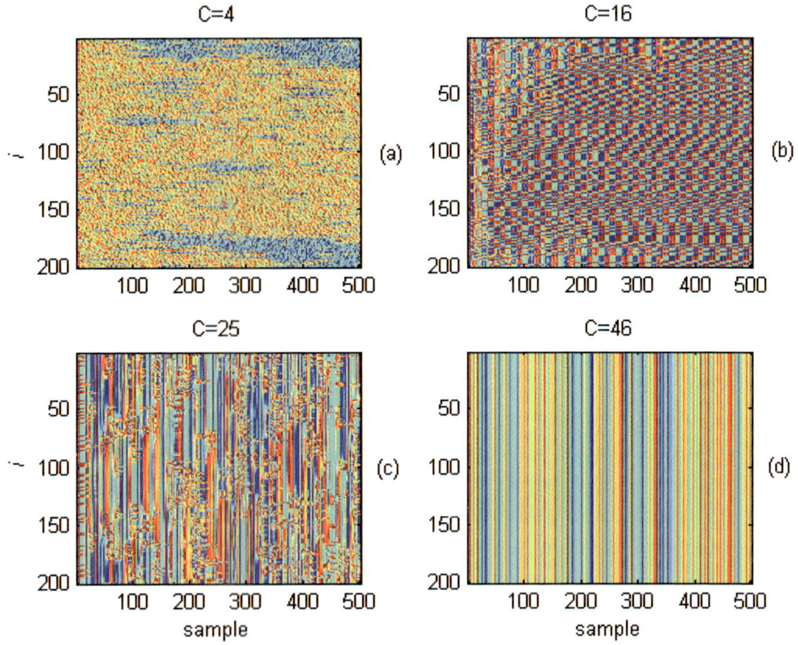

Fig. 13.3. Spatiotemporal patterns of the fuzzy chains ($l = 0.1$): (a) $C = 4$, (b) $C = 16$, (c) $C = 25$, (d) C = 46

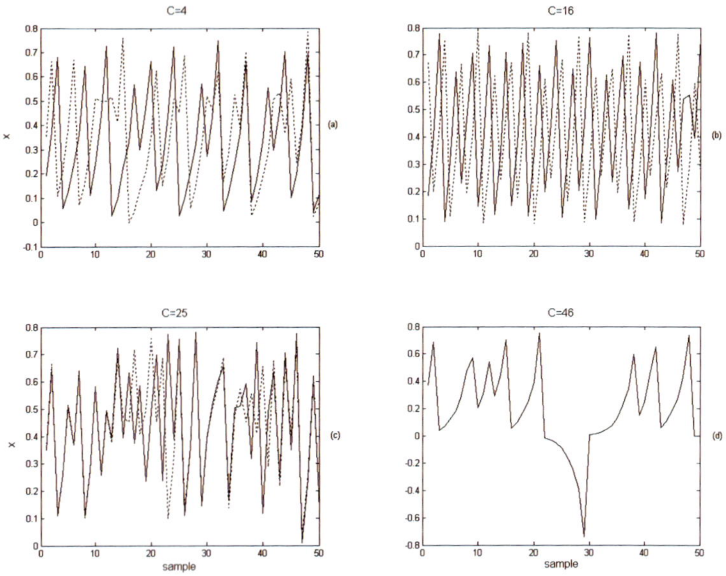

Fig. 13.4. Time series of two cells (x_{100} and x_{101}) in the fuzzy chains ($l = 0.1$): (a) $C = 4$, (b) $C = 16$, (c) $C = 25$, (d) $C = 46$

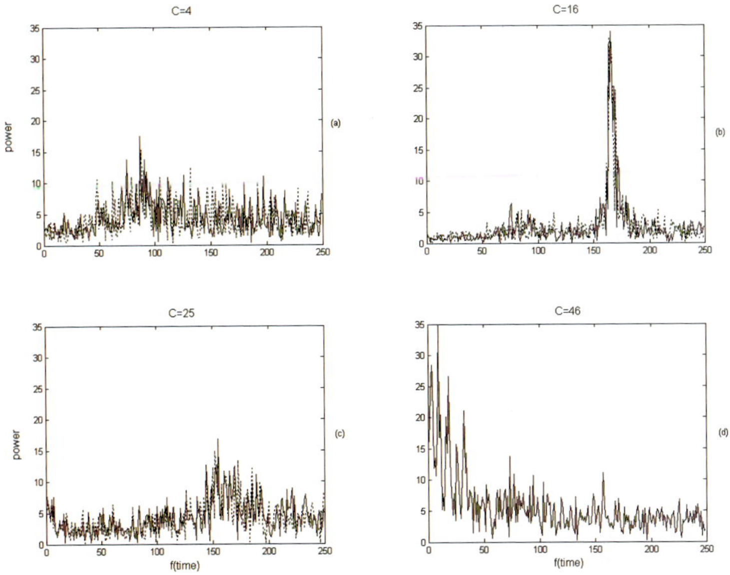

Fig. 13.5. Fourier spectrum of two cells (x_{100} and x_{101}) in the fuzzy chains ($l = 0.1$): (a) $C = 4$, (b) $C = 16$, (c) $C = 25$, (d) $C = 46$

When varying the chaotic dynamics of the fuzzy oscillator setting as $l = 0.9$, the same types of collective behavior as the one previously characterized for $l = 0.1$ can be obtained by maintaining the same chain structure as shown in Fig. 13.6.

Moreover, the time series of the two cells (x_{100}, x_{101}) and their Fourier spectra are shown in Figs. 13.7 and 13.8. The spatiotemporal chaos is recognizable in the independent chaotic evolution of the cells and in their spread frequency power distribution as shown in Fig. 13.7a–13.8a. The regular synchronized behavior is characterized by a sharp sequence of picks at specific frequency values (Figs. 13.7b–13.8b). The characteristics of the chaotic synchronized behavior are shown in Figs. 13.7d–13.8d. These results can qualitatively be compared with the ones obtained for $l = 0.1$ (Figs. 13.3, 13.4 and 13.5). As can be seen, the synchronization reached in the second experiment ($l = 0.9$, $C = 46$) is not as well defined as the one of the first experiment ($l = 0.1$). Increasing the number of connections, one will possibly obtain similar results.

13.3 Synchronization Index

To characterize various spatiotemporal patterns, a mathematical indicator, i.e., a synchronization index, is introduced in this section. Using this indica-

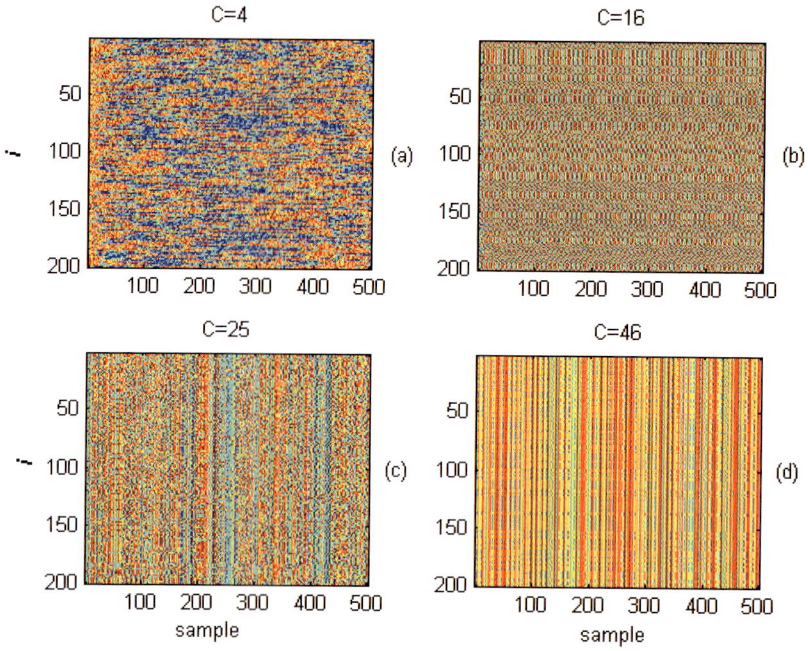

Fig. 13.6. Spatiotemporal patterns in the fuzzy chains ($l = 0.9$): (a) spatiotemporal chaos, (b) regular synchronization, (c) transition phase, (d) chaotic synchronization

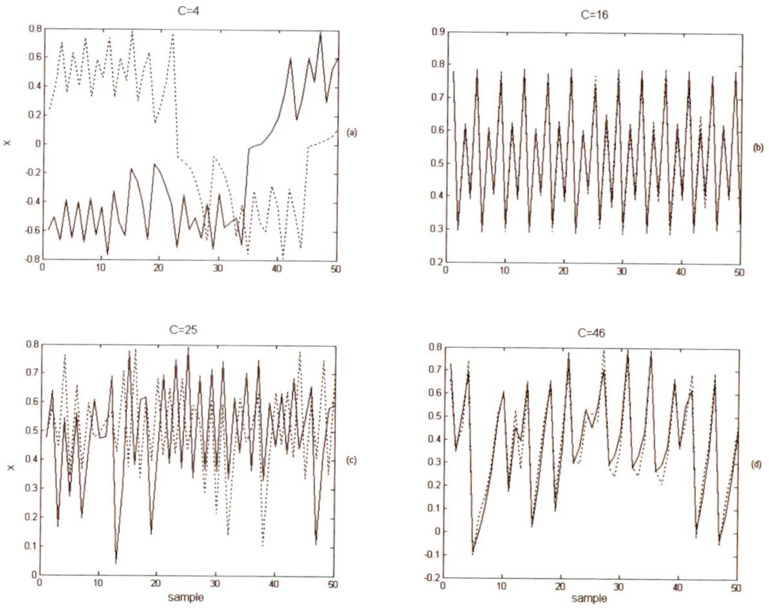

Fig. 13.7. Time series of two cells (x_{100}, x_{101}) in the fuzzy chains ($l = 0.9$)

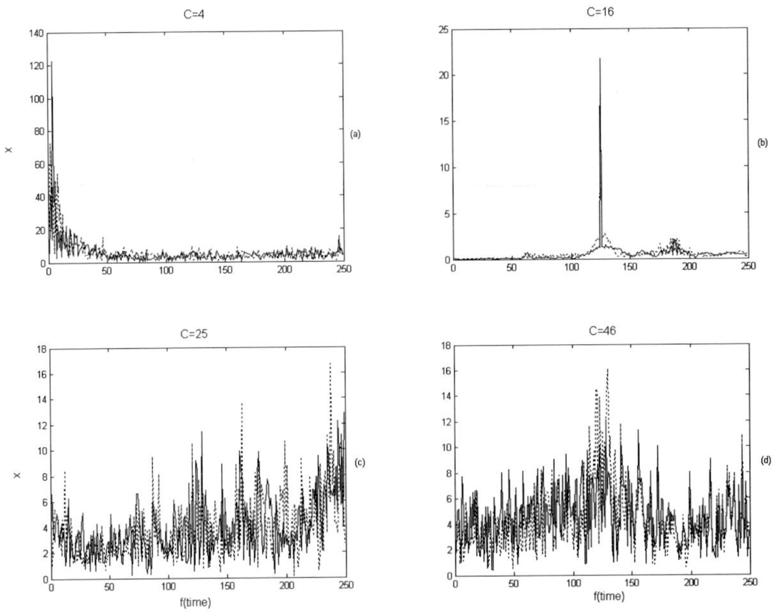

Fig. 13.8. Fourier spectrum of two cells (x_{100}, x_{101}) in the fuzzy chains ($l = 0.9$)

tor, the analysis of spatiotemporal chaos analysis as carried out in the previous section can be extended and detailed by tuning the Lyapunov exponent and the number of connections in suitable ranges. Furthermore, this evaluation index eases comparison and identification of complex network features versus adopted system parameters and, thus, avoids tedious computations in inspecting all spatiotemporal maps versus parameter variations.

The synchronization index is to evaluate the synchronization degree of a system consisting of coupled units [129, 270]. Involving all N entries in the rows of the matrix A generated by the N subunits of the system, the covariance matrix is derived as $W_{N\times N} = A \cdot A^T$. The synchronization index involves the eigenvalues of W, i.e., the square of the singular values of the matrix A. If all signals are uncorrelated, all singular values will be nonzero; if all signals are identical, there will be only one nonzero singular value and the rank of the matrix W is equal to 1; and if the signals are similar but not identical, very small but nonzero singular values can be derived. Therefore, the synchronization index is defined in regard to the minimum number m of eigenvalues, whose sum is greater than a percentage ξ of the trace of W, as

$$\sigma(\xi) = \min m \left| \sum_{i=1}^{m} \lambda_i^* \right| > \xi \cdot Tr(W), \tag{13.4}$$

where λ_i^* the i-th largest eigenvalue of the covariance matrix W. Starting from an N-coupled systems, this index ranges from 1 (full synchronization) to $[\xi N] + 1$ (no synchronization), and gives information about how many kinds of dynamics the system can exhibit. Here and throughout, the percentage ξ is assumed to be 98% and, accordingly, the index range is $1 \leq \xi(0.98) \leq 196$ due to $N = 200$.

Assume the Lyapunov exponent of each cell to be equal to 0.1, and the fuzzy array's structure to be regular. The synchronization index is evaluated to compare the collective dynamics by varying the number of connections from $C = 1$ to $C = 50$. The index σ versus the number of connections is plotted in Fig. 13.9, which implies a specific relation between the different ranges of connections and the previously defined spatiotemporal dynamics, detailed in the following:

a) For $1 < C < 15$, big values of σ characterize the spatiotemporal chaos;
b) For $15 < C < 20$, the system exhibits the increase of synchronization (regular synchronized behavior) quantified by the decrease of the synchronization index;
c) For $20 < C < 30$, the transition phase appears and the value of σ becomes larger;
d) For $C > 30$, the index decreases to 1, and chaotic synchronized behavior appears.

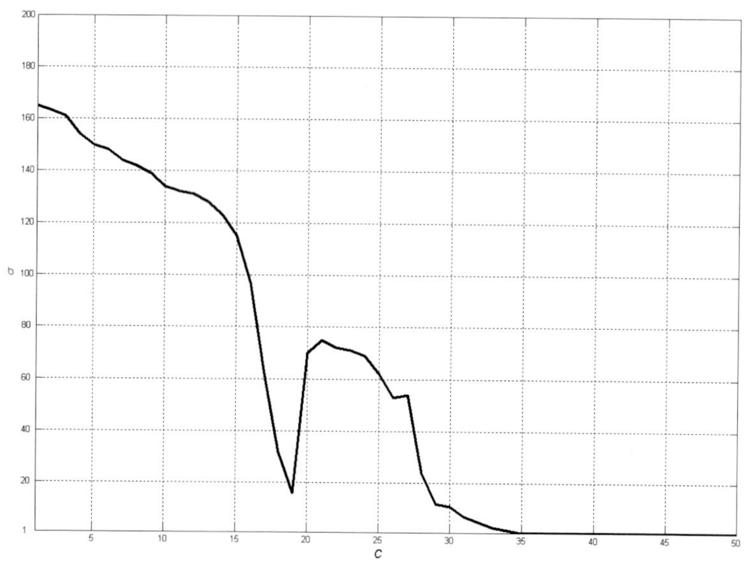

Fig. 13.9. Synchronization index σ versus the number of connections C in regular fuzzy chains at $l = 0.1$

Comparing these results to the Fourier two-dimensional transform, the feasibility of this index to analyse transition states in complex systems by varying network parameters can be verified.

A characterization of all spatiotemporal dynamic features of the array by evaluation of the two-dimensional Fourier transforms is shown in Fig. 13.10. Observing the three-dimensional graphs obtained in relation to the four C values as used previously, it is possible to recognize: the initial chaotic state, the transition phase and two main synchronized behaviors, the regular one and the chaotic one.

a) For $C = 4$, the spread number of picks both in space spectrum and time spectrum is visible (Fig. 13.10 (a));
b) For $C = 16$, the value of picks at a particular time spectrum becomes bigger (Fig. 13.10 (b));
c) For $C = 25$, the network dynamics exhibits a transition from the regular behavior to the chaotic synchronized one (Fig. 13.10 (c));
d) For $C = 46$, the distribution of picks, only in the time spectrum axis, shows perfect space synchronization of the cells with chaotic dynamics (Fig. 13.10 (d)).

Therefore, such a correspondence between the synchronization index and the two-dimensional Fourier transform opens a possibility to characterize a

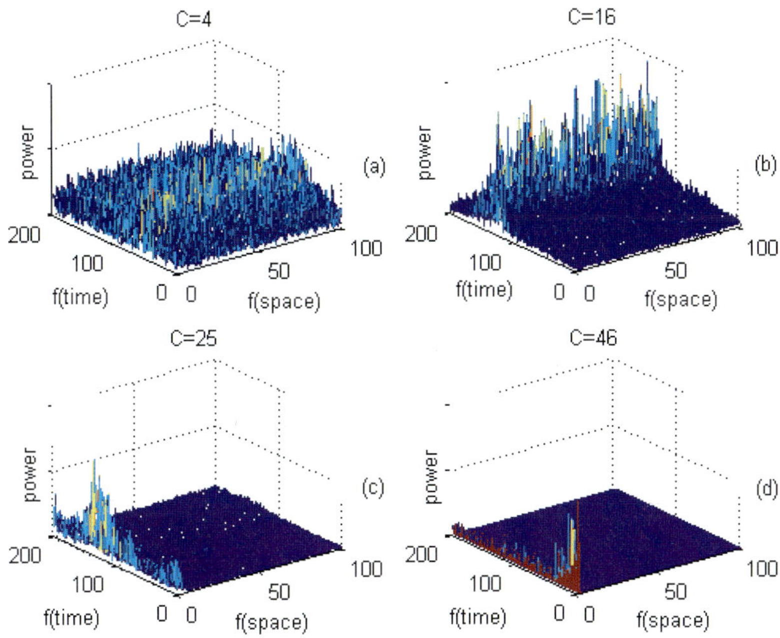

Fig. 13.10. Fourier 2-D transform of regular fuzzy chain with l=0.1: (a) $C = 4$, (b) $C = 16$, (c) $C = 25$, (d) $C = 46$

series of experiments quantifying the influence of parameters and network topologies on complex dynamics.

In order to enhance the lattices' self-synchronization properties in relation to the chaotic dynamics of the fundamental cell, the Lyapunov exponent is varied in steps of 0.1 from $l = 0.1$ to $l = 0.9$. According to the relation of the synchronization index in the regular chains for different Lyapunov exponents and the number of connections C as shown in Fig. 13.11, different collective behaviors are clearly distinguished. Increasing the value of the Lyapunov exponent, the transition from regularization to chaotic synchronization occurs with different trends, which are explained in the following:

a) Spatiotemporal chaos occurs with a relatively small C for all values of the Lyapunov exponent;
b) Regular synchronized behavior appears to different extents for different ranges of C, although perfect synchronization occurs at $l = 0.9$;
c) The transition phase is not always present; sometimes as C is varied, spatiotemporal chaos switches to synchronized dynamics following a smooth trend;
d) Chaotic synchronization is typically quantified for all values of the Lyapunov exponent when $C > 30$.

It is noted that although this index does not directly distinguish between chaotic and regular synchronization, by frequency analysis of the time series it is possible to gain this information.

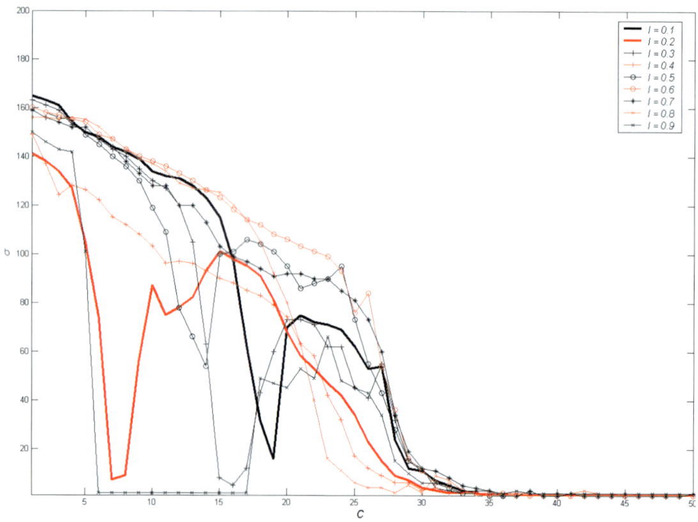

Fig. 13.11. Synchronization index σ versus the number of connections C in regular fuzzy chains by varying l

13.4 Complex Networks: Preliminaries

Many real-world systems of interest, like ecosystems, power grids or the Internet, can be modeled by complex dynamical networks. The overall dynamics of these models have two sources of complexity, the dynamics of their components and their interactions. Considering simple regular structures, the overall dynamics can be studied in terms of the dynamical behavior of their components, a scenario that has been investigated, and for which interesting results have been obtained. Simple architectures allow us to focus on the complexity caused by the nonlinear dynamics of the nodes, without considering additional complexity in the network structure itself. Recently, the focus has shifted to consider the complexity arising from the network structure. Along this line significant advances have been reported over the last few years, most significantly the small-world and scale-free complex network models. These models capture essential aspects common to most complex real-world systems. Furthermore, it can be shown that its structural characteristics determine the overall dynamics of a network. Another particularly interesting phenomenon of a network's overall behavior is synchronization. One way to break up synchronization by changing a deterministic protocol in, for instance, Internet traffic is likely to generate another synchrony. Thus, it suggests that a more efficient solution requires a better understanding of the nature of synchronization behavior in complex networks.

Watts and Strogatz have proposed in 1998 a new complex network architecture characterized by a regular lattice rewired in order to introduce increasing amounts of disorder: the "small-worlds" theory. Many real networks are characterized by topologies between order and randomness. A famous manifestation of small-world features is the so-called "six degrees of separation" principle by the psychologist Milgram [275]. This concept is based on the notion that everyone in the world is connected to anybody else through a chain of not more than six mutual connections. Such complex systems are realized to exhibit both highly clustered structures and reduced path lengths.

Two main concepts characterize the anatomy of a generic network and are fundamental to investigate the dynamic behavior of a whole system, i.e., *average path length* and *clustering coefficient.*

Definition 13.1 (Average path length). *The average path length L is defined as the mean of all distances between two cells in a network, and it gives information about the size of the whole network. Here, the distance between two cells in a generic network is the number of edges along the shortest path between them.*

It is remarked that the larger the average path length is, the bigger the separation between every pair of cells is.

Definition 13.2 (Clustering coefficient). *The clustering coefficient c_i for a vertex v_i is defined in the following way. Let N_i be a set of vertices that* *are connected to v_i, i.e., that share an edge with v_i (not including v_i). The maximum number of edges among the set N_i is $\frac{|N_i|(|N_i|-1)}{2}$. Thus, c_i can be defined, given that $|N_i| \geq 2$, as the actual number of edges among N_i divided by $\frac{|N|(|N|-1)}{2}$.*

The clustering coefficient c for the whole system is defined as the average of the clustering coefficient for each vertex,

$$c = \frac{1}{n} \sum_{i=1}^{n} c_i,$$

where n is the total number of the vertices.

It is evident that in a large network the cells connected to a particular cell can be connected with each other, too.

The structural characteristics of networks are fundamental to investigate the collective behavior in high-dimensional systems consisting of communicating nonlinear units to reach a common macro-behavior. In recent studies, network topologies are investigated focusing on four architectures that can be described by using a mathematical model: regular coupled networks, random graphs, "small worlds", and scale free (omitted here). The first three kinds of networks are roughly described in the following.

1. **Regular coupled networks:** Networks like chains, grids, and fully connected graphs are typical examples of regular structures. A regular configuration is often adopted to model high-dimensional systems to give an order structure to a complex behavior. Former studies focused on the dynamic features of a single node when it is either isolated or connected to others. A node could be a generic dynamical system with a stable point, a limit cycle or a chaotic behavior. Regular lattices are high-clustered and their path length grows linearly with the number of edges N.
2. **Random graphs:** Erdos and Renyi (ER) [276] have studied random graphs by varying the number of nodes and links. An ER random graph is obtained by tossing a set of buttons on the floor. With probability p each pair of buttons is tied with a thread. According to the value of probability p the graph properties change. The average path length of a random graph is $L = p(N-1)$. For small values of p, graphs are constituted by separated components. As p increases, the number of links increases and the network becomes more compact. When $p \approx \frac{\ln N}{N}$, the system can be considered as a unique entity. The clustering coefficient of a random graph is equal to $p = L/N \ll 1$. This means that in a friendship network the probability that two friends are friends themselves is not greater than the probability that two randomly chosen persons are connected.
   ER networks are homogeneous systems whose connectivity follows a Poisson distribution. Each node in a graph is connected to each other node through short paths and the maximum "degree of separation" grows with

log N. Random graphs are an ideal model, and due to their building simplicity it is often used to model gene networks, ecosystems and spread of diseases [277, 278].

3. **"Small worlds"** The introduction of randomness in regular lattices, obtained through a rewiring procedure, interpolates from a regular network to a random one with the same number of vertices and edges. This is the case of "small-world" systems in which in a regular topology are identifiable small amounts of irregular connections. Watts and Strogatz (WS) [274] have introduced a new class of networks based on "small-world" features that tune a graph between a regular lattice and a random network. "Small-world" networks have two fundamental features: short paths and high clustering. The starting configuration is a ring configuration with N vertices, with each connected to the same number $2C$ of neighbors. Rewiring is realized by choosing a vertex and an edge and reconnecting, with probability p, the end of the bond to a vertex chosen uniformly at random over the ring. All duplicate edges are forbidden and all vertices are processed in a clockwise direction. This procedure tunes the graph between regularity ($p = 0$) and disorder ($p = 1$) obtaining interesting structural properties as shown in Fig. 13.12. In "small-world" networks, small values of the rewiring probability p allow to introduce few far-connections that give birth to interesting nonlinear effects. The mean distance of each pair of vertices decreases, but the number of removed edges is small enough to maintain the same regular network clustering characteristics at local level. Average path length and clustering coefficient of "small worlds" versus rewiring probability are shown in Fig. 13.13.

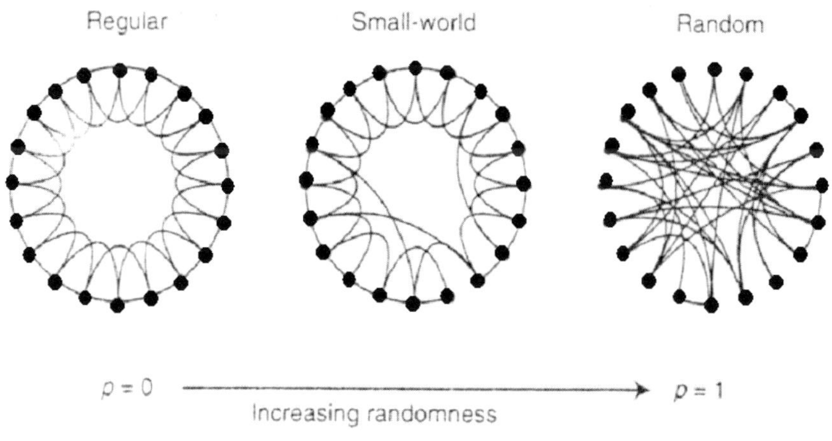

Fig. 13.12. "Small-world" rewiring scheme

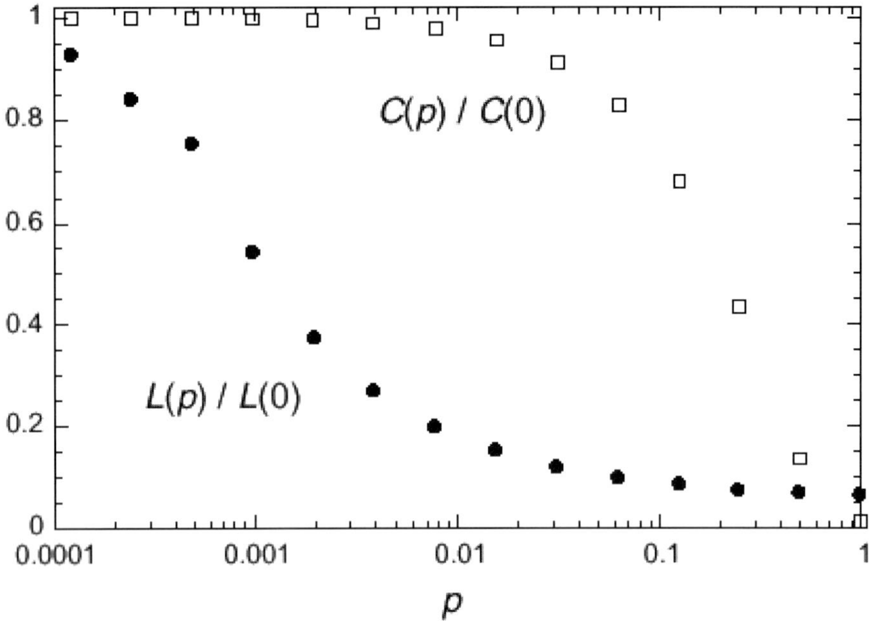

Fig. 13.13. Path length L and clustering coefficient c versus rewiring probability in "small-world" networks; the values are normalized by the values of a regular topology $L(0)$ and $c(0)$

The functional study of "small-world" connectivity for dynamic systems shows an enhancement of network speed and exchange of information.

13.5 Collective Behavior versus Network Topology

The evolution of high-order complex fuzzy systems has been studied characterizing the effects of the topology on the collective dynamics. The structure of the regular fuzzy chain has been modified varying the structure of the network connections continuously, starting from a regular topology and ending with a random one, passing through different "small-world" topologies.

Regular arrays are highly clustered and their path length grows linearly with the number of elements N, whereas networks with random connections are poorly clustered, but the mean distance between two vertices grows logarithmically with N. The introduction of randomness in a regular chain, obtained through the "small-world" rewiring procedure, interpolates from a regular network to a random one with the same number of vertices and edges. An example of a rewiring scheme of the "small world" is shown in Fig. 13.12.

The initial configuration is the previous regular array with N vertices, each connected to the same number $2C$ of neighbors. Rewiring is performed choosing a vertex and an edge, and reconnecting them with probability p; the end of the bond to a vertex is chosen uniformly at random. All duplicate edges are forbidden and all vertices are processed in clockwise direction. All "small-world" chains have many vertices with sparse connections, but the graphs never become disconnected guaranteeing the following relations:

$$N \gg 2C \gg \ln(N) \gg l. \tag{13.5}$$

This procedure moves the graph structure between regularity ($p = 0$) and disorder ($p = 1$) obtaining new structural properties.

Small values of the "small-world" rewiring probability p generate few rewired connections and give birth to interesting nonlinear effects: the mean distance of each pair of vertices decreases, but the small number of removed edges yields clustering characteristics at the local level as regular networks. The random topology with probability $p = 1$ modifies all connections.

Synchronization performances have been analysed by building chains of $N = 200$ fuzzy units, with the Lyapunov exponent ranging from $l = 0.1$ to $l = 0.9$ through different topologies. The probability of rewiring p has been investigated in the entire range from the regular topology value ($p = 0$) to the random topology one ($p = 1$). In Figs. 13.14–13.17, the synchronization index of these fuzzy chains with several values of the Lyapunov exponent is reported for four rewiring probabilities ($p = 0$, $p = 0.1$, $p = 0.5$, $p = 1$).

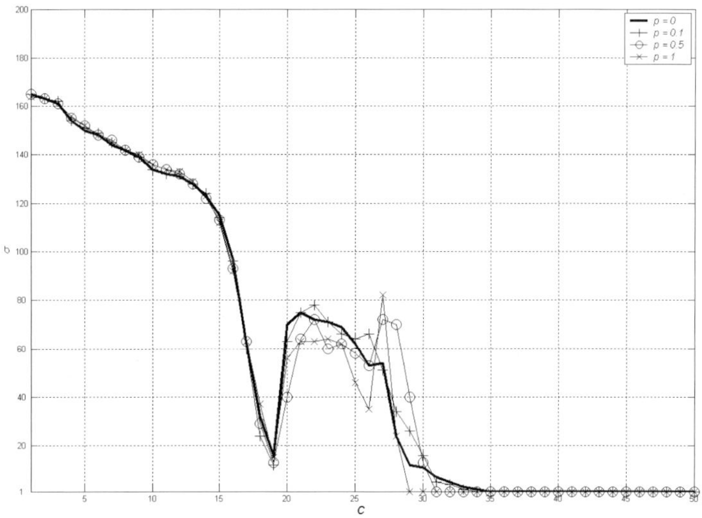

Fig. 13.14. Synchronization index σ versus the number of connections C in fuzzy chains with $l = 0.1$ by varying the topology: regular ($p = 0$), "small world" ($p = 0.1$, $p = 0.5$), random ($p = 1$)

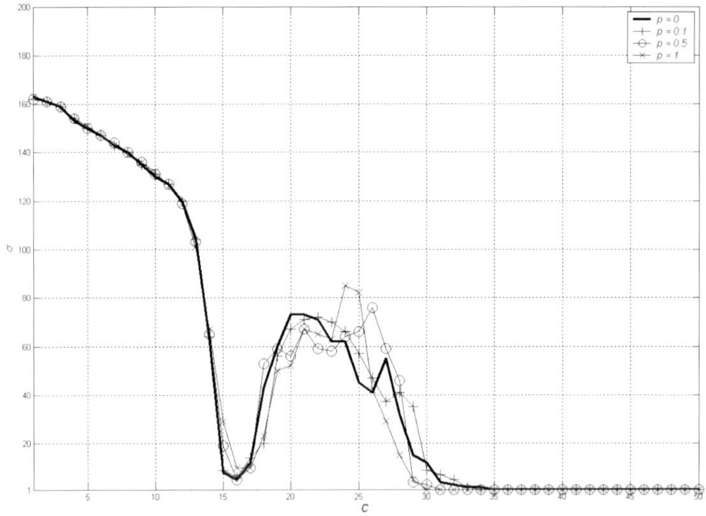

Fig. 13.15. Synchronization index σ versus the number of connections C in fuzzy chains with $l = 0.3$ by varying the topology: regular ($p = 0$), "small world" ($p = 0.1$, $p = 0.5$), random ($p = 1$)

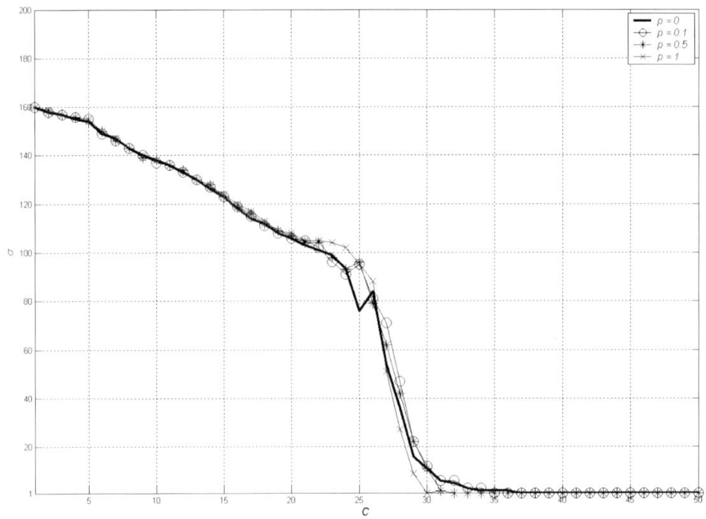

Fig. 13.16. Synchronization index σ versus the number of connections C in fuzzy chains with $l = 0.6$ by varying the topology: regular ($p = 0$), "small world" ($p = 0.1$, $p = 0.5$), random ($p = 1$)

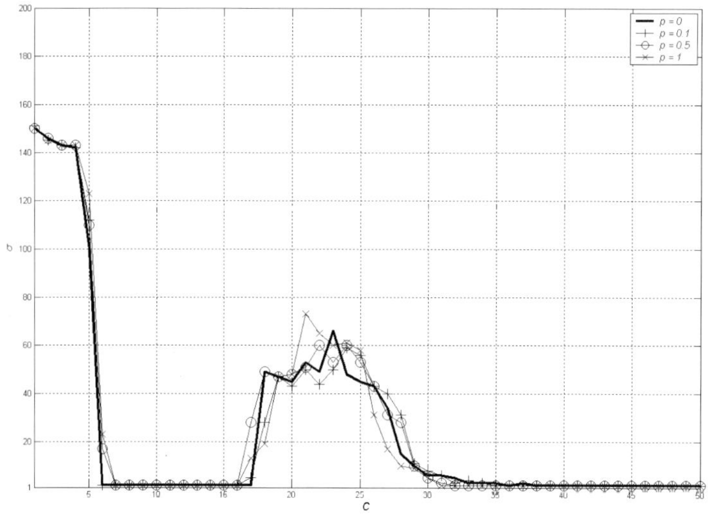

Fig. 13.17. Synchronization index σ versus the number of connections C in fuzzy chains with $l = 0.9$ by varying the topology: regular ($p = 0$), "small world" ($p = 0.1$, $p = 0.5$), random ($p = 1$)

The trends of the synchronization index versus the number of connections, for $l = 0.1$, $l = 0.3$, $l = 0.6$ and $l = 0.9$, are shown in Figs. 13.14–13.17, respectively, followed by some considerations:

a) The range of C related to the occurrence of spatiotemporal chaos is not influenced by the rewiring of the connections even if all are modified ($p = 1$);
b) The transition phase range of C from regular to chaotic synchronization is modified by the rewiring of the connections, with the exception of $l = 0.6$;
c) Chaotic synchronization occurs for the system with $p = 1$ considering a fewer number of connections.

The complexity introduced in the network architecture results in a decrease of the mean path length between two units, and enhances the transition to chaotic synchronization.

14

Fuzzy-chaos-based Cryptography

As an application example of integrating fuzzy logic and chaos theory, a fuzzy-model-based chaotic cryptosystem is introduced in this chapter.

14.1 Introduction

Cryptography, defined as the science and study of secret writing, concerns the ways in which communications and data can be encoded to prevent disclosure of their contents through eavesdropping or message interception, using codes, ciphers, or other methods, so that only certain people can see the real messages. The science of cryptography is very old, and can be traced to Ancient Egypt. From Julius Caesar to Mary, Queen of Scots, to Abraham Lincoln's Civil War ciphers, cryptography has been a part of history. At that time, cryptography was concerned only by those associated with the military, the diplomatic service and government in general, and was used as a tool to protect national secrets and strategies [279].

Nowadays, the Internet has become an indispensable part of our daily life. However, over the Internet various communications, such as E-mails, or the use of WWW browsers, are not secure for sending and receiving data. Therefore, varieties of cryptographic methods have been proposed to secure Internet communication. For instance, the Data Encryption Standard (DES) was adopted as a U.S. Federal Information Processing Standard for encrypting unclassified information. Others include IDEA (International Data Encryption Algorithm), and RSA (developed by Rivest, Shamir and Adleman). These encryption algorithms are based on number theory. However, none of them is absolutely secure.

Cryptography can be strong or weak. Cryptographic strength is measured in the time and resources, which would be required to recover the plaintext. A strong cryptography makes a ciphertext difficult to be deciphered without possession of the appropriate decoding tool. In other words, given all of today's computing power and available time – even a billion computers doing a

Zhong Li: *Fuzzy Chaotic Systems*, StudFuzz **199**, 275–283 (2006)
www.springerlink.com

billion checks in a second, it is still impossible to decipher the results of strong cryptography before the end of the universe. Straightforward, one would think that strong cryptography would hold up rather well against even an extremely determined cryptanalysis. Nevertheless, no one can prove that the strongest encryption obtainable today will hold up under tomorrow's computing power. Therefore, some emerging theories, such as chaos theory, can be adopted to strengthen existing cryptography.

The reason of applying chaos theory in cryptography lies in its intrinsic essential properties, such as sensitivity to initial conditions (or control parameters) and ergodicity, which meet Shannon's requirements of confusion and diffusion for cryptography. Shannon wrote in his seminal paper [280]: "In a good mixing transformation ... functions are complicated, involving all variables in a sensitive way. A small variation of any one (variable) changes (the outputs) considerably." An important difference between chaos and cryptography lies on the fact that systems used in chaos are defined on real numbers, while cryptography deals with systems defined on finite numbers of integers. Nevertheless, it is believed that the two disciplines can benefit from each other [281].

Chaotic cryptosystems can be analog or digital. The analog ones are based on a chaotic synchronization technique, which was proposed in [204], to design analog circuits for secure communication via noisy channels. However, this cannot be extended to design modern cryptographic algorithms implemented with digital techniques [282]. The digital chaotic ciphers can be categorized into stream ciphers and block ciphers. Stream ciphers employ chaotic systems to generate pseudo-random keystream to mask plaintext, while block ciphers use the plaintext and/or the secret keys multiple times to obtain ciphertext. In addition, some other chaotic encryption schemes have also been proposed and tested [283, 284, 285, 286, 287].

Since the pioneering work of Carroll and Pecora [204], which has led to many contributions regarding synchronization of two chaotic systems [25, 288], where two chaotic systems with suitable coupling produce identical oscillations, many theories [25, 288, 289, 290, 291] have been proposed to achieve synchronized behaviour of a master-slave configuration. This master-slave configuration consists of the original chaotic system as a *drive system* providing a *driving signal* to synchronize another system called the *response system*. In addition, chaotic signals are typically broadband, noise-like, and difficult to predict. Therefore, they can be used in various context for masking information-bearing waveforms. They can also be used as modulating waveforms in spread-spectrum systems. The idea of chaotic masking [292, 293] is to directly add the message in a noise-like chaotic signal at the end of the transmitter, while chaotic modulation [294, 295, 296, 297] is by injecting the message into a chaotic system as a spread-spectrum transmission. Later, at the receiver, a coherent detector with some signal processing is employed to recover the message. But the signal masking or parameter modulation approach to chaotic communication only provides a lower level of security as stated in [298]. Here, a fuzzy-model-based chaotic cryptosystem is proposed by the use of basic cryptosystem theory to provide a methodology with a higher level of security [299, 300, 301].

The proposed TS fuzzy-model-based chaotic cryptosystem has the following advantages [129]:

i) The fuzzy-model-based design methodology is suitable for most well-known Luré type discrete-time chaotic systems;
ii) The synchronization problem to be solved by using LMIs is systematic and straightforward, and may be solved by powerful software toolboxes;
iii) Superincreasing sequences generated by chaotic signals are time-varying;
iv) There exists a flexible choice for whether ciphertext is embedded in the state or output of the drive system;
v) It has multi-user capabilities.

In this chapter, Luré type discrete-time chaotic systems are first exactly represented by TS fuzzy models. Then, a superincreasing sequence is generated by using a chaotic signal, which can flexibly be used as an output of the TS fuzzy chaotic drive system, or any state in which the synchronization error approaches zero. In terms of a cryptosystem, the plaintext (message) is encrypted using the superincreasing sequence at the drive system side, which results in the ciphertext. The ciphertext may be added to the output or state of the drive system using the methodologies proposed in [207, 43, 302]. Further, this the ciphertext embedding scalar signal is sent to the response system end. Following the design of a response system, chaotic synchronization between drive and response system is achieved by solving LMIs. By synchronization, one can regenerate the same superincreasing sequence and recover the ciphertext at the response system end. Finally, using the regenerated superincreasing sequence, the ciphertext is decrypted into the plaintext.

14.2 Working Principle of the Cryptosystem

Some basic cryptosystem terminologies are first introduced. The composite message N to be transmitted is called *plaintext*, which is encoded by using the *superincreasing sequence* S_i. The encoded plaintext results in the *ciphertext* E. The process of recovering the plaintext from the ciphertext is executing the decryption function $\widehat{E}$. Encryption and decryption process use the keys K and $\widehat{K}$, respectively. The definition of a superincreasing sequence is given as follows.

Definition 14.1. *A real sequence* $\{S_i\}_{i=1}^{\ell}$ *is called a superincreasing sequence if the following is satisfied:*

$$S_j > \sum_{i=1}^{j-1} S_i, \ell \geq j > 1 \text{ and all } S_i > 0.$$

Note that in traditional superincreasing problems [303, 304], the sequence is a set of positive integers, while the superincreasing sequence used here will be a set of positive real numbers.

A fuzzy-model-based chaotic cryptosystem is shown in Fig. 14.1.

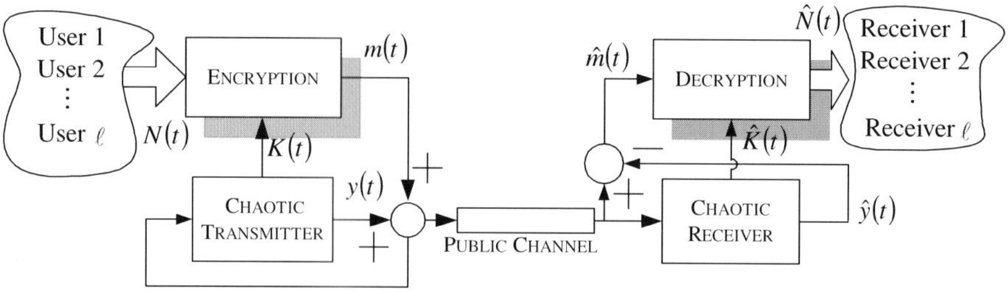

Fig. 14.1. Block diagram of chaotic encryption methodology

In the process of encrypting a plaintext, a discrete-time chaotic systems is first represented by a TS fuzzy model as a drive system, from the output of which a key stream $K(t-i)$, $i = 0, \ldots, \ell-1$ can be generated. Then, the key stream $K(t), K(t-1), \ldots, K(t-\ell+1)$ is used to construct a superincreasing sequence $\{S_i\}$, $i = 1, \cdots, \ell$, where ℓ is the number of messages, and $S_1(t) = |K(t)|+\tau$, $S_j(t) = \sum_{i=1}^{j-1} S_i(t) + |K(t-j+1)| + \tau$, in which $j = 2, \ldots, \ell$, and $\tau > 0$.

Combining the superincreasing sequence with the plaintext $N = [n_1\ n_2 \cdots n_\ell]$, $n_i \in \{0, 1\}$, results in the ciphertext:

$$E(N(t), K(t), K(t-1), \ldots, K(t-\ell+1)) = S(t)N(t)^T \equiv E(t),$$

where E is an encryption function. Further transform the encryption function $E(t)$ into $\xi(t) = (E - \frac{H(t)}{2})/(\gamma\frac{H(t)}{2})$, where $H = \sum_{i=1}^{\ell} S_i$ and γ is a scalar, such that $\xi(t) \in (-0.01, 0.01)$ is sufficiently small as not to destroy the chaotic characteristics of the masking signal. Add $\xi\ (t)$ to the masking signal and send a scalar coupling signal to response system.

In the process of decrypting the ciphertext, the key stream $\widehat{K}(t-i)$, $i = 0, \ldots, \ell-1$, and the signal $\hat{\xi}(t)$ are first recovered by the TS fuzzy response system via synchronization, which is to be described in the next section. From the key stream $\widehat{K}(t-i)$, one can obtain a superincreasing sequence $\{\widehat{S}_i(t)\}$, $i = 1, \cdots, \ell$, where $\widehat{S}_1(t) = |\widehat{K}(t)|+\tau$, $\widehat{S}_j(t) = \sum_{i=1}^{j-1} \widehat{S}_i(t)+|\widehat{K}(t-j+1)|+\tau$, in which $j = 2, \ldots, \ell$, and $\tau > 0$. Then, transform $\hat{\xi}(t)$ inversely to obtain $\widehat{E}(t) = \hat{\xi}(t)(\gamma\widehat{H}(t)/2)+\widehat{H}(t)/2$. Finally, the plaintext is recovered from decrypting the ciphertext $\widehat{E}(t)$ as follows:

$$\widehat{N}(t) = D\left(\widehat{K}(t), \widehat{E}\left(N(t), \widehat{K}(t), \widehat{K}(t-1), \ldots, \widehat{K}(t-\ell+1)\right)\right)$$

where D is an decryption function that makes use of the recovered key $\hat{K}(t-i)$, $i = 0, \ldots, \ell - 1$. The decryption algorithm is designed as:

$\widehat{V} = \widehat{E}$
For $i = \ell$ down to 1
Begin
If $\widehat{V} - \widehat{S}_i > -\varepsilon$
$\hat{n}_i = 1$
$\widehat{V} = \widehat{V} - \widehat{S}_i$
Else $\hat{n}_i = 0$
END

where $0 < \varepsilon < \tau$.

It is noted that if $n_\ell(t) = 1$ (or $n_\ell(t) = 0$), then $E(t) \geq S_\ell(t)$ or $(E(t) \leq S_\ell(t) - \tau)$. After the transient time, the response system is synchronized to the drive system, which means that $\widehat{E}(t) = E(t)$ and, therefore, $\widehat{S}_\ell(t) = S_\ell(t)$. Finally, $\hat{n}_\ell(t) = 1$ (or $\hat{n}_\ell(t) = 0$) from the above algorithm, and the other plaintexts $\hat{n}_{\ell-1}(t) \cdots \hat{n}_1(t)$ are recovered by the iteration loop.

It is further noted that when the cryptosystem simultaneously serves ℓ users, the plaintext N consists of multiuser messages, i.e., only one bit from each user is transmitted at a time, while for the case of only one user, ℓ bits of the user's message are simultaneously transmitted. This emphasizes this cryptosystem's multiuser capabilities and high data transmission rate for one user.

14.3 Decryption by Fuzzy-model-based Synchronization

In this section, the ciphertext is first masked by the output of a chaotic system, and the modulation process is carried out by injecting the masking signal into the fuzzy chaotic transmitter. Then, the masked signal is sent to the fuzzy chaotic receiver, where the ciphertext is extracted according to the masking methods.

Assume that the ciphertext $\xi(t)$ is directly added into the output of the chaotic system. The chaotic transmitter is expressed as a TS fuzzy model:

$$\begin{aligned} \textit{Transmitter Rule } i: \text{IF } \bar{y}(t) \text{ is } F_i \text{ THEN} & \\ x(t+1) &= A_i x(t) + b_i(t) + L_i M\xi(t) \\ \bar{y}(t) &= Cx(t) + M\xi(t),\ i = 1, 2, ..., r, \end{aligned}$$

where the gains L_i, $i = 1, 2, ..., r$, are to be determined; M is a public output masking key which masks the ciphertext xi; and $\bar{y}(t)$ is the coupling signal to be transmitted to a receiver through a public channel. The fuzzy-inferred result for the chaotic transmitter is derived as

$$
\begin{aligned}
x(t+1) &= \sum_{i=1}^{r} \mu_i(\bar{y}(t)) \left\{ \overline{A}_i x(t) + b_i(t) + L_i \bar{y}(t) \right\} \\
\bar{y}(t) &= Cx(t) + M\xi(t),
\end{aligned} \tag{14.1}
$$

where $\overline{A}_i = A_i - L_i C$.

To recover the ciphertext, the receiver is designed as

$$
\begin{aligned}
&\textit{Receiver Rule } i: \text{IF } \bar{y}(t) \text{ is } F_i \text{ THEN} \\
&\qquad \hat{x}(t+1) = A_i \hat{x}(t) + b_i(t) + L_i \left(\bar{y}(t) - \hat{y}(t) \right) \\
&\qquad \widehat{y}(t) = C\widehat{x}(t),\ i = 1, 2, ..., r,
\end{aligned}
$$

The overall receiver is inferred in the following

$$
\begin{aligned}
\hat{x}(t+1) &= \sum_{i=1}^{r} \mu_i(\bar{y}(t)) \left\{ A_i x(t) + b_i(t) + L_i(\bar{y}(t) - \hat{y}(t)) \right\} \\
\hat{y}(t) &= C\hat{x}(t).
\end{aligned} \tag{14.2}
$$

Letting error signals $e_x(t) \equiv x(t) - \hat{x}(t)$ and $e_y(t) \equiv \bar{y}(t) - \hat{y}(t)$ according to (14.1) and (14.2), the error dynamics of $e_x(t)$ and $e_y(t)$ are expressed as

$$
e_x(t+1) = \sum_{i=1}^{r} \mu_i(\bar{y}(t))(A_i - L_i C) e_x(t) \tag{14.3}
$$

$$
e_y(t) = Ce_x(t) + M\xi(t) \tag{14.4}
$$

The stability condition for (14.4) is derived using the Lyapunov method. The main result is addressed here.

Theorem 14.2. *Consider the chaotic transmitter (14.1) and receiver (14.2). Ciphertext can be recovered from* $\xi(t) = \frac{1}{M} e_y(t)$*, and all states of chaotic transmitter and receiver are synchronized in an asymptotic manner if there exist a common positive-definite matrix* P *and gains* L_i*, for* $i = 1, 2, ..., r$*, such that the following LMIs are satisfied:*

$$
\begin{bmatrix} P & (PA_i - W_i C)^T \\ PA_i - W_i C & P \end{bmatrix} > 0, \ \textit{for all } i, \tag{14.5}
$$

where $W_i \equiv PL_i$.

Proof. Given a Lyapunov function candidate as $V(\widetilde{x}(t)) = e_x^T(t) P e_x(t) > 0$. Taking difference of $V(t)$ along the error dynamics (14.3) yields

$$\begin{aligned}\triangle V(e_x(t)) &= V(e_x(t+1)) - V(e_x(t))\\ &= \sum_{i=1}^{r} \mu_i^2(\bar{y}(t)) e_x^T(t)[\overline{A}_i^T P \overline{A}_i - P] e_x(t)\\ &\quad + \sum_{i<j}^{r} \mu_i(\bar{y}(t))\mu_j(\bar{y}(t))\\ &\quad \cdot e_x^T(t)[\overline{A}_i^T P \overline{A}_j + \overline{A}_j^T P \overline{A}_i - 2P] e_x(t),\end{aligned} \tag{14.6}$$

where $P > 0$, and $\overline{A}_i = A_i - L_i C$. Notice that if $\overline{A}_i^T P \overline{A}_i - P < 0$, then $\overline{A}_i^T P \overline{A}_j + \overline{A}_j^T P \overline{A}_i - 2P < 0$. This means, if there are P and L_i such that the conditions (14.5) hold, then $\overline{A}_i^T P \overline{A}_i - P < 0$ by Schur complement. Let $-Q$ denote the maximum negative-definite matrix of $\overline{A}_i^T P \overline{A}_i - P$ for all i. Then $\triangle V(e_x(t)) \leq -e_x^T(t) Q e_x(t) < 0$. Thus, the synchronization error $e_x(t)$ converges to zero as $t \to \infty$. According to the equation (14.4),$e_y(t)$ converges to $M\xi(t)$ as $t \to \infty$.

Since the convergence rate of the synchronization error $e_x(t)$ affects the transmission performance, the decay rate design for chaotic cryptosystems can be carried out by solving LMIs problems as follows:

Chaotic Cryptosystem with decay rate:

$$\begin{aligned}&\underset{P,W_i}{minimize}\ \beta\\ &\text{subject to } P > 0,\ 0 < \beta < 1\\ &\begin{bmatrix} \beta P & (PA_i - W_i C)^{\mathrm{T}} \\ PA_i - W_i C & P \end{bmatrix} > 0, \text{ for all } i,\end{aligned}$$

where $W_i \equiv PL_i$. The equation (14.6) becomes $\Delta V(e_x(t)) \leq -(1-\beta)V(e_x(t))$ with parameter β tuning the decay rate.

Example 14.3. Consider the chaotic cryptosystem using the discrete-time Hénon map (14.7), which was described in Subsection 6.4.4 of the following form:

$$\begin{aligned}x_1(t+1) &= -x_1^2(t) + 0.3x_2(t) + 1.4\\ x_2(t+1) &= x_1(t)\\ y(t) &= x_1(t)\end{aligned} \tag{14.7}$$

Take $x_1(t)$ as the premise variable of the fuzzy rules, the Hénon map in the fuzzy rules consists of $x(t) = [x_1(t)\ x_2(t)]^T$, $C = [1\ 0]$, the fuzzy sets $F_1(y(t)) = \frac{1}{2}(1 + \frac{y(t)}{d})$ and $F_2(y(t)) = \frac{1}{2}(1 - \frac{y(t)}{d})$, and

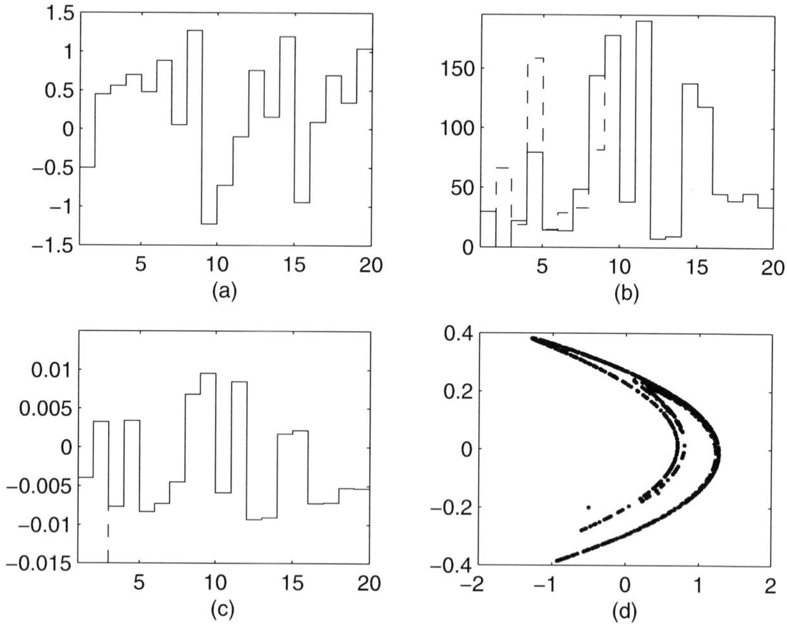

Fig. 14.2. (a) Chaotic coupling signal, (b) encryption function $E(\cdot)$, $\hat{E}(\cdot)$ (dotted line), (c) scaled ciphertext $m(t)$, $\hat{m}(t)$, and (d) phase portrait

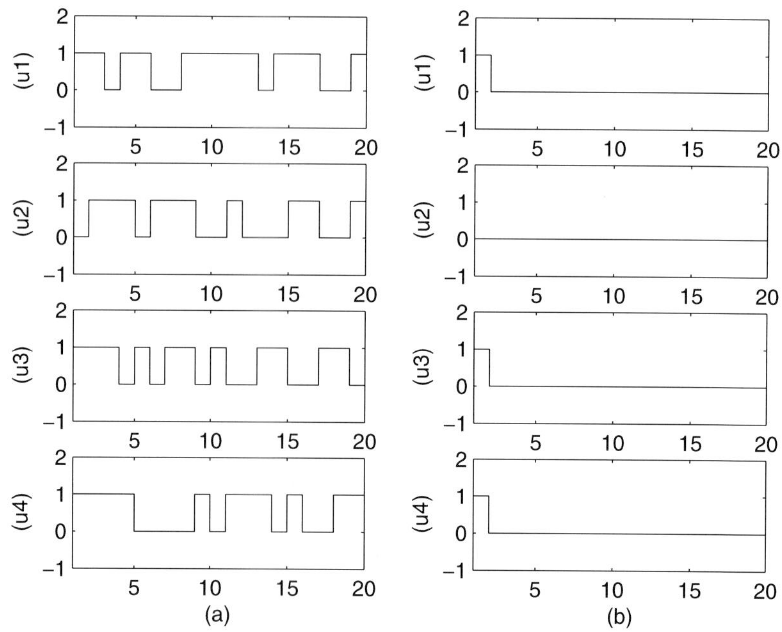

Fig. 14.3. (a) Plaintexts transmitted my users 1 through 4 (of 8 in total) and (b) errors between original message and recovered message of users 1 through 4

$$A_1 = \begin{bmatrix} -d & 0.3 \\ 1 & 0 \end{bmatrix}, \ A_2 = \begin{bmatrix} d & 0.3 \\ 1 & 0 \end{bmatrix}, \ b_1 = b_2 = \begin{bmatrix} 1.4 \\ 0 \end{bmatrix}$$

where $d = 2$ and $x_1(t) \in [-d \ d]$.

Assume that the cryptosystem serves eight users whose plaintexts are random binary sequences. Let the state $x_1(t)$ be the output which generates the key stream $K(t)$. The cyphertext, which is obtained from the encryption algorithm in Section 14.2, is added to the output of the drive system $\bar{y}(t)$. According to the fuzzy representation of the Hénon map and Theorem 14.2, the chaotic transmitter (14.1) and receiver (14.2) are designed with gain vectors $L_1 = [-2.1384, 5.6608]^T$ and $L_2 = [2.1384, 5.6608]^T$. Note that the parameter of the decay rate is solved as $\beta = 0.1$. For simplicity, we set the output masking key as $M = 1$. Thus, Fig. 14.2 shows the chaotic coupling signal scaled ciphertext m and $\hat{m}$ (dotted line), encrypting function E and $\widehat{E}$ (dotted line), and chaotic phase portrait, respectively. In Figs. 14.3 a–b, $u1 \sim u4$ are the first four user plaintexts transmitted (total of eight users) and errors between the recovered plaintext and original plaintexts of users $u1 \sim u4$.

References

1. S. Korner, "Laws of thought," *Encyclopedia of Philosophy*, Vol. 4, MacMillan: New York, 1967, pp. 414–417.
2. C. Lejewski, "Jan Lukasiewicz," *Encyclopedia of Philosophy*, Vol. 5, MacMillan: New York, 1967, pp. 104–107.
3. L.A. Zadeh, "Fuzzy sets," *Inform. Contr.*, Vol. 8, 1965, pp. 338–353.
4. L.A. Zadeh, "Fuzzy algorithms," *Inform. Contr.*, Vol. 12, 1968, pp. 94–102.
5. L.A. Zadeh, "Outline of a new approach to the analysis of complex systems and decision processes," *IEEE Trans. Syst. Man, and Cyber., SMC*, Vol. 3, 1973, pp. 28–44.
6. R.M. Tong, "The construction and evaluation of fuzzy models," in *Advances in Fuzzy Sets Theory and Applications*, M.M. Gupta, R.K. Ragade and R.R. Yager (eds.), North-Holland: Amsterdam, 1979, pp. 559–576.
7. W. Pedrycz, *Fuzzy Control and Fuzzy Systems*, Wiley: New York, 1989.
8. M. Sugeno and T. Yasukawa, "A fuzzy-logic-based approach to qualitative modeling," *IEEE Trans. Fuzzy Systems*, Vol. 1, 1993, pp. 7–31.
9. T. Takagi and M. Sugeno, "Derivation of fuzzy control rules from human operators actions," *Proc. IFAC Symposium on Fuzzy Information*, Marseille, 1983, pp. 55–60.
10. T. Takagi and M. Sugeno, "Fuzzy identification of systems and its application to modeling and contorl," *IEEE Trans. Syst., Man, and Cyber.*, Vol. 15, pp. 116–132, 1985.
11. M. Sugeno and G.T. Kang, "Fuzzy modeling and control of multilayer incinerator," *Fuzzy Sets and Systems*, Vol. 18, 1986, pp. 329–346.
12. H. Ying, "Fuzzy systems technology: a brief overview," *IEEE Circ. Syst. Soc. Newsletter*, Vol. 11, No. 3, 2000, pp. 28–37.
13. H. Ying, M. McEachern, D. Eddleman and L.C. Sheppard, "Fuzzy control of mean arterial perssure in postsurgical patients with sodium nitroprusside infusion," *IEEE Trans. Biomedical Eng.*, Vol. 39, 1992, pp. 1060–1070.
14. H. Ying, *Fuzzy Control and Modeling: Analytical Foundations and Applications*, IEEE Press, 2000.
15. *The New Encyclopaedia Britannica* (Ready Reference and Index), Micropaedia, Vol. 2, Encyclopaedia Britannica, Inc., Chicago, IL, 1984, p. 743.
16. A.B. Cambel, *Applied Chaos Theory: A Paradigm for Complexity*, Academic Press: San Diego, CA, p. 15, 1993.

17. M. Hénon and C. Heiles, "The applicability of the third integral of motion: some numerical experiments," *Astronomical Journal*, Vol. 69, pp. 73–79, 1964.
18. B.V. Chirikov, "Resonance processes in magnetic traps," *J. Nucl. Ener.-Part C*, Vol. 1, pp. 253–260, 1960.
19. T.Y. Li and J.A. Yorke, "Period three implies chaos," *Amer. Math. Monthly*, Vol. 82, pp. 481–485, 1975.
20. R.L. Devaney, *An Introduction to chaotic Dynamical Systems*, Addison-Wesley: New York, 1987.
21. G. Chen, "Control and anticontrol of chaos", *Proc. Int'l. Conf. on Control of Oscillations and Chaos*, St. Petersburg, Russia, pp. 181–186, 1997.
22. M. Vellekoop and R. Berglund, "On Intervals, Transitivity = Chaos", *Amer. Math. Monthly*, Vol. 101, No. 4, pp. 353–355, 1994.
23. D.A. McAndrew, "Chaos, complexity, and fuzziness: science looks at teaching English," *English Journal*, Vol. 86, No. 7, pp. 37–43, 1997.
24. J.M. Ottino, "The mixing of fluids," *Scientific American*, Vol. 260, pp. 56–67, 1989.
25. G. Chen and X. Dong, *From Chaos to Order: Perspectives and methodologies, and Applications*, World Scientific: Singapore, 1997.
26. S. Neil Rsband, *Chaotic dynamics of nonlinear systems*, John Wiley & Sons, Inc. 1990.
27. Y. Chen and A.Y.T. Leung, *Bifurcation and Chaos in Engineering*, Springer, 1998.
28. A.L. Fradkov and A.Yu. Pogromsky, *Introduction to Control of Oscillations and Chaos*, World Scientific: Singapore, 1997.
29. S. Boccaletti, C. Grebogi, Y.-C. Lai, H. Mancini and D. Maza, "The control of chaos: theory and applications," *Physics Reports*, Vol. 329, 2000, pp. 103–197.
30. E. Ott, C. Grebogi and J.A. Yorke, "Controlling chaos," *Physical Review Letters*, Vol. 64, 1990, pp. 1196–1199.
31. G. Chen, "Chaos: its control and generation for engineering applications," *Dynamics of Continuous, Discrete and Impulsive Systems*, Vol. 10, Nos. 1–3, 2003, pp. 235–245.
32. H.O. Wang, K. Kanaka and T. Ideda, "Fuzzy modeling and control of chaotic systems", *Proc. 1996 IEEE Int. Symp. on Circ. Syst., ISCAS'96*, pp. 209–212, 1996.
33. O. Calvo and J.L. Cartwright, "Fuzzy control of chaos," *Int. J. Bifurc. Chaos*, Vol. 8, No. 8, pp. 1743–1747, 1998.
34. H.L. Hiew, "An adaptive fuzzy system for modeling chaos," *Information Science*, Vol. 81, No. 3, pp. 193–212, 1994.
35. P.E. Kloeden, "Chaotic iterations of fuzzy sets," *Fuzzy Sets and Systems*, Vol. 42, No. 1, pp. 37–42, 1991.
36. P. Grim, "Self-reference and chaos in fuzzy logic," *IEEE Trans. Fuzzy Systems*, Vol. 1, No. 4, pp. 237–253, 1993.
37. S. Baglio, L. Fortuna and G. Manganaro, "Design of fuzzy iterators to generate chaotic time series with assigned Lyapunov exponent," *Electronics Letters*, Vol. 32, pp. 292–293, 1996.
38. M. Porto and P. Amato, "A fuzzy approach for modeling chaotic dynamics with assigned properties," *Proc. 9th IEEE Intl. Conf. on Fuzzy Systems*, Vol. 1, pp. 435–440, 2000.

39. Z. Li, J.B. Park, G. Chen, Y.H. Joo and Y.H. Choi, "Generating chaos via feedback control from a stable TS fuzzy system through a sinusoidal non-linearity," *Int. J. Bifur. Chaos*, Vol. 12, No. 10, pp. 2283–2291, 2002.
40. Z. Li, J.B. Park, Y.H. Joo, G. Chen and Y.H. Choi, "Anticontrol of chaos for discrete TS fuzzy systems," *IEEE Trans. Circ. Syst.-I*, Vol. 49, No. 2, pp. 249–253, 2002.
41. Z. Li, J.B. Park and Y.H. Joo, "Chaotifying continuous-time TS fuzzy systems via discretization," *IEEE Trans. Circ. Syst.-I*, Vol. 48, No. 10, pp. 1237–1243, 2001.
42. Z. Li, W. Halang, G. Chen and L.F. Tian "Chaotifying a continuous-time TS fuzzy system via time-delay feedback," *J. of Dynamic of Disc, Cont., Impul. Syst.*, Vol. 10, No. 6, pp. 813–832, 2003.
43. K.Y. Lian, P. Liu and C.S. Chiu, "Fuzzy model-based application to chaotic encryption using synchronization," *Int. J. Bifurc. Chaos*, Vol. 13, No. 1, pp. 215–225, 2003.
44. E.N. Lorenz, "Deterministic nonperiodic flow," *J. Atmos. Sci.*, Vol. 20, pp. 130–141, 1963.
45. P. Ivars, *Newton's Clock: Chaos in the Solar System*, W.H. Freeman, 1993.
46. W.J. Freeman, "The physiology of perception," *Scientific American*, Feb., pp. 78–85, 1991.
47. A.L. Goldberger, "Applications of chaos to physiology and medicine," in *Applied Chaos*, J. Kim and J. Stringer (eds.), Academic Press: New York, pp. 321–331, 1992.
48. V. Novak, *Fuzzy sets and their applications*, Adam Hilger: Bristol, 1989.
49. H.-J. Zimmermann, *Fuzzy set theory and its applications*, 2nd ed., Kluwer, 1991.
50. G.J. Klir and B. Yuan (eds.), *Fuzzy sets, fuzzy logic and fuzzy system: Selected papers by Lotfi A. Zadeh*, World Scientific: Singapore, 1996.
51. H.T. Nguyen and A. Walker, *First course in fuzzy logic*, CRC Press, 1999.
52. P. Hajek, *Metamathematics of fuzzy logic.*, Kluwer, 1998.
53. E. Turunen, *Mathematics behind fuzzy logic*, Physica Verlag, 1999.
54. V. Novak, I. Perfilieva and J. Mockor, *Mathematical principles of fuzzy logic*, Kluwer, 2000.
55. S. Gottwald, *A treatise on many-valued logic*, Research Studies Press Ltd., 2001.
56. G.J. Klir and T.A. Folger, *Fuzzy sets, uncertainty, and information*, Prentice-Hall International, 1998.
57. E. Cox, *The fuzzy systems handbook: A practitioner's guide to building, using, and maintaining fuzzy systems*, 2nd ed., AP Professional, 1998.
58. D. Dubois and H. Prade, *Fuzzy sets and systems: theory and applications*, Academic Press, 1980.
59. L.X. Wang, *Adaptive fuzzy systems and control: design and stability analysis*, PTR Prentice Hall, 1994.
60. K. Tanaka, *An introduction to fuzzy logic for practical applications*, Springer, 1996.
61. R. Yager and D.P. Filev, *Essentials of fuzzy modeling and control*, John Wily & Sons, 1994.
62. P.Y. Liu, "Mamdani fuzzy system: universal approximator to a class of random processes," *IEEE Trans. Fuzzy Systems*, Vol. 10, No. 6, pp. 756–766, 2002.
63. G. Chen, J.L. Moiola and H.O. Wang, "Bifurcation control: theories, methods, and application," *Int. J. Bifurcation Chaos*, Vol. 10, No. 3, pp. 511–548, 2000.

64. D.K. Arrowsmith and C.M. Place, *An Introduction to Dynamical Systems*, Cambridge University Press, 1990.
65. P. Glendinning, *Stability, Instability and Chaos*, Cambridge University Press, 1994.
66. H.-O. Peitgen, H. Jürgens and D. Saupe, *Chaos and fractals – New frontiers of science*, 2nd ed., Springer, 2004.
67. R. Gilmore and M. Lefranc, *The topology of chaos – Alice in stretch and squeezeland*, John Wiley & Sons: New York, 2002.
68. E.A. Jackson, *Perspectives of nonlinear dynamics*, Vols. 1 and 2, Cambridge University Press, 1990.
69. A.F. Ranada, "Phenomenology of chaotic motion," in *Methods and applications of nonlinear dynamics*, A.W. Saenz (ed.), World Scientific: Singapore, pp. 1–93, 1986.
70. M. Tabor, *Chaos and integrability in nonlinear dynamics*, Wiley: New York, 1989.
71. J. Carr, *Applications of center manifold theory*, Springer-Verlag: New York, 1981.
72. S. Wiggins, *Introduction to applied nonlinear dynamical systems and chaos*, Springer-Verlag: New York, 1990.
73. A.D. Bruno, *Local methods in nonlinear differential equations*, Springer-Verlag: New York, 1989.
74. T. Kapitaniak, *Controlling chaos – Theoretical and practical methods in nonlinear dynamics*, Academic Press, 1996.
75. S. Boccaletti, C. Grebogi, Y.-C. Lai, H. Mancini and D. Maza, "The control of chaos: theory and applications," *Physics reports*, Vol. 39, pp. 103–197, 2000.
76. B.R. Andrievskii and A.L. Fradkov, "Control of chaos: methods and applications. I. Method," *Automation and remote control*, Vol. 65, No. 5, pp. 673–713, 2003.
77. B.R. Andrievskii and A.L. Fradkov, "Control of chaos: methods and applications. I. Applications," *Automation and remote control*, Vol. 65, No. 4, pp. 505–533, 2004.
78. E. Ott, C. Grebogi and J.A. Yorke, "Controlling chaos," *Phys. Rev. Lett.*, Vol. 64, pp. 1196–1199, 1990.
79. F.J. Romeiras, C. Grebogi, E. Ott and W.P. Dayawansa, "Controlling chaotic dynamical system," *Physica D*, Vol. 58, pp. 165–192, 1992.
80. F. Takens, "Detecting strange attractors in turbulence," in *Dynamical systems and turbulence, Lecture notes in mathematics*, Vol. 898, D.A. Rand and L.-S. Young (eds.), Springer: New York, pp. 366–381, 1980.
81. C. Grebogi, Y.C. Lai and S. Hayes, "Control and applications of chaos," *Int. J. Bifurcat. Chaos*, Vol. 7, pp. 2175–2197, 1997.
82. C. Grebogi, Y.C. Lai and S. Hayes, "Control and applications of chaos," *J. Franklin Inst.*, Vol. 334B, pp. 1115–1146, 1997.
83. C. Grebogi and Y.C. Lai, "'Controlling chaos in high dimensions," *IEEE Trans. Circ. Syst.-I*, Vol. 44, pp. 971–975, 1997.
84. C. Grebogi and Y.C. Lai, "Controlling chaotic dynamical systems," *Syst. Control Lett.*, Vol. 31, No. 3, pp. 307–312, 1997.
85. E.A. Jackson, "The entrainment and migration controls of multiple attractor systems," *Phys. Lett. A*, Vol. 151, pp. 478–484, 1990.
86. E.A. Jackson, "On the control of complex dynamic systems," *Physica D*, Vol. 50, pp. 341–366, 1991.

87. E.A. Jackson, *Perspectives of nonlinear dynamics*, Vol. 1 and 2, Cambridge University Press, New York, 2nd ed., 1991.
88. E.A. Jackson, "Controls of dynamic flows with attractors," *Phys. Rev. A*, Vol. 44, pp. 4839–4853, 1991.
89. E.A. Jackson, "An open-plus-closed-loop (OPCL) control of complex dynamic system," *Physica D*, Vol. 85, pp. 1–9, 1995.
90. E.A. Jackson and A. Hübler, "Periodic entrainment of chaotic logistic map dynamics," *Physica D*, Vol. 44, pp. 407–420, 1990.
91. Y. Braiman and I. Goldhirsch, "Taming chaotic dynamics with weak periodic perturbations," *Phys. Rev. Lett.*, Vol. 66, pp. 2545–2548, 1991.
92. R. Chacón, "Geometrical resonance as a chaos eliminating mechanism," *Phys. Rev. Lett.*, Vol. 77, pp. 482–485, 1996.
93. T. Kapitaniak, "Controlling chaotic oscillators without feedback,", *Chaos, Solitons and fractals*, Vol. 2, pp. 519–530, 1992.
94. T. Kapitaniak, L. Kocarev and L.O. Chua, "Controlling chaos without feedback and control signals," *Int. J. Bifurcat. Chaos*, Vol. 3, pp. 459–468, 1993.
95. M. Pettini, "Controlling chaos through parametric excitations," in *Dynamics and stochastic processes*, R. Lima, L. Streit and R.V. Mendes (eds.), Springer-Verlag: New York, pp. 242–250, 1989.
96. W.-H. Steeb, J.A. Louw and T. Kapitaniak, "Chaotic behavior of an anharmonic oscillator with two external periodic forces," *J. Phys. Soc. Japan*, Vol. 55, pp. 3279–3281, 1986.
97. K. Pyragas, "Continuous control of chaos by self-controlling feedback," *Phys. Lett. A*, Vol. 170, pp. 421–428, 1992.
98. K. Pyragas, "Control of chaos via an unstable delayed feedback controller," *Phys. Rev. Lett.*, Vol. 86, pp. 2265–2268, 2001.
99. J.E.S. Socolar, D.W. Sukow and D.J. Gauthier, "Stabilizing unstable periodic orbits in fast dynamical systems," *Phys. Rev. E*, Vol. 50, pp. 3245–3248, 1994.
100. M.E. Bleich, D. Hochheiser, J.V. Moloney, "Controlling extended systems with spatially filtered, time-delayed feedback," *Phys. Rev. E*, Vol. 55, pp. 2119–2126, 1997.
101. A.V. Naumenko, N.A. Loiko, S.I. Turovets *et al.*, "Bias current impulsive feedback control of nonlinear dynamics in external cavity laser diodes," *Electron. Lett.*, Vol. 34, pp. 181–182, 1998.
102. A.V. Naumenko, N.A. Loiko, S.I. Turovets *et al.*, "Chaos control in external cavity laser diodes using electronic impulsive delayed feedback," *Int. J. Bifurcat. Chaos*, Vol. 8, pp. 1791–1799, 1998.
103. W.H. Hai, Y.W. Duan and L.X. Pan, "An analytical study for controlling unstable periodic motion in magneto-elastic chaos," *Phys. Lett. A*, Vol. 234, pp. 198–204, 1997.
104. T. Hikihara, M. Touno and T. Kawagoshi, "Experimental stabilization of unstable periodic orbit in magneto-elastic chaos by delayed feedback control," *Int. J. Bifurcat. Chaos*, Vol. 7, pp. 2837–2846, 1997.
105. M.E. Brandt, H.T. Shih and G. Chen, "Linear time-delay feedback control of a pathological rhythm in a cardiac conduction model," *Phys. Rev. E*, Vol. 56, pp. 1334–1337, 1997.
106. F.J. Elmer, "Controlling friction," *Phys. Rev. E*, Vol. 57, pp. 4903–4906, 1998.
107. K. Konishi, H. Kokame and K. Hirata, "Decentralized delayed-feedback control of an optimal velocity traffic model," *Europ. Phys. J.- B*, Vol. 15, pp. 715–722, 2000.

108. K. Konishi, H. Kokame and K. Hirata, "Coupled map car-following model and its delayed-feedback control," *Phys. Rev. E*, Vol. 60, pp. 4000–4007, 1999.
109. C. Batlle, E. Fossas and G. Olivar, "Stabilization of periodic orbits of the Buck converter by time-delayed feedback," *Int. J. Circ. Theory Appl.*, Vol. 27, pp. 617–631, 1999.
110. J. Banks, J. Brooks, G. Cairns, G. Davis and P. Stacey, "On the Devaney's Definition of Chaos", *Amer. Math. Monthly*, Vol. 99, No. 4, pp. 332–334, 1992.
111. B. Barna, "Über die Iterationen reeller Funktionen III," *Publ. Math. Debrecen*, Vol. 22, pp. 269–278, 1975.
112. G. Chen, S. Hsu and J. Zhou, "Snap-back repellers as a cause of chaotic vibration of hte wave equation with a Van de Pol boundary condition and energy injection at the middle of the span," *J. Math. Phys*, Vol. 39, No. 12, pp. 6459–6489, 1998.
113. R. Devaney, *An introduction to chaotic dynamical systems*, New York: Addison-Wesley Publishing Co., 1989.
114. R. Devaney, *A first course in chaotic dynamical systems: Theory and experiment*, Cambridge, MA: Perseus Books, 1992.
115. P. Diamond, "Chaotic behavior of systems of difference equations," *Int. J. Systems Sci.*, Vol. 7, pp. 953–956, 1976.
116. P. Diamond and P.E. Kloeden, "Characterization of compact subsets of fuzzy sets," *Fuzzy Sets and Systems*, Vol. 35, pp. 241–249, 1990.
117. P. Diamond und P.E. Kloeden, *Metric Spaces of Fuzzy Sets: Theory and Applications,* World Scientific: Singapore, 1994
118. J. Guckenheimer, G. Oster and A. Ipaktachi, "The dynamics of density dependent population models," *J. Math. Biol.*, Vol. 4, pp. 101–147, 1977.
119. M. Hénnon, "A two-dimensional mapping with a strange attractor," *Commun. Math. Phys.*, Vol. 50, pp. 69–77, 1976.
120. W. Huang and X. Ye, "Devaney's chaos or 2-scattering implies Li-Yorke's chaos," *Topology and Its Applications*, Vol. 117, pp. 259–272, 2002.
121. O. Kaleva, "On the convergence of fuzzy sets," *Fuzzy Sets and Systems*, Vol. 17, pp. 53–65, 1985.
122. P.E. Kloeden, "On Sharkovsky's cycle coexistence ordering," *Bull. Austral. Math. Soc.*, Vol. 20, pp. 171–177, 1979.
123. P.E. Kloeden, "Chaotic difference equations in R^n," *J. Austral. Math. Soc. (Series A)*, Vol. 31, pp. 217–225, 1981.
124. P.E. Kloeden, "Cycles and chaos in higher dimensional difference equations," *Proc. 9th Int. Conf. Nonlinear Oscillations*, Vol. 2, pp. 184–187, (Kiev Naukova Dumka) 1984.
125. P.E. Kloeden, "Chaotic iterations of fuzzy sets," *Fuzzy Sets and Systems*, Vol. 42, No. 1, pp. 37–42, 1991.
126. P.E. Kloeden, M. Deakin and A. Tirkel, "A precise definition of chaos", *Nature*, Vol. 264, p. 295, 1976.
127. P.E. Kloeden and A.I. Mees, "Chaotic phenomena," *Bull. Math. Biol.*, Vol. 47, pp. 697–738, 1985.
128. C.P. Li and G. Chen, "An improved version of the Marotto theorem," *Chaos, Solitons and Fractals*, Vol. 18, pp. 69–77, Sept. 2003.
129. Z. Li, W. Halang, and G. Chen (Eds.), *Integration of Fuzzy Logic and Chaos Theory*, Springer-Verlag, Heidelberg, 2006.
130. F.R. Marotto, "Snap-back repellers imply chaos in R^n," *J. of Math. Anal. Appl.*, Vol. 63, pp. 199–223, 1978.

131. R.M. May, "Simple mathematical models with very complicated dynamics," *Nature*, Vol. 261, pp. 459–467, 1976.
132. A.N. Sharkovsky, "Coexistence of cycles of a continuous transformation of a line into itself," *Ukrain. Math. Zh.*, Vol. 16, No. 1, 1964.
133. A.N. Sharkovsky, "On cycles and the structure of a continuous transformation," *Ukrain. Math. Zh.*, Vol. 17, No. 3, pp. 104–111, 1965.
134. Y.M. Shi and G. Chen, "Chaos of discrete dynamical systems in complete metric spaces," *Chaos, Solitons and Fractals*, Vol. 22, pp. 555–571, Nov. 2004.
135. Y.M. Shi and G. Chen, "Discrete chaos in Banach spaces," *Science in China Ser. A: Mathematics*, Vol. 48, No. 2, pp. 222–238, 2005.
136. K. Shiraiwa and M. Kurata, "A generalization of a theorem of Marroto," *Nagoya Math J.*, **82**, pp. 83–97, 1981.
137. C.P. Silva, "Shil'nikov theorem – a tutorial," *IEEE Trans. Circuits Syst.-I*, Vol. 40, pp. 675–682, 1993.
138. S. Smale, "Differentiable dynamical systems," *Bull. Amer. Math. Soc.*, Vol. 73, pp. 747–817, 1967.
139. P.R. Stein and S.M. Ulam, "Nonlinear transformation studies on electronic computers," *Rozprawy Matematyczne XXXIX*, pp. 1–65, 1964.
140. I. Stewart, "The Lorenz attractor exists," *Nature*, Vol. 406, pp. 948–949, 31 Aug. 2000.
141. W. Tucker, "The Lorenz attractor exists," *C.R. Acad. Sci. Paris*, Vol. 328, pp. 1197–1202, 1999.
142. T. Zhou, Y. Tang and G. Chen, "Chen's attractor exists," *Int. J. Bifurc. chaos*, Vol. 14, No. 9, pp. 3167–3177, 2004.
143. S. Čelikovský and G. Chen, "On a generalized Lorenz canonical form of chaotic systems," *Int. J. Bifurc. Chaos*, Vol. 12, No. 8, pp. 1789–1812, 2002.
144. T. Zhou, G. Chen, and S. Čelikovský, "Ši'lnikov chaos in the generalized Lorenz canonical form of dynamical systems," *Nonlinear Dynamics*, Vol. 39, pp. 319–334, 2005.
145. R. Madan, *Chua's Circuit: A Paradigm for Chaos*, World Scientific: Singapore, 1993.
146. P. Arena, R. Caponetto, L. Fortuna and D. Porto, "Chaos in fractional order Duffing system," *Proc. ECCTD'97*, pp. 1259–1272, 1997.
147. D. Driankov, H. Hellendoorn and M. Reinfrank, *An Introduction to Fuzzy Control*, Springer-Verlag: Berlin, 1993.
148. S. Baglio, L. Fortuna and G. Manganaro, "Uncertainty analysis to generate chaotic time series wiht assigned Lyapunov exponent," *Proc. NOLTA'95*, pp. 861–864, 1995.
149. J.P. Eckmann, S. Oliffson Kamphorst, D. Ruelle and S. Ciliberto, "Lyapunov exponents from time series," in *Coping with chaos*, E. Ott, T. Savor and J.A. Yorke (eds.), New York: Wiley-Interscience, 1994.
150. Y. H. Joo, H.S. Hwang, K.B. Kim, and K.B. Woo, "Fuzzy system modeling by fuzzy partition and GA hybrid schemes," *Fuzzy Sets and Systems*, Vol. 86, No. 3, pp. 279–288, 1997.
151. G. Chen and X. Dong, "Identification and control of chaotic sustems: an artificial neural network approach," *Proc. IEEE Int. Symp. on Circ. Sys.*, Seattle, WA, pp. 1177–1182, 1995.
152. K. Tanaka, T. Ikeda and H.O. Wang, "Robust stabilization of a class of uncertain nonlinear systems via fuzzy control: quadratic stabilizability, H^{∞} control

theory, and linear matrix inequalities," *IEEE Trans. Fuzzy Systems,* Vol. 4, No. 1, pp. 1–13, 1996.

153. K. Tanaka, T. Ikeda and H.O. Wang, "Fuzzy regulators and fuzzy observers: relaxed stability conditions and LMI-based designs," *IEEE Trans. Fuzzy Systems*, Vol. 6, No. 2, pp. 250–265, 1998.
154. W.L. Brogan, *Modern Control Theory*, Englewood Cliffs, NJ: Prentice Hall, 1991.
155. H.J. Lee, *Robust Fuzzy-Model-Based Controller Design for Nonlinear Systems with Parametric Uncertainties*, Master thesis, Yonsei University, Seoul, Korea, 2000.
156. S.H. Strogatz, *Nonlinear Dynamics and Chaos*, New York: Addison Wesley, 1994.
157. R. Marino and M.W. Spong, "Nonlinear control techniques for flexible joint manipulators: a single link case study," *IEEE Conf. on Robotics and Automation*, pp. 1026–1030, 1986.
158. H.A. Malki, D. Misir, D. Feigenspan and G. Chen, "Fuzzy PID control of a flexible-joint robot arm with uncertainties from time-varing loads," *IEEE Trans. on Control Syst. Tech.*, Vol. 5, No. 3, pp. 371–378, 1997.
159. K.S. Tang, K.F. Man, G.Q. Zhong and G. Chen, "Generating chaos via $x|x|$," *IEEE Trans. Circuits Syst. I,* Vol. 48, No. 5, pp. 636–641, 2001.
160. E.H. Abed, H.O. Wang and A. Tesi, "Control of bifurcations and chaos", in W.S. Levine (ed.) *The Control Handbook*, IEEE Press, 1996.
161. P. Glendinning, *Stability, instability and chaos.* Cambridge University Press, 1994.
162. J. Guckenheimer and P. Holmes, *Nonlinear Oscillations, Dynamical Systems and Bifurcations of Vector Fields.* Springer-Verlag, 1983.
163. J.K. Hale and H. Koçak, *Dynamics and Bifurcations.* Springer-Verlag, 1991.
164. Y.A. Kuznetsov, *Elements of Applied Bifurcation Theory.* Springer-Verlag, 1995.
165. J. Aracil, A. Ollero and A. Garcia-Cerezo, "Stability indices for the global analysis of expert control systems", *IEEE Trans. Systems, Man, and Cybernetics*, Vol. 19, pp. 988–1007, May 1989.
166. J. Aracil, F. Gordillo and T. Álamo, "Global Stability Analysis of Second-Order Fuzzy Control Systems", in *Advances in Fuzzy Control*, R. Palm, D. Driankov and H. Hellendorn (eds.), pp. 11–31, Springer-Verlag, 1997.
167. F. Gordillo, J. Aracil and T. Álamo, "Determining Limit Cycles in Fuzzy Control Systems", *Proc. of the 6th IEEE Int. Conf. on Fuzzy Systems (IEEE-FUZZ'97)*, Barcelona, pp. 193–198, 1997.
168. J. Moiola and G. Chen, *Hopf Bifurcation Analysis: A Frequency Domain Approach.* Singapore: World Scientific, 1996.
169. J. Llibre and E. Ponce, "Bifurcation of a periodic orbit from infinity in planar piecewise linear vector fields", *Nonlinear Analysis*, Vol. 36, pp. 623–653, 1999.
170. M. Basso, R. Genesio and A. Tesi, "A frequency method for predicting limit cycle bifurcations", *Nonlinear Dynamics*, 13, pp. 339–360, 1997.
171. F. Cuesta, F. Gordillo, J. Aracil and A. Ollero, "Stability Analysis of Nonlinear Multivariable Takagi-Sugeno Fuzzy Control Systems", *IEEE Transactions on Fuzzy Systems*, Vol. 7, No. 5, pp. 508–520, 1999.
172. J. Llibre and E. Ponce, "Global first harmonic bifurcation diagram for odd piecewise linear control systems", *Dynamics and Stability of Systems*, Vol. 11, No. 1, pp. 49–88, 1996.

173. M. Sugeno, "On Stability of Fuzzy Systems Expressed by Fuzzy Rules with Singleton Consequents", *IEEE Trans. on Fuzzy Systems*, Vol. 7, No. 2, April 1999.
174. E. Freire, E. Ponce, F. Rodrigo and F. Torres, "Bifurcation sets of continuous piecewise linear systems with two zones", *Int. Journal of Bifurcation and Chaos*, Vol. 8, No. 11, pp. 2073–2097, 1998.
175. E. Freire, E. Ponce and J. Ros, "Limit cycle bifurcation form center in symmetric piecewise-linear systems", *Int. Journal of Bifurcation and Chaos*, Vol. 9, No. 5, pp. 895–907, 1999.
176. V. Carmona, E. Freire, E. Ponce and F. Torres, "Instability in the simplest class of continuous switched linear systems with stable component", *Proc. 16th IFAC WORLD CONGRESS*, Prague, July 2005.
177. R. Abraham and J. Marsden, *Foundations of mechanics.* Benjamin: New York, 1978.
178. L.A. Cherkas, "Estimation of the number of limit cycles of autonomous systems", *Differential Equations*, Vol. 13, pp. 529–547, 1977.
179. Ch. Lee, "Fuzzy logic control system: Fuzzy logic controller – Part I and II," *IEEE Trans. Syst., Man, and Cyber.*, Vol. 20, No. 2, pp. 404–435, 1990.
180. M. Sugeno, *Industrial Applications of Fuzzy Control*, Elsevier Science, 1985.
181. E. Mamdani and S. Assilian, "An experiment in linguistic synthesis with a fuzzy logic controller", *Intl. Journal on Man-Machine Studies*, Vol. 7, No. 1, pp. 1–13, 1975.
182. T.J. Procyk and E. Mamdani, "A linguistic self-organizing process controller", *Automatica*, Vol. 15, pp. 15–30, 1979.
183. R. Jang and N. Gulley, "Fuzzy Logic Toolbox for use with Matlab", The MathWorks Inc., 1996.
184. M.P. Kennedy, "Three steps to chaos – part II: A Chua's circuit primer," *IEEE Trans. Circ. Syst.* Vol. 40, pp. 657–674, 1993.
185. T. Matsumoto, "A chaotic attractor from Chua's circuit," *IEEE Trans. Circ. Syst.*, Vol. 31, pp. 1055–1058, 1984.
186. R.N. Madan (ed.), *Chua's Circuit: A Paradigm for Chaos*, World Scientific, 1993.
187. G.A. Johnson, T.E. Tigner and E.R. Hunt, "Controlling chaos in Chua's circuit," *J. Circ. Syst. Comput.*, Vol. 3, pp. 109–117, 1993.
188. L. Chen, G. Chen and Y.W. Lee, "Fuzzy modeling and adaptive control of uncertain chaotic systems," *Information Sciences*, Vol. 121, pp. 27–37, 1999.
189. Y.Z. Tang, N.Y. Zhang and Y.D. Li, "Stable fuzzy adaptive control for a class of nonlinear systems," *Fuzzy Sets and Systems*, Vol. 104, pp. 279–288, 1999.
190. L.X. Wang, "Stable adaptive fuzzy control of nonlinear systems," *IEEE Trans. Fuzzy Systems*, Vol. 1, No. 2, pp. 146–155, 1993.
191. S. Boyd, L. El Ghaoui, E. Feron and V. Balakrishnan, *Linear Matrix Inequalities in Systems and Control Theory*, SIAM Books: Philadelphia, 1994.
192. L. Xie, "Output feedback H_∞ control of systems with parameter uncertainties," *Int. J. Contr.*, Vol. 63, No. 4, pp. 741–750, 1996.
193. K. Tanaka, T. Ikeda and H.O. Wang, "A unified approach to controlling chaos via an LMI-based fuzzy control system design," *IEEE Trans. Circ. Syst. - I*, Vol. 45, No. 10, pp. 1021–1040, 1998.
194. K. Tanaka and M. Sugeno, "Stability analysis of fuzzy systems using Lyapunov's direct method," *Proc. NAFIPS'90*, pp. 133–136, 1990.

195. K. Tanaka and M. Sugeno, "Stability analysis and design of fuzzy control systems," *Fuzzy Sets and Systems*, Vol. 45, No. 2, pp. 135–156, 1992.
196. K. Tanaka and M. Sano, "Concept of stability margin or fuzzy systems and design of robust fuzzy controllers," *Proc. 2nd IEEE Int. Conf. Fuzzy Syst. Eng.*, Vol. 1, pp. 29–34, 1993.
197. H. Wang, K. Tanaka and M. Griffin, "Paralle distributed compensation of nonlinear systems by Takagi and Sugeno's fuzzy model," *Proc. FUZZ-IEEE'95*, pp. 531–538, 1995.
198. H. Wang, K. Tanaka and M. Griffin, "An approach to fuzzy control of nonlinear systems: stability and design issues," *IEEE Trans. Fuzzy Syst.*, Vol. 4, pp. 14–23, 1996.
199. C.-W. Park, C.-H. Lee, and M. Park, "Design of an adaptive fuzzy model based controller for chaotic dynamics in Lorenz systems with uncertainty," *Information Sciences*, Vol. 147, pp. 245–266, 2002.
200. G. Feng and G. Chen, "Adaptive control of discrete-time chaotic systems: a fuzzy control approach," *Chaos, Solitons & Fractals*, Vol. 23, pp. 459–467, 2005.
201. J.F. Heagy, T.L. Carroll and L.M. Pecora, "Synchronous chaos in coupled oscillator systems," *Phys. Rev. Lett.*, Vol. 50, pp. 1874–1885, 1994.
202. Y.S. Tang, A.I. Mees and L.O. Chua, "Synchronization and chaos," *IEEE Trans. Circ. Syst.*, Vol. 30, pp. 620–626, 1983.
203. S. Boccaletti, J. Kurths, G. Osipov, D.L. Valladares and C.S. Zhou, "The synchronization of chaotic systems," *Physics Reports*, Vol. 366, pp. 1–101, 2002.
204. L.M. Pecora and T.L. Carroll, "Synchronization in chaotic systems," *Phys. Rev. Lett.*, Vol. 64, pp. 821–824, 1990.
205. P. Ashwin, "Synchronization from chaos," *Nature*, Vol. 422, pp. 384–385, 27 March 2003.
206. L. Callenbach, S. Linz, and P. Hänggi, "Synchronization of simple chaotic flows," *Phys. Lett. A*, Vol. 287, pp. 90–98, 2001.
207. K.-Y. Lian, C.-S. Chiu, T.-S. Chiang and P. Liu, "LMI-based fuzzy chaotic synchronization and communiations," *IEEE Trans. Fuzzy Systems*, Vol. 9, No. 4, pp. 539–553, 2001.
208. K.-Y. Lian, T.-S. Chiang, P. Liu, and C.-S. Chiu, "Synchronization and secure communication for chaotic systems: the modulated fuzzy system design," *Int. J. Bifur. Chaos*, Vol. 11, No. 5, pp. 1397–1410, 2001.
209. K.-Y. Lian, T.-S. Chiang, C.-S. Chiu and P. Liu, "Synthesis of fuzzy model-based designs to synchronization and secure communications for chaotic systems," *IEEE Trans. Syst., Man, and Cybern. – Part B: Cybernetics*, Vol. 31, No. 1, pp. 66–83, 2001.
210. G. Chen and T. Ueta, "Yet another chaotic attractor," *Int. J. of Bifur. Chaos*, Vol. 9, pp. 1465–1466, 1999.
211. Y.W. Wang, Z.H. Guan and H.O. Wang, "LMI-based fuzzy stability and synchronization of Chen's system," *Phys. Lett. A*, Vol. 320, pp. 154–159, 2003.
212. O.E. Rössler, "An equation for hyperchaos," *Phys. Lett. A*, Vol. 71, pp. 155–159, 1979.
213. G. Chen and X. Dong, "From chaos to order – perspectives and methodologies in controlling chaotic nonlinear dynamical systems," *Int. J. of Bifurcation and Chaos*, Vol. 3, pp. 1363–1409, 1993.
214. T. Shinbrot, C. Grebogi, E. Ott and J.A. Yorke, "Using small perturbations to control chaos," *Nature*, Vol. 363, pp. 411–417, 1993.

215. G. Chen and X. Dong, *From Chaos to Order: Methodologies, Perspectives and Applications*, World Scientific: Singapore, 1998.
216. G. Chen and X. Dong, "On feedback control of chaotic continuous-time systems," *IEEE Trans. Circuits and Systems-I*, Vol. 40, No. 9, pp. 591–601, 1993.
217. H. Wang, K. Tanaka and T. Ikeda, "Fuzzy modeling and control of chaotic systems," *IEEE International Symposium on Circuits and Systems*, Vol. 3, pp. 209–212, 1996.
218. G. Chen, "Chaos: Control and anticontrol," *IEEE Circuits Syst. Soc. Newsletter*, Vol. 9, No. 1, pp. 1–5, 1998.
219. S.J. Schiff, K. Jerger, D.H. Duong, T. Chang, M.L. Spano and W.L. Ditto, "Controlling chaos in the brain," *Nature*, Vol. 370, pp. 615–620, 1994.
220. W. Yang, M. Ding, A.J. Mandell and E.Ott, "Preserving chaos: Control strategies to preserve complex dynamics with potential relevance to biological disorders," *Phys. Rev. E*, Vol. 51, pp. 102–110, 1995.
221. I. Triandaf and I.B. Schwartz, "Tracking sustained chaos: A segmentation method," *Phys. Rev. E*, Vol. 62, pp. 3529–2534, 2000.
222. G. Chen and D. Lai, "Feedback control of Lyapunov exponents for discrete-time dynamical systems," *Int. J. of Bifurcation and Chaos*, Vol. 6, pp. 1341–1349, 1996.
223. G. Chen and D. Lai, "Anticontrol of chaos via feedback," *Proc. IEEE Conf. Decision and Control*, San Diego, CA, pp. 367–372, 1997.
224. G. Chen and D. Lai, "Feedback control of chaos," *Int. J. of Bifurcation and Chaos*, Vol. 8, pp. 1585–1590, 1998.
225. Y. Shi and G. Chen, "Chaotification of discrete dynamical systems governed by continuous maps," *Int. J. of Bifurcation and Chaos*, Vol.15, pp.547–556, 2005.
226. X.F. Wang and G. Chen, "Chaotifying a stable LTI system by tiny feedback control," *IEEE Trans. Circuits and Systems-I*, Vol. 47, No. 3, pp. 410–415, 2000.
227. X.F. Wang, G. Chen and X. Yu, "Anticontrol of chaos in continuous-time systems via time-delay feedback," *Chaos*, Vol. 10, No. 4, pp. 1–9, 2000.
228. T.S. Zhou, G. Chen and Q.G. Yang, "A simple time-delay feedback anticontrol method made rigorous," *Chaos*, Vol. 14, pp. 662–668, 2004.
229. H.G. Zhang, Z.L. Wang and D.R. Liu, "Chaotifying fuzzy hyperbolic model using adaptive inverse optimal control approach," *Intl. J. of Bifurcation and Chaos*, Vol. 14, No. 10, pp. 3505–3517, 2004.
230. Y.H. Joo, L.S. Shieh and G. Chen, "Hybrid state-space fuzzy model-based controller with dual-rate sampling for digital control of chaotic systems," *IEEE Trans. Fuzzy Systems*, Vol. 7, No. 4, pp. 394–408, 1999.
231. H.J. Lee, J.B. Park and G. Chen, "Robust fuzzy control of nonlinear systems with parametric uncertainties," *IEEE Trans. Fuzzy Systems*, Vol. 9, No. 2, pp. 369–379, 2001.
232. G.H. Golub and C.F. Van Loan, *Matrix Computations*, Baltimore, MD: Johns Hopkins Univ. Press, 1983.
233. B.C. Kuo, *Digital Control Systems*, New York: Holt, Rinehart and Winston, 1980.
234. L.S. Shieh, J.L. Zhang and J.W. Sunkel, "A new approach to the digital redesign of continuous-time controllers," *Control Theory and Advanced Technology*, Vol. 8, pp. 37–57, 1992.

235. L.S. Shieh, J.L. Zhang and W. Wang, "Digital modeling and digital redesign of analogue uncertain systems using genetic algorithms," *J. of Guidance, Control and Dynamics*, Vol. 20, No. 4, pp. 721–728, 1997.
236. W. Chang, J.B. Park, Y.H. Joo and G. Chen, "Design of sampled-data fuzzy-model-based control systems by using intelligent digital redesign," *IEEE Trans. Circ. Syst.-I*, Vol. 49, No. 4, pp. 509–517, 2002.
237. K. Ogata, *Modern Control Engineering*, Prentice-Hall: Englewood Cliffs, NJ, 1990.
238. S. Sastry, *Nonlinear Systems: Analysis, Stability, and Control*, Springer, 1999.
239. A. Isidori, *Nonlinear Control Systems*, Springer: Berlin, 1995.
240. H. Lu and Z. He, "Chaotic behaviors in first-order autonomous continuous-time systems with delay," *IEEE Trans. Circ. Syst. Video Tech.*, Vol. 43, pp. 700–702, 1996.
241. J. Huffman, "Dealing with chaos: Neural nets and fuzzy logic — tools for building products in the information age," *WESCON/94. Idea/Microelectronics. Conference Record*, pp. 332–336, 1994.
242. K. Hirota, "Fuzzy-neuro-chaos: Research and industrial applications in Japan," *IEEE Int. Conf. on Syst., Man and Cyber.*, Vol. 3, pp. 2446–2459, 1995.
243. T. Akagi and M. Sugeno, "Fuzzy identification of systems and its applications to modeling and control," *IEEE Trans. Syst., Man, Cyber*, Vol. 15, No. 1, pp. 116–132, 1985.
244. H.O. Wang, K. Tanaka and M.F. Griffin, "Parallel distributed compensation of nonlinear systems by Takagi-Sugeno fuzzy model," *Proc. FUZZ IEEE/IFES '95*, pp. 531–538, 1995.
245. C.I. Byrnes and A. Isidori, "Exact linearization and zero dynamics," *Proc. 29th Conference on Decision and Control*, Honolulu, pp. 2080–2084, 1990.
246. S.S. Sastry and A. Isidori, "Adaptive control of linearization systems," *IEEE Trans. Auto. Contr.*, Vol. 34, No. 11, pp. 1123–1131, 1989.
247. P. Celka, "Delay-differential equation versus 1D-map: Application to chaos control," *Physica D*, Vol. 104, pp. 127–147, 1997.
248. J.D. Farmer, "Chaotic attractors of an infinite-dimensional dynamical systems," *Physica D*, Vol. 4, pp. 366–393, 1982.
249. H.J. Lee, H. Kim, Y.H. Joo, W. Chang and J.B. Park, "A new intelligent digital redesign: Global approach," *IEEE Trans. Fuzzy Syst.*, Vol. 12, No. 2, pp. 274–284, 2004.
250. L.S. Shieh, W.M. Wang and M.K. Appu Panicker, "Design of PAM and PWM digital controllers for cascaded analog systems," *ISA Trans.*, Vol. 37, pp. 201–213, 1998.
251. C.A. Rabbath and N. Hori, "Reduced-order PIM methods for digital redesign," *IEE Proc. Control Theory Appl.*, Vol. 150, No. 4, pp. 335–346, 2003.
252. W. Chang, J.B. Park, H.J. Lee and Y.H. Joo, "LMI approach to digital redesign of linear time-invariant systems," *IEE Proc. Control Theory Appl.*, Vol. 149, No. 4, pp. 297–302, 2002.
253. J.W. Sunkel, L.S. Shieh and J.L. Zhang, "Digital redesign of an optimal momentum management controller for the space station," *J. Guid. Control Dyn.*, Vol. 14, No. 4, pp. 712–723, 1991.
254. J. Xu, G. Chen and L.S. Shieh, "Digital redesign for controlling chaotic Chua's circuit," *IEEE Trans. Aero. Electr.*, Vol. 32, No. 8, pp. 1488–1499, 1996.

255. L.S. Shieh, Y.J. Wang and J.W. Sunkel, "Hybrid state-space self-tuning control of uncertain linear systems," *IEE Proc. Control Theory Appl.*, Vol. 140, No. 2, pp. 99–110, 1993.
256. L.S. Shieh, W.M. Wang and J.B. Zheng, "Robust control of sampled-data uncertain systems using digitally redesigned observer-based controllers," *Int. J. Control*, Vol. 66, No. 1, pp. 43–64, 1997.
257. L.S. Shieh, W.M. Wang and J.S.H. Tsai, "Digital redesign of H_∞ controller via bilinear approximation method for state-delayed systems," *Int. J. Control*, Vol. 70, No. 5, pp. 665–683, 1998.
258. L.S. Shieh, W.M. Wang, J. Bain and J.W. Sunkel, "Design of lifted dual-rate digital controller for X-38 vehicle," *J. Guid. Control Dyn.*, Vol. 23, pp. 629–639, 2000.
259. H.J. Lee, J.B. Park and Y.H. Joo, "An effecient observer-based sampled-data control: Digital redesign approach," *IEEE Trans. Circuits Syst. I*, Vol. 50, No. 12, pp. 1595–1601, 2003.
260. S.M. Guo, L.S. Shieh, G. Chen and C.F. Lin, "Effective chaotic orbit tracker: a prediction-based digital redesign approach," *IEEE Trans. Circuits Syst. I*, Vol. 47, No. 11, pp. 1557–1570, 2000.
261. S.M. Guo, L.S. Shieh, C.F. Lin and J. Chandra, "State-space self-tuning control for nonlinear stochastic and chaotic hybrid systems," *Int. J. Birfurcation Chaos*, Vol. 11, No. 4, pp. 1079–1113, 2001.
262. W. Chang, J.B. Park, Y.H. Joo and G. Chen, "Design of sampled-data fuzzy-model-based control systems by using intelligent digital redesign," *IEEE Trans. Circuits Syst. I*, Vol. 49, No. 4, pp. 509–517, 2002.
263. S.M. Guo, W. Wang and L.S. Shieh, "Discretization of two degree-of-freedom controller and system with state, input and output delays," *IEE Proc. Control Theory Appl.*, Vol. 147, No. 1, pp. 87–96, 2000.
264. L.S. Shieh, J.L. Zhang and J.W. Sunkel, "A new approach to digital redesign of continuous-time controllers," *Control Theory and Advanced Technology*, Vol. 8, pp. 37–57, 1992.
265. D.A. Lawrence, "Analysis and design of gain scheduled sampled-data control systems," *Automatica*, Vol. 37, pp. 1041–1048, 2001.
266. W.T. Baumann, "Discrete-time control of continuous-time nonlinear systems," *Proc. Conf. on Infor. Sci. and Syst.*, pp. 119–124, Baltimore, MD, 1989.
267. K. Takatsuka, "Nonlinear dynamics in coupled fuzzy control systems I. Coherence and chaos-frustration in triangle configuration," *Physica D*, Vol. 82, pp. 95–116, 1995.
268. Y. Tomonaga and K. Takatsuka, "Strange attractors of infinitesimal widths in the bifurcation diagram with an unusual mechanism of onset nonlinear dynamics in coupled fuzzy control systems II," *Physica D*, Vol. 111, pp. 51–80, 1998.
269. M. Bucolo, L. Fortuna and M. La Rosa, "Complex dynamics through fuzzy chains," *IEEE Trans. Fuzzy Systems*, Vol. 12, No. 3, pp. 289–295, 2004.
270. L. Fortuna, M. La Rosa, D. Nicolosi and G. Sicurella, "Spatio-temporal dynamics towards self-synchronization index", *Proc. 12th Intl. IEEE Workshop on Nonlinear Dynamics of Electronic Syst.*, Evora, Portugal, 2004.
271. Y. Bar-Yam, *Dynamics of Complex Systems*, Addison-Wesley, 1997.
272. S.H. Strogatz, "Exploring complex networks," *Nature*, Vol. 410, pp. 268–276, 2001.

273. D.J. Watts, *Small Worlds*, Princeton University Press: Princeton, NJ, 1999.
274. J. Watts and S.H. Strogatz, "Collective dynamics of "small-world" networks," *Nature*, Vol. 393, pp. 440–442, 1998.
275. S. Milgram, "The small-world problem", *Psychology Today*, Vol. 2, pp. 60–67, 1967.
276. P. Erdos and A. Rényi, "On the evolution of random graphs," *Publ. Math. Inst. Hung. Acad. Sci.*, Vol. 5, pp. 17–61, 1960.
277. R. May, *Stability an dcomplexity in model ecosystems*, Princeton Universtiy Press, 1973.
278. S.A. Kauffmann, "Metabolic stability and epigenesis in randomly constructed genetic nets," *J. Theor. Biol*, Vol. 22, pp. 437–467, 1969.
279. Z. Li, P. Li 1, Y. Mao and W.A. Halang, "Chaos-based pseudo-random number generators and chip implementation," *Proc. 16th IFAC World Congress*, Prague, July 2005.
280. C.E. Shannon, "Communication theory of secrecy systems," *Bell Syst. Tech. J.*, Vol. 28, pp. 656–715, 1949.
281. M.S. Baptista, "Cryptography with chaos," *Physics Letters A*, Vol. 240, pp. 50–54, 1998.
282. D.R. Frey, "Chaotic digital encoding: an approach to secure communication," *IEEE Trans. Circuits and Systems – II*, Vol. 40, No. 10, pp. 660–666, 1993.
283. Z. Kotulski and J. Szczepanski, "On constructive approach to chaotic pseudorandom number generators," *RCMCIS'2000*, pp. 191–203, 2000.
284. B. Schneier, *Applied Cryptography – Protocols, algorithms, and source code in C*, 2nd ed., John Wiley & Sons: New York, 1996.
285. J. Szczepanski and Z. Kotulski, "Chaotic pseudorandom numbers generators based on chaotic dynamical systems," *Open Sys. And Information Dyn.*, Vol. 7, pp. 1–10, 2000.
286. W.K. Wong, L.P. Lee and K.W. Wong, "A modified chaotic cryptographic method," *Computer Physics Communications*, Vol. 138, pp. 1932–1934, 2001.
287. H. Zhou and X.T. Ling, "Problems with the chaotic inverse system encryption approach," *IEEE Trans. Circuits and Systems – I*, Vol. 44, No. 3, pp. 268–271, 1997.
288. M. Lakshmanan and K. Murali, "Chaos in nonlinear oscillators: controlling and synchronization", World Scientific: Singapore, 1996.
289. O. Morgül and E. Solak, "Observer based synchronization of chaotic systems," *Phys. Rev. E*, Vol. 54, pp. 4803–4811, 1996.
290. H.J.C. Huijberts, H. Nijmeijer and A.Yu. Pogromsky, "An observer point of view on synchronization of discrete-time systems," *Proc. ISCAS 2000*, pp. 491–494, 2000.
291. G. Grassi and S. Mascolo, "Synchronizing hyperchaotic systems by observer design," *IEEE Trans Circuits Syst – II*, Vol. 46, pp. 478–483, 1999.
292. K.M. Cuomo, A.V. Oppenheim and S.H. Strogatz, "Synchronization of Lorenz-based chaotic circuits with applicatons to communications," *IEEE Trans Circuits Syst – II*, Vol. 40, pp. 626–633, 1993.
293. L.J. Kocarev, K.D. Halle, K. Eckert, L.O. Chua and U. Parlitz, "Experimental demonstration of secure communications via chaotic synchronization," *Int. J. Bifurcation Chaos*, Vol. 2, pp. 709–713, 1992.
294. T.L. Liao and N.S. Huang, "An observer-based approach for chaotic synchronization with applications to secure communications," *IEEE Trans Circuits Syst I*, Vol. 46, pp. 1144–1150, 1999.

295. C.W. Wu and L.O. Chua, "A simple way to synchronize chaotic systems with applications to secure communication systems," *Int. J. Bifurcation Chaos*, Vol. 3, pp. 1619–1627, 1993.
296. K.S. Halle, C.W. Wu, M. Itoh and L.O. Chua, "Spread spectrum communication through modulation of chaos," *Int. J. Bifurcation Chaos*, Vol. 3, pp. 469–477, 1993.
297. K.-Y. Lian, T.-S. Chiang and P. Liu, "Discrete-time chaotic systems: applications in secure communications," *Int. J. Bifurcation Chaos*, Vol. 10, pp. 2193–2206, 2000.
298. K. Short, "Steps toward unmasking secure communications," *Int. J. Bifurcation Chaos*, Vol. 4, pp. 959–977, 1994.
299. M. Brucoli, D. Cafagna, L. Carnimeo and G. Grassi, "Design of a hyperchaotic cryptosystem based on identical and generalized synchronization," *Int. J. Bifur. Chaos*, Vol. 9, pp. 2027–2037, 1999.
300. G. Grassi and S. Mascolo, "Synchronizing hyperchaotic systems by observer design," *IEEE Trans Circuits Syst II*, Vol. 46, pp. 478–483, 1999.
301. T. Yang, C.W. Wu and L.O. Chua, "Crytpography based on chaotic system," *IEEE Trans. Circuits Syst. I*, Vol. 44, pp. 469–472, 1997.
302. K.Y. Lian, P. Liu, C.S. Chiu and T.S. Chiang, "Robust Dead-Beat Synchronization and Communication for Discrete-time Chaotic Systems," *Int. J. Bifur. Chaos*, Vol. 12, pp. 835–846, 2002.
303. M.R. Garey and D.S. Johnson, *Computers and intractability: a guide to the theory of NP-completeness*, WH Freeman & Co: San Francisco, 1976.
304. D.E.R. Denning, *Cryptography and data security*, Addison Wesley: New York, 1982.

Printing: Krips bv, Meppel
Binding: Stürtz, Würzburg